CRC SERIES IN BIOCOMPATIBILITY

Series Editor-in-Chief

David F. Williams, Ph.D.
Senior Lecturer in Dental and Medical Materials
Department of Dental Sciences
University of Liverpool
Liverpool, ENGLAND

FUNDAMENTAL ASPECTS OF BIOCOMPATIBILITY
Editor: David F. Williams, Ph.D.

BIOCOMPATIBILITY OF ORTHOPEDIC IMPLANTS
Editor: David F. Williams, Ph.D.

BLOOD COMPATIBILITY
Editors: David F. Williams, Ph.D. and Donald J. Lyman, Ph.D.

BIOCOMPATIBILITY OF DENTAL MATERIALS
Editors: D. C. Smith, Ph.D. and David F. Williams, Ph.D.

SYSTEMIC ASPECTS OF BIOCOMPATIBILITY
Editor: David F. Williams, Ph.D.

BIOCOMPATIBILITY IN CLINICAL PRACTICE
Editor: David F. Williams, Ph.D.

BIOCOMPATIBILITY OF CLINICAL IMPLANT MATERIALS
Editor: David F. Williams, Ph.D.

Systemic Aspects of Biocompatibility

Volume II

Editor

David F. Williams, Ph.D.

Senior Lecturer in Dental and Medical Materials
Department of Dental Sciences
University of Liverpool
Liverpool, England

CRC Series in Biocompatibility
Series Editor-in-Chief

David F. Williams, Ph.D.
Senior Lecturer in Dental and Medical Materials
Department of Dental Sciences
University of Liverpool
Liverpool, England

RA1231
M52
S97
v.2
1981

CRC Press, Inc.
Boca Raton, Florida

Library of Congress Cataloging in Publication Data
Main entry under title:

Systemic aspects of biocompatibility.

(CRC uniscience series on biocompatibility)
Bibliography: p.
Includes index.
1. Metals—Toxicology. 2. Metals in the body.
3. Metals in medicine. 4. Polymers and polymerization
—Toxicology. 5. Polymers in medicine. I. Williams,
David Franklyn. II. Series. [DNLM: 1. Biocompatible materials. QT34 S996]
RA1231.M52S97   615.9'02   80-17499
ISBN 0-8493-6621-6 (v. 1.)
ISBN 0-8493-6622-4 (v. 2.)

This book represents information obtained from authentic and highly regarded sources. Reprinted material is quoted with permission, and sources are indicated. A wide variety of references are listed. Every reasonable effort has been made to give reliable data and information, but the author and the publisher cannot assume responsibility for the validity of all materials or for the consequences of their use.

All rights reserved. This book, or any parts thereof, may not be reproduced in any form without written consent from the publisher.

Direct all inquiries to CRC Press, Inc., 2000 N.W. 24th Street, Boca Raton, Florida 33431.

© 1981 by CRC Press, Inc.

International Standard Book Number 0-8493-6621-6 (Volume I)
International Standard Book Number 0-8493-6622-4 (Volume II)

Library of Congress Card Number 80-17499
Printed in the United States

SERIES PREFACE

One of the most noticeable and beneficial aspects of the recent developments in medicine has been the exploitation of technological advances and innovations. Modern hospitals and clinics contain many pieces of hardware that are used for diagnostic and treatment purposes and which testify to this progress. While it is difficult to single out any one particular branch of biomedical engineering, as this subject is now called, as giving the most benefit to the patient, the tremendous advances that have been made in surgery through the use of implanted devices must certainly number among the more significant. Anyone who has witnessed the transformation that is produced in an arthritic patient by treatment with a total joint replacement prosthesis can be left with little doubt as to the clinical importance of this type of development.

Although a wide spectrum of implanted devices now exists, fulfilling needs in such diverse surgical disciplines as ophthalmology, cardiology, neurosurgery, dentistry, and orthopedics, they all have one thing in common; they must have intimate contact with the patient's tissues, providing a real, physical interface. It is the existence of this interface that poses the most intriguing questions, provides the most interesting challenges, and restricts developments most seriously. This series of volumes is about this interface, the reactions which occur there, and the effects these reactions have on prostheses and, especially, on patients. The term biocompatibility has been introduced with reference to these interactions. It is the term used to describe the state of affairs when a material exists within a physiological environment without either the material adversely and significantly affecting the body, or the environment of the body adversely and significantly affecting the material. Absolute biocompatibility may be regarded as utopia, while in a more realistic way we have to consider the various degrees of biocompatibility that we find in practice. As discussed in the opening chapter to *Fundamental Aspects of Biocompatibility* there are many ways of looking at biocompatibility and the literature is replete with observation, comment and argument over what constitutes "adverse" and "significant" interactions and what "biocompatibility" really means.

It is partly because of the somewhat confused state that exists at this moment and partly because of the tremendous importance of this subject that this series of volumes on biocompatibility has been written and compiled. The planning of the series was based on two premises, first that a better understanding of biocompatibility could be achieved by considering both the fundamental aspects of these interactions and their clinical effects, and secondly that a collection of papers within one series derived from all branches of surgery would, through a process of cross-fertilization of ideas, serve to enhance this understanding.

The series starts, therefore, in the first volume, with a discussion of the fundamental aspects of biocompatibility, covering the principles of materials degradation in the physiological environment and the basic features of the tissue response to implanted foreign bodies. *Biocompatibility of Orthopedic Implants*, *Blood Compatibility*, and *Biocompatibility of Dental Materials* are volumes which deal with the major clinical areas that utilize implants: orthopaedic, cardiovascular, and dental surgery. In each of these volumes the clinical features of biocompatibility are emphasized, using descriptions of the devices, materials, and clinical procedures as a basis for discussion where necessary and appropriate.

Systemic Aspects of Biocompatibility returns to the theme of the fundamental interactions and explores the question of the systemic effects that may arise after biomaterial-tissue contact and especially reviews the relationship between biocompatibility and toxicology. *Biocompatibility in Clinical Practice* then brings together a wide range of surgical disciplines, the clinical features of biocompatibility associated with diverse procedures such as the treatment of hydrocephalus, the use of contact lenses, bladder replacements, and cardiac pacemakers being reviewed and compared. In *Biocompatibility of Clinical Implant Materials,* biocompatibility is discussed from the materials point of view, each major material being reviewed in terms of its properties, clinical use, and biocompatibility.

These volumes are not merely independent publications on biocompatibility, but rather provide a series of complementary sources of instruction and reference, providing what is believed to be the first comprehensive review of biocompatibility.

The series could not, of course, have been completed without the assistance and patience of the many contributing authors, to whom I am deeply grateful.

David Williams

PREFACE

Systemic Aspects of Biocompatibility is basically concerned with the effects that biomaterials may have on tissues remote from the site of their application. An examination of the literature on this subject will reveal that very little factual data exists for, in the main, attention has been primarily focused at the tissue-material interface and the local effects of any interaction. Since many substances are readily transported in the tissues, it must be expected that products of any interaction may be dispersed systemically and it is highly relevant to consider their fate and their potential for tissue damage at these distant sites. An attempt is made in these two volumes, therefore, to correlate systemic biocompatibility with the known toxicology of the materials and, indeed, many of the chapters have been contributed by acknowledged authorities on the toxicology of the specific biomaterials considered.

As with the other parts in the series, the rationale and planning are discussed in the first introductory chapter, and there is little need to preempt that discussion in a preface. It is hoped that, as a sequel to *Fundamental Aspects of Biocompatibility,* this work will provide a further basis for the understanding of the fundamentals of biocompatibility.

These volumes could not have been completed without the expertise of either the contributing authors or the editing staff at CRC Press. To all these I wish to express my gratitude.

David F. Williams

THE EDITOR

David F. Williams, Ph.D. is Senior Lecturer in Dental and Medical Materials in the School of Dental Surgery, University of Liverpool, England.

Dr. Williams received a B.Sc. with 1st class Honors in 1965 and a Ph.D. in 1969, both in Physical Metallurgy at the University of Birmingham. He has been on the staff at Liverpool University since 1968, apart from a year as Visiting Associate Professor in Bioengineering at Clemson University, South Carolina in 1975.

Dr. Williams is a Chartered Engineer and a Fellow of the Institution of Metallurgists. He has been a council member of the Biological Engineering Society and is currently chairman of the Biomaterials Group of that Society.

Dr. Williams has major research interests in the interactions between metals and tissues, polymer degradation and the clinical performance of biomaterials. He has authored many research papers and reviews in these areas and has previously written two books and edited two volumes.

CONTRIBUTORS

R. D. Bagnall, Ph.D.
Senior Lecturer
Dental Materials Science
Edinburgh University Dental School
Edinburgh, Scotland

C. Allen Bradley, Ph.D.
Professor and Chairman
Department of Biopharmaceutical
 Sciences
University of Arkansas for Medical
 Sciences
Little Rock, Arkansas

Stanley A. Brown, D. Eng.
Associate Professor of Bioengineering
 in Orthopedic Surgery
University of California
Davis, California

Ulrich Borchard, Ph.D.
Professor of Pharmacology
University of Dusseldorf
Dusseldorf, West Germany

Thomas Ming Swi Chang, M.D., F.R.C.P.(c)
Director, Artificial Cells and Organs
 Research Center
Professor of Medicine and Physiology
McGill University
Montreal, Quebec
Canada

G. C. Clark
Professor of Dental Sciences
University of Liverpool
Liverpool, England

Michael G. Crews, Ph.D.
Assistant Professor of Food and
 Nutrition
Texas Tech University
Lubbock, Texas

Jeanine Grisvard, Ph.D.
Assistant Professor of Cellular Biology
University of Paris XI
Orsay, France

Etienne Guille, Ph.D.
Assistant Professor of Molecular Plant
 Biology
University of Paris XI
Orsay, France

Thomas J. Haley, Ph.D.
Science Advisor to the Director
National Center for Toxicological
 Research
Jefferson, Arkansas

Stephen P. Halloran, B.Sc., M.Sc.
Senior Biochemist
Department of Clinical Biochemistry
St. Peter's Hospital
Chertsey
Surrey, England

Arne Hensten-Pettersen, Dr. Odont.
Research Associate
NIOM, Scandinavian Institute of
 Dental Materials
Oslo, Norway

Hogne Hofsøy, D.D.S.
Assistant Professor of Microbiology
University of Oslo
Oslo, Norway

Leon L. Hopkins, Ph.D.
Professor and Chairman
Department of Food and Nutrition
Texas Tech University
Lubbock, Texas

Robert A. Jacob, Ph.D.
Chief Clinical Nutrition Laboratory
Jamaica Plain, Massachusetts

Nils Jacobsen, Dr. Odont.
Associate Professor of Dental
 Technology
University of Oslo
Oslo, Norway

Jindřich Kopeček, Ph.D.
Head, Laboratory of Medical Polymers
Institute of Macromolecular Chemistry
Prague, Czechoslavakia

Sverre Langård, M.D.
Head, Department of Occupational
 Medicine
Telemark Sentralsjukehus
Porsgrunn, Norway

Laszlo Magos, M.D.
Toxicologist
Medical Research Council Toxicology
 Unit
Carshalton
Surrey, United Kingdom

Katharine Merritt, Ph.D.
Associate Professor of Microbiology in
 Orthopedic Surgery
University of California
Davis, California

Robert J. Pariser, M.D.
Assistant Professor of Medicine
 (Dermatology)
Eastern Virginia Medical School
Norfolk, Virginia

Trevor Rae, Ph.D.
Research Associate
University of Cambridge Clinical
 School
Cambridge, England

Igor Sissoeff, Ph.D.
Assistant Professor of Molecular
 Biology
University of Paris XI
Orsay, France

Gail K. Smith, Ph.D., V.M.D.
Assistant Professor of Orthopedic
 Surgery
School of Veterinary Medicine
University of Pennsylvania
Philadelphia, Pennsylvania

Michael K. Ward, M.D.
Consultant Physician
Royal Victoria Hospital
Newcastle, England

David F. Williams, Ph.D.
Senior Lecturer in Dental and Medical
 Materials
University of Liverpool
Liverpool, England

TABLE OF CONTENTS

Volume I

METALLIC MATERIALS

Chapter 1
Introduction 1
David F. Williams

Chapter 2
Low Molecular Weight Complexes of Transition Metal Ions in Biological Fluids 7
G. C. F. Clark

Chapter 3
Metal-Enzyme Interactions 21
Trevor Rae

Chapter 4
Structure and Function of Metallo-DNA in the Living Cell 39
Etienne Guille, Jeanine Grisvard, and Igor Sissoeff

Chapter 5
Synergism and Antagonium in Metal Toxicology 87
Laszlo Magos

Chapter 6
Hair as a Biopsy Material 101
Robert A. Jacob

Chapter 7
Some Biological Aspects of Nickel 115
Nils Jacobsen, Arne Hensten-Pettersen, and H. Hofsøy

Chapter 8
Biological Properties of Cobalt 133
G. C. F. Clark

Chapter 9
Chromium Toxicology 143
Sverre Langård and Arne Hensten-Pettersen

Chapter 10
Biological Properties of Molybdenum 163
David F. Williams

Chapter 11
Biological Effects of Titanium 169
David F. Williams

Chapter 12
Metabolism and Toxicity of Vanadium..179
Michael G. Crews and Leon C. Hopkins

Chapter 13
Biological Properties of Aluminum..187
C. Allen Bradley

Chapter 14
Copper...211
Stephen P. Halloran

Chapter 15
Mercury..237
David F. Williams

Chapter 16
Argyria: Silver in Biological Tissues..251
Robert J. Pariser

Index..259

Volume II

Chapter 1
Systemic Biocompatibility of Metallic Surgical Implants........................1
Gail K. Smith

Chapter 2
Metal Toxicology in Hemodialysis..23
Michael K. Ward

Chapter 3
Hypersensitivity to Metallic Biomaterials.....................................33
Katharine Merritt and Stanley A. Brown

POLYMER-BASED MATERIALS

Chapter 4
Introduction to the Toxicology of Polymer-Based Materials.....................49
David F. Williams

Chapter 5
Biocompatibility of Monomers..57
Thomas J. Haley

Chapter 6
Pharmacology, Toxicology, and Clinical Actions of Monomeric and Polymeric Methyl
Methacrylate...105
Ulrich Borchard

Chapter 7
The Toxicology of Additives in Medical Plastics 143
David F. Williams

Chapter 8
Soluble Polymers in Medicine 157
Jindřich Kopeček

Chapter 9
Biocompatibility and Experimental Therapy of Immobilized Enzymes and
Proteins.. 179
Thomas Ming Swi Chang

Chapter 10
The Principles of Controlled Drug Release................................. 187
R. D. Bagnall

Index 197

Chapter 1

SYSTEMIC BIOCOMPATIBILITY OF METALLIC SURGICAL IMPLANTS

Gail K. Smith

TABLE OF CONTENTS

I.	Introduction	2
II.	Corrosion: Cause for Concern	2
III.	Systemic Effects	4
	A. General Principles	4
	B. Role of Trace Elements in Normal Metabolism	5
	C. Pharmacotoxicological Considerations	6
	D. Systemic Effects of Implant Corrosion (Real and Potential)	7
	1. Immunologic Considerations	8
	a. Iron and Susceptibility to Infectious Disease	8
	b. Metal Hypersensitivity	9
	2. Carcinogenic Effects	10
	a. Industrial Exposure and Experimental Investigations	10
	b. Clinical Carcinogenicity	11
	3. Metabolic Effects	12
	a. Iron Metabolism	13
	1. Absorption	14
	2. Transport	14
	3. Storage	15
	4. Excretion	15
	b. Iron Overload	15
	c. Transport and Disribution of Corrosion Products from 316L Stainless Steel	17
	d. Metal-Induced Systemic Sequellae	17
IV.	Conclusions	18
	References	20

I. INTRODUCTION

Metals used as surgical implants have traditionally been selected for optimum strength and corrosion resistance in the biological environment. The metals and alloys of current popularity are those which, over the years, have performed acceptably in the clinical situation. It should be recognized that acceptable performance implies only a relative degree of implant-tissue biocompatibility. The two major alloy systems in use today in the U.S., 316L stainless steel and cobalt chromium, although corrosion resistant by conventional engineering standards, nevertheless, have finite corrosion rates in vivo, releasing ions and complexes of their constituent elements upon implantation. This phenomenon, as predicted from thermodynamic theory, is supported in practice by both experimental and clinical experience. Mechanisms of corrosion, as discussed in a previous volume, range from uniform attack to accelerated modes such as pitting, fretting, or crevice corrosion. Irrespective of mode, however, the net effect is the unregulated introduction of quantities of metal ions or complexes into the physiological system, both locally and systemically.

Corrosion products have been strongly implicated in the local tissue changes in the vicinity of a metal implant and in this context have been extensively investigated. In sharp contrast, however, there exists few reports even remotely addressing the question of potential systemic consequences of metallic implantation. This situation exists despite information from toxicological, nutritional, biochemical, and industrial exposure literature indicating a strong dependence of normal body function on precise trace element homeostasis. Accordingly, the fact that metallic implant consititiuents such as iron, chromium, nickel, and cobalt are themselves normal, trace elements should raise concern regarding the act of metallic implantation and long-term patient safety. It is this concern which prompted the author to undertake an investigation of the systemic transport and distribution of corrosion products of 316L stainless steel implants in rabbits. When germane, this work will be discussed in the course of the chapter to follow.

II. CORROSION: CAUSE FOR CONCERN

The process of corrosion from metallic implants occasions the release of foreign material, both soluble and insoluble, into the surrounding tissue. This has been made abundantly clear through the early studies of Ferguson, Laing, and Hodge[1,2] and reinforced by several investigators since. The classic work of Ferguson, et al.[2] in 1962 demonstrated the potential for corrosion products from various surgical metals to be deposited at remote sites systemically. Their work, although lacking in statistical treatment, raises many questions regarding the interaction of corrosion products with normal trace element metabolism.

Von Ludinghausen, et al.,[3] using neutron activation analytical techniques, demonstrated the enrichment of chromium and manganese in tissues surrounding 316L stainless steel implants in clinical cases. Employing similar methodology, Coleman, Herrington, and Scales[4] found in nine patients who had received metal-to-metal cast cobalt chromium total hip arthroplasties, a 10 to 20-fold increase in blood cobalt, a 3-fold increase in blood chromium concentration, a 10 to 50-fold elevation in urine cobalt, and a 15-fold increase in urine chromium concentration when compared with respective preoperative values. The authors suggested the need for further studies to elucidate chromium's half-life and distribution in the body.

Lux and Zeisler[5] investigated the concentration profile of various trace elements and corrosion products in tissue surrounding 316 stainless steel implants as a function of distance from the implant. Results of extensive analysis of metallosis tissue from 38 patients including neutron activation, electron diffraction, and electron microscopy led the investigators to some interesting observations. Ostensibly, 316 stainless steel implant corrosion occurs in a transcrystalline manner, resulting in the deposition of microcrystallites in the surrounding tissue where further degradation continues. All components of 316 stainless steel were evident in the tissue immediately adjacent to the implant and in proportion, generally consistent with the bulk 316 stainless steel composition. Analyses of tissue, however, at various distances (1 to 4 cm) from the implant revealed that nickel and molybdenum are selectively and rapidly transported from the site, that chromium is removed at a lower rate, and that iron forms an intracellular biological compound, thus being removed still slower than chromium. (Surgical hemorrhage and hemosiderin formation could, of course, be influencing the last observation.) Formation of a biological chromium compound was also postulated. In addition, an inverse relationship (interaction) between tissue zinc and biologically bound iron was thought to exist. It should be emphasized that this study was the first to consider the dynamic distribution, albeit locally, of implant corrosion products in the biological system. Apparently there exist selective transport or diffusion processes which influence, at relative rates, the systemic distribution of these corrosion complexes.

In summary, then, evidence thus far suggests that (1) corrosion is a phenomenon common to all currently used implant alloys, (2) that the process results in contamination of the local environment with metal constituents, and (3) that these corrosion products diffuse or are transported by mechanisms and to sites as yet undetermined.

Taylor,[6] on a purely theoretical basis, has predicted the total body (or organ) accumulation over time, $Q(t)$, of human total hip recipients. Using the expression:

$$Q(t) = Q_o + R(1 - e^{-kt})/k \tag{1}$$

where R = corrosion rate, Q_o = normal body (or organ) metal concentration, k = fractional rate of excretion of the metal,* and t = time, it was demonstrated that for long times at the steady state when the rate of metallic release into the circulation is balanced by excretion rate or

$$\frac{dQ(t)}{dt} = 0 \tag{2}$$

the total body content of the metal, Q_E, becomes:

$$Q_E = Q_o + R/k \tag{3}$$

Choosing a high corrosion rate, R, and assuming an implant surface area of 200 cm², Taylor predicted for cobalt chromium devices at steady state, an 18-fold increase in body cobalt and a 273-fold increase in body chromium. For 316L stainless steel implants a 3.4-fold increase is predicted in nonheme iron, a 90-fold increase in body nickel, and 183-fold increase in body chromium. Unquestionably, the estimated corrosion rates, R, used in this calculation are far in excess of actual in vivo rates (approximately, a few hundred ng/cm²/day). Nevertheless, even for low corrosion rates, the predictions become biologically significant when the calculation is applied to the case of a porous sintered metallic implant having an estimated surface area 50 to 100 times that used above (i.e., 10,000 to 20,000 cm²).

---

* Publications 2 and 10 of the Internal Commission on Radiological Protection.

It is unfortunate that apart from these few cited investigations, (and their rather peripheral significance), little else is known regarding corrosion-mediated systemic phenomena. The need for continued research in this vein is clear, particularly when considering some of the more recent developments and trends in orthopedics. For example, the application of metallic devices for orthopedic procedures continues to expand at a rapid pace. Estimates based on discussions with surgeons and industrial representatives indicate that more than 100 million surgical implants have been used in orthopedic practice in the U.S. alone. By design or omission, nearly 10 million remain in place. The growing selection of permanent prostheses and their use in younger patient populations raises concern as to chronic exposure to implant corrosion for periods as long as 50 years. Moreover, the release of wear products accompanying total joint replacement devices, particularly the metal-on-metal devices, further complicates the problem. Last, the allure of porous-surfaced devices promoting bony ingrowth and rigid implant-to-bone fixation ensures increased development and utilization. The attendant high surface area amplifies many-fold these considerations.

Several factors are perhaps responsible for the apparent neglect in evaluating implant-mediated systemic effects. First, and probably most important, such investigations appear superfluous judging from past clinical experience. That is, clinically, metals have been applied internally for many years, yet only very recently with the suspicion of ion-mediated hypersensitivity or latent tumor formation have systemic effects been implicated. Obviously, the untoward effects (if indeed there are untoward effects) do not manifest as overt toxicity clinically, and/or have very long latency periods (> 20 years), or produce effects not distinguishable from other clinical problems.

Second, regarding corrosion product measurement, technology has lagged behind the desire to accurately and reproducibly measure trace ion concentrations typically in the parts per billion (ng/mℓ) range. Coincident, therefore, with advances in atomic absorption spectroscopy, neutron activation, emission spectrography, and reagent purity, the importance of trace elements in normal and disease states is only recently gaining appreciation. An additional deterrent to investigation is the inherently large biological variation in normal trace element levels which often necessitates costly and time consuming studies to demonstrate trace element aberrations on a statistically significant basis.

III. SYSTEMIC EFFECTS

A. General Principles

In the absence of specific literature on systemic toxicity of corrosion products, one is forced to investigate general aspects of trace element function and toxicity with the hope that such information will yield insight into potential systemic effects of implanted metals. Information is typically derived from (1) industrial exposure studies, (2) experimental investigations to determine deficiency or excess states of various nutritional or toxic metals, or (3) incidental clinical cases (human or veterinary) relating the presence of an implant with either allergy or tumor formation. In general, potential systemic effects of implant corrosion (either in theory or practice) fit into three main categories: metabolic, carcinogenic, or immunologic. Carcinogenic or immune phenomena are usually recognizable clinical entities. Metabolic effects, however, may be subtle and clinically undetected. It can, perhaps, be argued that tumors and allergies are themselves products of altered cellular metabolism and should be included under metabolic effects. However, as used in this context, metabolic effects constitute abrupt changes in normal trace element homeostasis as a direct consequence of implant corrosion. This category and related potential systemic sequellae have not as yet been addressed in the literature. All three categories will be discussed in sections to follow.

B. Role of Trace Elements in Normal Metabolism

"There probably does not exist a single enzyme-catalyzed reaction in which either substrate, product, enzyme or some combination within this triad is not influenced in a very direct and highly specific manner by the precise nature of the inorganic ions which surround or modify it."[7] Green[8] almost 40 years ago had postulated that "enzyme catalysis is the only rational explanation of how a trace of some substance can produce profound biochemical effects." More specifically, Harper[13] has listed the mechanisms for this activity: (1) direct participation in catalysis, (2) combination with substrate to form a metal-substrate complex upon which the enzyme acts, (3) formation of a metalloenzyme that binds substrates in an enzyme-metal-substrate (or enzyme-metal-coenzyme-substrate) complex, (4) combinaton of metal with a reaction product to alter equilibrium, and (5) maintenance of quaternary (or tertiary) structure. In recent years with advances in physiological and biochemical research, increasing emphasis is being placed on the importance of trace elements and their roles in normal and disease states. For a more comprehensive review of this information, the works of Underwood[9] and Reinhold,[10] and References 57 and 58 should be consulted.

As the term trace implies, these elements are evident in all biological systems in minute concentrations (1×10^{-6} to $1 \times 10^{-12}/g/g$ wet tissue) often at or below the limit of analytical detection. For a given organism, concentrations of trace elements may vary from one location or tissue to another or at the same location at differing times. The trace elements are grouped into three general categories: essential, possibly essential, and nonessential. Cotzias[11] has formally defined an essential trace element by the following criteria: (1) it is present in all healthy tissues of all living things, (2) its concentration from one animal to the next is fairly constant, (3) its withdrawal from the body induces, reproducibly, the same structural and physiological abnormalities regardless of species studied, (4) its addition either prevents or reverses these abnormalities, (5) the abnormalities induced by deficiency are always accompanied by pertinent, specific, biochemical changes, and (6) these biochemical changes can be prevented or cured when the deficiency is prevented or cured.

Essential elements include *iron*, iodine, *copper*, zinc, *manganese*, *cobalt*, *molybdenum*, selenium, *chromium*, *nickel*, tin, silicon, and *vanadium*. The *italicized* metals are all constituents of orthopedic implant alloys. The category possibly essential encompasses those elements which deprivation from an organism causes no harm, but addition results in noticeable benefits, e.g., the fluoride ion in caries resistance.

Twenty or thirty nonessential trace elements including *aluminum*, antimony, mercury, cadmium, and *titanium* appear in highly variable concentrations in the body and are thought to represent contaminants reflecting the contact of the organism with its environment.[9] It has been observed that concentrations of these elements in human organs have a skewed (log-normal) distribution, whereas essential elements have a normal distribution. Liebscher and Smith[12] have suggested that the difference between essential and nonessential distribution patterns reflect the system's ability or inability, respectively, to control incorporation and retention of these trace elements. That is, for essential elements physiological control mechanisms are postulated which standardize the level of an element throughout a population, thereby leading to a normal or symmetrical distribution. Nonessential elements, on the other hand, are distributed throughout a population relative to their ambient levels in each particular environment, such as food, water, and air.

Considering the low magnitude of biological trace element concentrations and the profound biochemical activity attributable to these levels, it becomes clear that the presence of a chronically corroding metallic implant may represent a threat to implant recipient health. Indeed, within the context of systemic biocompatibility, it is not illog-

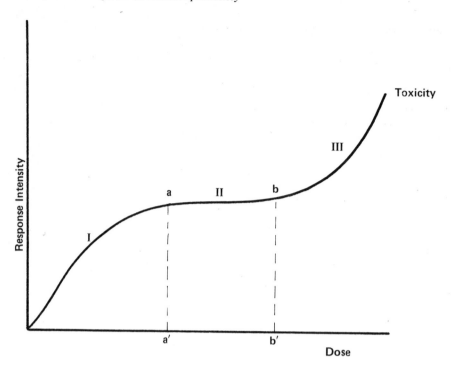

FIGURE 1. Hypothetical dose-response curve.

ical to regard implants as repositories of slowly and continuously released drugs. Although only a few elements, namely As, Pb, Cd, and Hg, have received widespread attention owing to their profound toxicity at low concentrations, it should be emphasized that all trace elements are potentially toxic if present at high enough levels for long enough time periods. The biological response to dose increments of any substance can be illustrated hypothetically on a dose-response curve (see Figure 1).

Part I of the dose-response curve expresses the biological action of an element and illustrates that with each increment of dose there results a proportional change in response. At point (a) the optimum physiological response is obtained; for essential elements any dose less than (a') represents a deficiency state. Part II, the plateaued portion of the dose-response curve (a to b), expresses the dose range for which optimal biological function (response) is achieved. The width of the plateau reflects the organism's homeostatic capacity. For doses higher than b' (Part III) the pharmacotoxicological action of the substance is observed. The element at this dose acts as a foreign drug often independent of its biological action either stimulating, irritating, or depressing some function of the organism.

Unfortunately, information is not available to construct dose-response curves for each individual constituent of metallic implants. If such were the case, it would be a simple exercise to superimpose corrosion-mediated local or systemic levels of corrosion products on their respective curves and predict the sequellae.

C. Pharmacotoxicological Considerations

The foregoing discussion has emphasized the profound biochemical activity of trace elements and the importance of maintaining levels of these elements within the homeostatic capacity of the system. Corrosion from metallic implants may disrupt this finely tuned equilibrium. Whether or not trace metals accumulate to toxic levels depends upon the inherent toxicity of the metal (or metal-protein complex) involved, the amount absorbed, and the rate at which the body detoxifies or excretes them. Through selective processes, i.e., absorption, sequestration, metabolism, and excretion, the body can maintain physiological levels of trace elements or foreign substances.

Absorption refers to a process requiring cell membrane penetration, which for the case of trace element entry into the system occurs in the duodenum and proximal ileum. Skin and lung epithelium are also potential sites of absorption, although minimal. Absorption, whether in the gut or within any cell or organelle, is facilitated by those substances having small size, high concentrations, high lipid solubility, and nonionic and nonpolar characteristics. Membrane transport, whether passive or active, constitutes an important factor in the toxic potential of any substance by permitting access of an element into those intracellular areas where its toxicity is ultimately manifested upon some particularly sensitive enzyme system, organelle, or genetic material. In the case of metallic surgical implants, the selective membrane permeability of the gut epithelium is bypassed, forcing the biological system to contend with quantities of substances, particularly metal ions, which would have otherwise undergone regulatory absorption.

Once absorbed, a toxic or charged moiety can be rendered inactive by sequestration. This refers to the ability of various protein components of the milieu interne to form ionic, hydrogen, or Van der Waal-type bonds with substances. The phenomenon can be relatively nonspecific as in the weak binding of albumin to an assortment of free, diffusible chemicals, or more specific as in the binding of iron by transferrin, a β_1 globulin (protein). Organ storage is a form of sequestration. Bone, liver, and occasionally other tissues can function in binding and storing toxic compounds, e.g., DDT in fat, tetracycline in bone and skin, lead in bone, and iron in reticulo-endothelial cells.[9]

Many active compounds, toxic or otherwise, may be metabolized to less active or inactive products. The liver, for instance, has the capacity to mobilize specific enzymic systems which function to neutralize circulating toxic entities. Two common examples in this regard are ethyl alcohol and barbiturates. In general, however, trace elements in excess are not detoxified in this manner.

Excretion represents a potent mechanism to eliminate toxic substances. It is mainly a function of the kidney with less important mechanisms existing for gastrointestinal and to a still lesser extent, respiratory excretion. The kidneys, by selective excretion and reabsorption, control the concentrations of most of the consitituents of body fluids and, as well, eliminate most of the end products of bodily metabolism. Excretion rate of a substance is determined by an interplay of glomerular filtration rate, tubular reabsorption, and tubular excretion of this substance. Glomerular filtration is dependant on the pressure inside the glomerular capillaries, the colloid osmotic pressure of the blood, the pressure in Bowman's capsule, and the size of the substance filtered (molecules must be less than 70,000 mol wt). Tubular absorption and excretion mechanisms are essentially identical with those for transport through any other membrane in the body. In the tubular epithelium there exist active and passive transport processes. The active mechanisms are highly selective and depend upon ATP to function, while the passive processes (diffusion mediated) are determined by the degree of lipid solubility, relative concentration, degree of ionization, and the polarity of the substance. Interestingly, since trace elements are typically 95% protein bound, they are not readily excreted by renal mechanisms. Of course, in trace element excess states, the percentage of free ionic or unbound elements is elevated and, therefore, excretion is correspondingly increased.

D. Systemic Effects of Implant Corrosion (Real and Potential)

As mentioned, systemic effects of implant corrosion, whether real or potential, fall into three main categories: immunologic, carcinogenic, and metabolic. These groups will be discussed in order. Little factual information is known regarding metabolic effects of corrosion products. In the absence of such information, one must speculate as to potential systemic effects from a knowledge of a corrosion product's metabolism as a trace element. Because iron is both widely studied and a large component of 316L stainless steel implants ($\approx$ 60%), it will receive the thrust of the following discussion. Other constituents of surgical metallic devices and their related systemic effects will also be mentioned. The iron system affords a convenient model to evaluate the toxic potential of iron corrosion products. When sufficient information on other implant constituents, e.g., Co, Cr. Ni, Mo, etc., become known these systems should be evaluated in a similar fashion.

In general, experiments to reveal the toxic effects of trace element overdoses necessitate cautious interpretation. Often such investigations are of short-term duration employing rapid per os, i.v., or s.c. administration of metal ions or complexes. They fail to approximate the unique situation created by implanted metals and their attendant slow and continuous release of metal ions into the system. Similarly, results of industrial exposure surveys, although often of a more long-term nature, typically relate to changes in the exposed skin or respiratory epithelium, subsequent to chronic contact or inhalation of noxious gases or particulate material. Here again the discrepancies in routes of administration and means of dissemination are obvious.[59]

1. Immunologic Considerations
a. Iron and Susceptibility to Infectious Disease

For three decades it has been recognized that a host's response to bacterial infection is a reduction in the iron content of the blood.[14] The mechanism of this reduction has been identified as a suppression of intestinal assimilation of iron together with an increased storage of iron in the liver. The net effect is to make growth-essential iron less available to microbial invaders, thus manifesting a so-called nutritional immunity for the host. To illustrate the strength of this phenomenon, patients enduring an infection are unable to mobilize iron from reticuloendothelial cell storage sites, often to the extent that transient iron deficient-erythropoiesis may develop despite normal body iron content.[15] This suggests that the physiological system has evolved to accept a short period of iron deficiency rather than risk a microbial invasion. Conversely, it has been demonstrated clinically that patients with iron deficiency anemia have a lowered susceptibility to infections when compared to patients with other types of anemia.[16] Weinberg[14] has summarized the role of iron in nutritional immunity: "a very consistent finding is that the intricate checks and balances between the iron chelators of microbes and of hosts are readily and markedly upset by changes in the environmental concentration of iron. If the metal is added, microbial growth is enhanced, if the metal is deleted, host defense is strengthened. This situation applies not only in experimental systems in vitro and in vivo but also in clinical disease situations."

As an example, Kochan[17] has demonstrated the microbiostatic action of various mammalian sera in culture media with respect to bacterial growth of tubercle bacilli (Table 1). Normal human serum containing transferrin (iron binding protein), saturated 30% with iron, inhibited bacterial multiplication of the BCG strain of *M. tuberculosis*. The addition, however, of 36μM of iron saturated the available transferrin and neutralized its bacteriostatic activity. Similar results were recorded for bovine, mouse, and rabbit sera. Guinea pig serum, however, owing to its normally high transferrin saturation prior to addition of iron, was equally susceptible to tubercle bacillus proliferation before and after iron addition. Other microorganisms demonstrating similar behavior include species of *Candida, Clostridium, Escherichia, Pasteurella, Shi-*

Table 1
CORRELATION OF LEVEL OF IRON SATURATION OF
TRANSFERRIN (TR) WITH BACTERIAL GROWTH IN
MAMMALIAN SERUMS IN VITRO (BACTERIAL GROWTH
IS EXPRESSED AS THE NUMBER OF GENERATIONS OF
TUBERCLE BACILLI IN 14 DAYS)

Source of serums	Number of samples	Fe concentration in serum (μM)	Saturation of Tr (%)	Bacterial growth	
				No added Fe	36 μM added Fe
Human	10	17	30.0	0—1	10—14
Bovine	4	34	39.0	0—1	10—14
Mouse	10	41	60.2	1—5	9—15
Rabbit	8	35	64.3	1—5	10—15
Guinea pig	20	49	84.4	9—14	9—14

From Weinberg, D. E., *Science*, 184, 952, 1974. Copyright 1974 by the American Association for the Advancement of Science. With permission.

gella, and *Staphylococcus*. In vivo studies using strains of *Pseudomonas aeruginosa*, *S. typhimurium*, *Listeria monocytogenes*, and *E. coli* have corroborated the in vitro results.[14]

Smith[18] has exposed rabbits to various surface areas of stainless steel 316L implants. Implanted surface area ranged from 2.9 cm²/kg body weight (surface area/body weight ratio equivalent to a 70 kg man having a standard total hip prosthesis) to 100 times this amount (290 cm²/kg). This was accomplished using very small (55 μm) 316L stainless steel microspheres. Results indicated that rabbits having implanted 316L stainless steel, irrespective of surface area, demonstrated a significantly higher incidence of obvious clinical infection ($P \ll 0.01$) than control populations, either unoperated or sham operated. Although apparently supporting the theory of nutritional immunity, the evidence, of course, is only circumstantial. Any or all of the constituents of 316L stainless steel could potentially act to suppress immunity, either by direct effects on lymphocyte or neutrophil function or by altering the reticuloendothelial system as a whole. Nevertheless, it can be concluded that the presence of a 316L stainless steel implant contributes in some way to the diminution in infectious disease resistance in rabbits. Whether the net effect is to promote greater numbers of infections within a population or to simply increase the virulence and, therefore, pathogenicity of these infections which randomly occur, could not be established in the study. Whether a similar phenomenon can be observed in populations of human implant recipients has not been examined.

b. Metal Hypersensitivity

Numerous investigators[19-21] in evaluating local implant acceptance in the musculoskeletal system, have long alluded to the possibility that metal ions or corrosion products in combination with complex host chemistry could become antigenic and thus precipitate some form of immunological response. The actual discovery of the clinical incidence of implant hypersensitivity, however did not occur until the mid-1960s, demonstrating that "incidence is often proportional to interest". Since 1965 a plethora of anecdotal clinical cases and retrospective surveys have been published which verify the existence of implant-mediated hypersensitivity reactions.

In general, metal implant hypersensitivity is observed clinically in two forms: (1) skin hypersensitivity manifesting as a generalized urticaria or eczematous-type skin reaction[22-24,26] and (2) local implant hypersensitivity resulting in implant loosening, pain, and failure.[25,27-30] The first type of reaction is undoubtedly a product of the implanted metallic device, in that patients exhibit positive patch or scratch tests to one or more of the implant's constituents and removal of the offending implant, in general, results in rapid remission of signs (24 to 72 hr).

The second type (implant loosening), however, has uncertain etiology. Reactions more commonly occur in patients having metal-on-metal (cobalt chromium) prostheses which, of course, generate large amounts of metallic wear products in situ. Whether the loosening then is caused by an actual hypersensitivity reaction or by local toxicity from wear and/or corrosion products elliciting bone necrosis is unclear. A few cases of failed metal-to-plastic prostheses associated with hypersensitivity reactions have been reported, suggesting that perhaps hypersensitivity alone is the mechanism of failure.[25,29,30]

Summarizing the literature concerning implant loosening and hypersensitivity reactions, clinical evidence so far indicates that: (1) patients having stable total joint devices demonstrate no sensitivity on random patch or scratch testing to constituents of the metallic device, (2) of the patients with unstable prosthetic devices, a percentage (ranging from 13%[30] to 74%[29]) are found to be sensitive to either nickel, chromium, or cobalt, (3) the presence of a metallic device can sensitize some individuals ($\sim 6\%$)[30] to constituent elements, although it is unclear whether this represents a new sensitivity or merely an anamnestic response from a pre-existing undetected hypersensitivity, and (4) a 6% incidence of hypersensitivity to cobalt, chromium, or nickel in the random population would warrant the routine patch and scratch testing of all arthroplasty candidates.

2. Carcinogenic Effects

The potential for metallic implants to incite tumor formation has long been a topic of concern relative to implant biocompatibility.[59] Implant tumorigenesis, even of low morbidity, would represent an unacceptable risk, particularly for the younger implant candidate. It can be argued that tumors are local phenomena and not logically included within a chapter on systemic biocompatibility. It should be recognized, however, that the same corrosion-related tumorigenic conditions at the implant site could hypothetically be duplicated at a remote storage site, such as the liver, where corrosion products eventually accumulate. To date, no retrospective studies correlating the presence of a metallic implant with the incidence of tumor formation regardless of site, have appeared in the literature. An abundance of experimental investigations, however, in laboratory animals and several industrial exposure surveys have demonstrated, although often under circumstances unrelated to implantation, the carcinogenic potential of many of the consitituents of surgical alloys. These investigations and the clinical incidence of implant tumorigenesis will be briefly discussed in the section to follow.

a. Industrial Exposure and Experimental Investigations

Heuper,[31-33] by parenteral administration of nickel in rats, rabbits, and guinea pigs, has implicated this metal as a carcinogen producing sarcomas of bone, connective tissue, nerve tissue, and muscle in rats and rabbits and benign and malignant lung tumors in guinea pigs and rats. Heath and Daniel[34-36] have demonstrated the carcinogenicity of pure metallic cobalt, cadmium, and nickel when suspended in serum and injected into rat skeletal muscle. Malignant metastasizing tumors including fibrosarcomas, rhabdomyosarcomas, and cellular sarcomas developed rapidly (3 months or longer) with an incidence of 50 to 75%. Doll[37] has indicated the carcinogenic hazards to humans directly involved in nickel industries. The role of Ni and Co in carcinogenesis is central to the work of Greene[38] who has shown that infusion of Ni in animal models results in the prediction of considerably elevated blood ion levels in patients with stainless steel implants due to low clearance rates. The predicted ion levels are similar to those observed in workers using nickel carbonyl compounds who experience a statistically significant increase in soft tissue tumors. Bech et al.[39] suspect cobalt as a possible industrial carcinogen. Metallic chromium has so far exhibited no carcinogenic tendencies for either rats or humans, although chromates have been demonstrated carcinogenic for both rats[33] and humans.[40]

Oppenheimer et al.[41] using Vitallium® and stainless steel foils implanted subcutaneously have induced fibrosarcomas in rats. More recently, Heath, et al.[42] have shown that wear particles from a cobalt-chromium alloy total joint replacement prosthesis, when suspended in serum and injected into the thigh muscle of female rats, have elicited tumor formation as early as 4½ months postinjection with an incidence of 19% at the time of publication. They have incriminated cobalt more so than chromium as the primary carcinogenic agent. Recently Gaechter, et al.[43] using Sprague-Dawley rats investigated the carcinogenicity of seven surgical metal alloys implanted in rod form intramuscularly. Their results after 2 years revealed no significant differences in tumor incidence between rats having metallic implants and those having none. The rats, however, on a metabolic basis, were extremely underdosed relative to the surface area of implanted metal.

b. Clinical Carcinogenicity

Human and veterinary literature contain a few scattered reports of tumor formation adjacent to sites of metallic implantation. Although often direct causality cannot be established, recent reports from the veterinary literature have stimulated renewed interest in the carcinogenic potential of currently used implant alloys.

The first documented human case of a tumor in the proximity of a metallic implant was reported by McDougall[44] in 1956. A malignant neoplasm of uncertain type, closely resembling a Ewing-type sarcoma, had arisen at the site of a stainless steel plate and screws implanted 30 years previously to stabilize a fractured humerus. Plate and screws were of differing composition resulting in extensive galvanic corrosion and discoloration of surrounding tissue. Precise etiology, whether from 30 years of chronic inflammation, or electrolytic phenomena, or a singular metallic constituent could not be determined. Dube and Fisher[45] in 1972 observed a similar phenomenon in an 84-year-old man treated in 1944 with a stainless steel 316 plate for a tibial nonunion. After a 26-year latency period, a hemangioendothelioma developed adjacent to plate and screws which later were demonstrated to have differing compositions. Again, no precise etiology could be identified. The latency period of 26 years is consistent with the findings of Siddon and MacArthur[46] who reported on seven cases of pulmonary malignancy adjacent to metallic foreign bodies averaging 24 years *in situ*. One final human case report as documented by Delgado,[47] was the formation of an osteosarcoma adjacent to a bone plate on the tibia 3 years postimplantation. Case presentation was unclear as to the types of metal involved.

The veterinary literature contains comparatively more cases of implant-related tumor incidence despite the relatively infrequent application of metallic devices in animal clinics. Brodey et al.[48] in 1959 were the first to document a tumor at the site of a metallic implant in an animal: an osteosarcoma adjacent to a Jonas splint, a multicomponent intramedullary fracture fixation device. Banks et al.[49] in 1975 described two cases of osteogenic sarcoma in dogs; one arising after 2½ years in the midshaft radius adjacent to a bone plate (of undetermined composition) and the other in the tibia associated with a Jonas intramedullary pin, 6 years *in situ*.

Records at the Animal Medical Center in New York during the 6-year period from 1968 to 1974 revealed eight cases of malignant tumors associated with metallic devices: seven dogs and one cat.[50] Latency periods varied from 6 months to 6 years, while tumor types and incidences were five osteosarcomas, one fibrosarcoma, and two undifferentiated sarcomas. Again, the Jonas splint was implicated in five of the tumor cases while two cases had intramedullary Steinmann pins and one, an AO plate and screws. Six tumors arose in the midshaft femur, one in the humerus, and one in the radius. The atypical location of the observed tumor types suggested a direct causality between metallic implant and tumor formation in these instances.

Harrison et al.[51] in 1976 presented two cases of osteosarcoma associated with stainless steel 316L implants in dogs. One case in a 12-year-old Doberman Pinscher involved the distal humerus 11 years after implantation of a Steinmann intramedullary pin for fracture fixation. The other was a 12-year-old Irish Wolfhound who 6 years previously received a plate and screws, type stainless steel 316L, for a fractured tibia.

At the University of Pennsylvania Veterinary Clinic, four implant-related tumor cases in dogs have been recorded.[52] Three osteosarcomas developed at the site of bone plating with stainless steel 316L, one Richards and two AO plates. In the fourth case a fibrosarcoma occurred subsequent to intramedullary pinning of the tibia with stainless steel 316L. The latency period for tumor formation varied from 3 to 8 years. Interestingly, three of the four dogs developed neoplasms despite surgical removal of the bulk of the implant several years previously, although some screws or small cerclage wires were left in place by virtue of uncertain or inaccessible location.

The possibility that tumor formation in these cases represents random occurrence is highly unlikely. Osteosarcoma as a clinical entity occurs infrequently in the random population, and when observed, has definite breed and site predilection. The fact that several of the osteosarcomas described have appeared in atypical breeds and at atypical sites, suggests a causal relationship between metal implant and neoplasia.

It is interesting to compare latency periods for implant-related tumor formation in the dog vs. that in man. As mentioned two of the three implant-associated tumors in man were found to occur after long latency periods, 26 to 30 years. Similarly, Siddon[46] observed that lung neoplasms, secondary to metallic foreign bodies, had a mean latency of 24 years. Dogs, on the other hand, demonstrated tumors developing 6 months to 11 years postimplantation with a mean of 4.6 years. If one accepts the adage that dogs age at a rate approximately seven times that of man, a prediction of the mean latency period in humans would calculate to be 32 years; a figure in reasonably good agreement with the few human cases on record.

Why then are not more cases of tumors in the vicinity of an implant appearing in the human medical literature? Probably the most plausible explanation is that only a relatively small population of individuals presently have surgical devices implanted for greater than 25 years. This group is certain to grow corresponding to the trend of implanting total joint prosthesis in younger patient populations. In addition, a contributing factor may be the current medical-legal atmosphere which could deter the publication of this type of information. The final possibility is of course, that no additional incidence of implant-associated tumors has occurred in man since 1972. Judging from the three documented cases and the increasing frequency of incidence in the veterinary literature, this is possible but unlikely.

3. Metabolic Effects

Iron in nutrition, encompassing both deficiency and excess states, has been investigated more thoroughly than any other trace element including other constituents of metallic implants. Its relatively high concentration in biological tissues (ppm range)

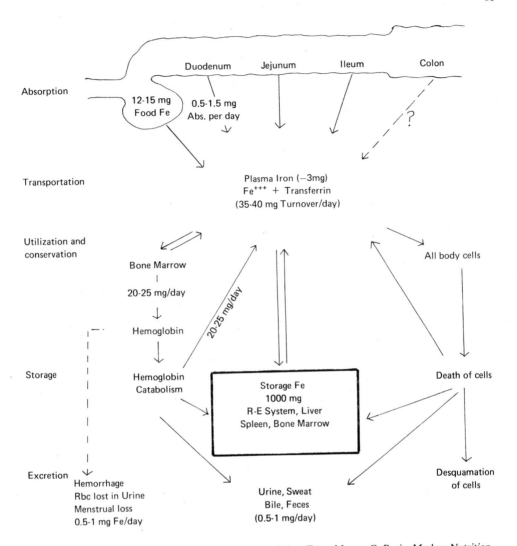

FIGURE 2. Schematic outline of iron metabolism in adults. (From Morre, C. B., in *Modern Nutrition in Health and Disease*, 5th ed., Goodhart, R. and Shils, M., Eds., Lea & Febiger, Philadelphia, 1973, 325. With permission.)

has long afforded reproducible analysis. Therefore, of all implant constituents, iron provides perhaps the best model to study the effects of implant corrosion on known metabolic pathways. Accordingly, iron metabolism in general will be briefly reviewed.

a. *Iron Metabolism*[9,15,53,54] *(see Figure 2)*

Iron, a biologically ubiquitous metal, is essential to all higher forms of life owing to its central role in the heme molecule facilitating oxygen and electron transport. The capacity of porphyrin-Fe-protein complexes to reversibly bind large quantities of oxygen makes hemoglobin and myoglobin well suited to the transport and storage of oxygen in higher organisms. Iron-containing enzymes include the cytochromes, catalase, cytochrome c reductase, succinic dehydrogenase, and fumaric dehydrogenase. Total body iron in an average adult ranges from 2 to 6 g depending on the weight, hemoglobin concentration, age, sex, and size of the storage compartment.

On the basis of function, two iron compartments are recognized: an essential compartment, containing 70% of the total body iron, is composed of hemoglobin, myoglobin, heme enzymes, cofactor and transport iron; a nonessential storage compartment accounts for 30% of the total body iron in a normal individual and consists of iron storage in the form of ferritin and hemosiderin, primarily in the liver, spleen, and bone marrow. The essential compartment can be further subdivided into the following distributions: 85% hemoglobin, 5% myoglobin, 10% in heme enzymes and iron cofactors in other enzyme systems, and 0.012% as transport iron bound to transferrin.

1. Absorption

Absorption of iron represents the single most important factor maintaining the normal balance of iron in the body. It is influenced by age, state of health, current iron status, conditions within the gastrointestinal tract, the amount and chemical form of iron ingested, and the relative iron chelating nature of other dietary constituents, including phosphates, phytates, ascorbic acid, and amino acids. Normally, only 5 to 15% of the total ingested iron is absorbed by the gastric mucosa. The actual mechanisms of absorption and transport of iron or iron chelates at the mucosal cell level are to date unclear. However, it appears that iron enters the mucosal brush border by a passive diffusion process and exists on the serosal surface, binding to the plasma transferrin occurring by an energy-requiring step. Intracellular absorbed iron in excess of immediate physiological needs is combined with a protein, apoferritin, to form ferritin, a water-soluble iron storage complex. As the mucosal cells become laden with ferritin, further absorption is impeded (consistent with diffusion kinetics) until ferritin iron is released to the plasma in response to body needs. Most absorbed iron in the form of intracellular ferritin is lost into the intestinal lumen when the crypt cells complete their 2- to 3-day maturation and migration to the tips of the villi and are sloughed. Intraluminal factors which decrease iron absorption include rapid transit time, achylia, malabsorption syndrome, precipitation by alkalinization, phosphates, phytates, and ingested alkaline clays and antacid preparations. Despite intensive research, the systemic factors regulating iron absorption have not been identified. In general, iron absorption increases whenever erythropoiesis is stimulated, during pregnancy, and in patients with hemochromatosis, while decreased absorption is associated with depressed erythropoiesis and with iron overload.

2. Transport

After an iron atom enters the physiological system it is virtually trapped, cycling almost endlessly from plasma to developing rubriblasts, into the circulation for 100 to 160 days then to the phagocytic R.E. cells where the iron is cleaved from hemoglobin, and finally released into the plasma to repeat the cycle. From the standpoint of distribution of total body iron, the transport compartment comprises the smallest percentage (0.008%). However, kinetically, it is by far the most active, turning over as many as ten times every 24 hr. The vehicle of this rapid transport and turnover of iron is transferrin, a $beta_1$-globulin of approximately 86,000 mol. wt and half-life of 8 to 10.5 days. Synthesized in the liver, the total body transferrin of 7 to 15 g is equally distributed in the intra- and extravascular space. It functions both to accept iron from gut absorption, storage sites, and R.E. cells and to deliver iron to erythroid marrow for hemoglobin systhesis, to R.E. cells for storage, to the developing fetus, and to all cells for incorporation into iron metalloenzymes. Normally, approximately one third of the total body transferrin (termed total iron-binding capacity, TIBC = 300 − 360 μg/100 ml) is saturated with iron; the remainder represents a latent or unbound reserve (unbound iron-binding capacity, UIBC).

The degree of saturation (%) and the total iron-binding capacity are important parameters in the study of iron metabolism and related disease syndromes. For example, increased TIBC is characteristically found in iron deficiency, the third trimester of

Table 2
DAILY IRON EXCRETION

	External iron loss (mg)
Gastrointestinal	
Blood	0.35
Mucosal	0.10
Biliary	0.20
Urinary	0.80
Skin (hair, sweat)	0.20

pregnancy, and in response to hypoxic states, while decreased TIBC is evident in infection, protein malnutrition, iron overload conditions, malignancy, cirrhosis of the liver, nephrosis, and protein-losing exteropathies. Figure 3 illustrates the relationships between plasma iron and transferrin in disease. In general, the level of plasma iron and saturation of TIBC are determined by the sum of factors extracting iron from the blood for storage and utilization balanced against those factors releasing iron into the blood, e.g., absorption, hemolysis, and storage site release.

3. Storage

Iron in excess of metabolic need is stored intracellularly as either ferritin or hemosiderin in various tissues of the body. Ferritin is normally found in many tissues of the body, although the liver, R.E. system, and intestinal mucosa are the storage sites metabolically most significant.

Hemosiderin, a granular, water-soluble compound is thought to be composed of an aggregation of ferritin molecules. It can be seen microscopically in unstained tissue sections of bone marrow films as clumps or granules of golden refractile pigment. Primary storage sites are the hepatic parenchymal cells and R.E. cells of the bone marrow, liver, and spleen. Although the relationship between ferritin and hemosiderin regarding iron distribution is unclear, it is postulated that the relative content of iron in either storage form is a function of the total storage iron concentration. Both ferritin and hemosiderin, by chemically binding or shielding iron from the surrounding intracellular environment, serve to reduce its inherent toxicity. Iron in its ionic form, in excess of the total iron binding capacity of the blood, is extremely and instantaneously toxic as demonstrated by Gitlow and Beyers.[55] Slow i.v. injection of only 10 mg of ferric ammonium citrate totally saturated the blood's iron-binding capacity of all patients tested, the excess randomly diffusing into all body tissues and manifesting toxicity in the forms of coughing, sneezing, nausea, and occasionally vomiting.

4. Excretion

Owing to a very limited capacity to excrete iron, the body tenaciously conserves its endogenous content of iron, excreting less than 1/1,000 of the total amount daily (see Figure 2.) The distribution of external iron loss is shown in Table 2.

b. Iron Overload

Considering the precisely regulated intestinal absorption of iron, coupled with such limited physiological excretory capabilities, one can easily envisage the development of an iron overload state should iron, by whatever route, pathological or iatrogenic, gain access to the endogenous system. For example, unregulated absorption of iron occurs in idiopathic hemochromatosis, a situation of excessive intestinal absorption in which iron is continually deposited in the parenchymal cells of various organs, often resulting in liver disease, cardiac failure, and diabetes (Figure 3). Excessive intake of

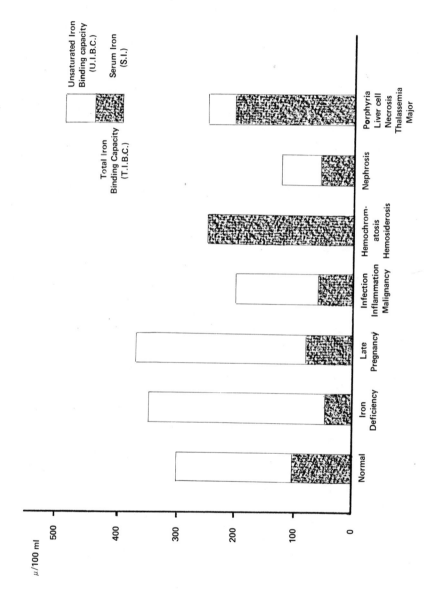

FIGURE 3. Relationships of serum iron, unsaturated iron-binding capacity, and total iron-binding capacity in various clinical conditions. (From Moore, C. V., in *Modern Nutritio in Health and Disease*, 5th ed., Goodhart, R. and Shils, M., Eds. Lea & Febiger, Philadelphia, 1973, 334. With permission.)

iron is demonstrated by the Bantu of Africa, who by virtue of their consumption of large quantities of food cooked and stored in iron pots, develop hepatic and reticuloendothelial involvement manifesting as portal cirrhosis. Additionally, a few reported instances of parenteral iron administration to treat misdiagnosed anemias have resulted in iatrogenic iron overload states with consequent toxic signs indistinguishable from hemochromatosis. Evidence therefore indicates that excessive amounts of intracellular iron stemming from an iron overload condition may cause progressive destruction of parenchymal cells and subsequent fibrotic replacement. The positive relationship of cardiomyopathies and diabetes mellitus in iron overload to the deposition of hemosiderin in the myocardium and the pancreas, further substantiate the toxicity of iron even in its bound storage form. In an iron overload state, the serum iron and transferrin saturation are usually increased, while the total iron-binding capacity is somewhat depressed (See Figure 3).

c. Transport and Distribution of Corrosion Products from 316L Stainless Steel

Implants made of 316L stainless steel afford yet another means of unregulated iron entry into the biological system as they contain about 60% iron. Smith[18] implanted high surface areas of stainless steel in rabbits (1 to 100 times surface area per body weight of a total hip prosthesis in man) and demonstrated significantly elevated plasma iron concentrations from 20 weeks postimplantation onward. A parallel pattern of diminished TIBC and elevated percent transferrin saturation was also observed, indicating that the situation of iron overload as shown in Figure 3 was being approached, albeit minimally. Not surprisingly, the rabbits also exhibited an 85% elevation in mean liver content of iron. Although the implant period was only 28 weeks, it was assumed that the corrosion process and iron accumulation in the liver would have continued for the duration of implantation. The long term effects of this situation are unknown, but, on a statistically significant basis, rabbits with implanted 316L stainless steel were found to be more susceptible to infectious disease, as discussed in a previous section.

In the same study plasma chromium concentrations (composition 316L stainless steel ≈ 18% Cr) were found to be elevated as early as 4 weeks postimplantation, the magnitude of which appeared to reflect the implanted surface area of 316L stainless steel. The liver also demonstrated accumulations of chromium in relation to implanted 316L stainless steel surface area. Interestingly, the kidney exhibited no elevations of either iron or chromium, consistent with its lack of reticuloendothelial capacity. As discussed in the iron metabolism section, the confinement of corrosion products to R.E. depots alone would indicate that the specific iron-binding capacity of the plasma is not exceeded. Otherwise, uniform distribution of corrosion products into all parenchymatous organs nonspecifically would have been observed. This finding confirms that an excessive overload state as in hemochromatoses is not manifesting.

d. Metal-Induced Systemic Sequellae

At sacrifice, (7 months postimplantation) blood samples were drawn and submitted for biochemical profile, measuring 20 blood or serum parameters. A number of interesting changes were revealed. Moderately, but significantly elevated SGOT values in four of the five experimental rabbit populations suggested an ongoing process of cellular damage manifesting in increased cell membrane permeability and/or frank cellular necrosis. Similar changes were noted in SAP and total bilirubin. The triad of elevated SGOT, SAP, and total bilirubin would implicate the liver as a primary site of sytemic toxicity. However, a causal association between liver necrois and iron or chromium accumulations could not be established. Such low-level biochemical changes (approximately twofold) if occurring in human implant recipients would likely be obscured by the large biological variation inherent in human populations. The long-term effects of these chronically altered parameters are unknown.

Smith[18] also observed a slight but significant depression in mean corpuscular hemoglobin concentration, MCHC, in a majority of the experimental rabbit populations. Reduced MCHC is a sign of hypochromic anemia irrespective of etiology and means that the ratio of hemoglobin on a weight basis to the volume in which it is contained is reduced. Although admittedly, many causes are possible, this finding hypothetically could reflect corrosion product interaction working to the detriment of the implant recipient. Specifically, Tan and Woodworth[56] have investigated the relative stabilities of transferrin when bound to other metals. The chromium-transferrin complex was found to be second only in stability to the iron-transferrin complex. In fact, it is thought that chromium binds with transferrin irreversibly. This is in contrast to iron which, although as yet unproven, is believed to utilize ATP to promote iron reduction, $Fe^{3+} \rightarrow Fe^{2+}$, and thus facilitate release from transferrin into red cell precursors.[53] Chromium's reported inability to dissociate probably stems from its high oxidation potential between Cr^{2+} and Cr^{3+}, making a reversible transition, as with iron, energetically unfavorable. Therefore, although chromium behaves like iron in its potential to bind to transferrin, it in theory may form a defective complex which acts as a competitive inhibitor of iron utilization at the red cell membrane. Figure 4 schematically illustrates this point.

Changes in serum chemistries and blood parameters were found to be independent of implanted stainless steel surface area and appear to be a previously unappreciated result of the implantation of stainless steel. Minimally, the results should raise concern regarding the systemic toxic potential of 316L stainless steel corrosion products.

Because the two largest constituents of 316L stainless steel are iron and chromium, the investigation was systematically directed toward the elaboration of these two elements relative to their known physiology and toxicology. Nickel, however, is also a primary constituent of stainless steel ($\approx 12\%$) and known under certain circumstances to be highly toxic. It should be emphasized therefore that the effects observed in this investigation may be attributable to any one or a combination of the constituents of 316L stainless steel. It was concluded that the significance of these findings to human populations is as yet indeterminable. There is clear need to extend these studies to longer term animal models and to investigate thoroughly the pathophysiology of the corrosion-mediated disease process.

IV. CONCLUSIONS

This chapter has attempted to outline, either specifically or by example, the systemic biocompatibility of metallic implants. Real or potential systemic effects, carcinogenic, immunologic, or metabolic were reviewed based on pertinent information presently available. Accordingly, not all constituents of surgical alloys received equal coverage. Indeed, a comprehensive review was neither intended nor possible within the content of this chapter. To appreciate the enormity and diversity of potential implant-related systemic sequellae, based on more periphrally related material, one has only to consult one of the many surveys of trace element research.[9,53,57,58] It quickly becomes obvious that continued, well-planned research specifically addressing actual implant conditions is necessary to satisfy the concern regarding systemic biocompatibility and long term patient safety. Before concluding this chapter, it is perhaps appropriate to place this chapter and indeed the topic of system implant biocompatibility into proper perspective.

It has been expressed that the singular motivating factor for concern regarding metallic implants is the safety and well-being of the implant recipient. All too commonly, the adverse effects of a drug or environmental pollutant are realized only in retrospect.

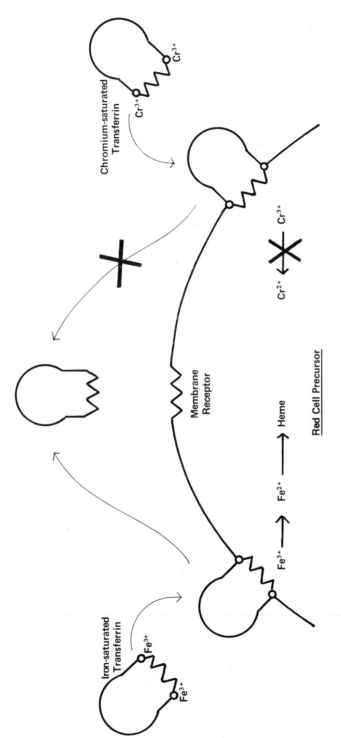

FIGURE 4. Hypothetical competetive inhibition of chromium at iron absorption sites on the red cell membrane.

Although untoward, implant-related systemic effect appear to have rare occurrence, in actual fact they have never been prospectively studied. In evaluating the risks vs benefits of applying a metallic surgical device, the surgeon is presently not cognizant of all the risks. Clearly the parameters of *Time* and implanted surface area or *Dose* must be factored into the risk/benefit formula. For example, there is general consensus that a 65-year-old man with severely debilitating degenerative joint disease of the hip would optimally benefit with a total joint arthroplasty. However, what are the clinical criteria in the case of a 20-year-old Viet Nam veteran in need of a total knee replacement? More specifically, what are the limitations of the currently used implant metals? What is the maximum safe duration of implantation and the maximum allowable dose of implantable metal?

For the present these questions remain unanswered. Obviously the overwhelming clinical need for (and benefit of) metallic surgical devices far outweigh the side effects, real or potential. This situation, however, should not foster complacency on the part of the surgeon or the device manufacturer. It is only through healthy circumspection and constructive criticism of the currently used surgical device armamentarium that sound improvements can be implemented in the future.

REFERENCES

1. **Ferguson, A. B., Laing, P. G., and Hodge, E. S.,** The ionization of metal implants in living tissue, *J. Bone Jt. Surg. Am. Vol.,* 42, 66, 1960.
2a. **Ferguson, A. B., Akahoshi, Y., Laing, P. G., and Hodge, E. S.,** Trace metal ion concentration in the liver, kidney, spleen, and lung of normal rabbits, *J. Bone Jt. Surg. Am. Vol.,* 44, 317, 1962.
2b. **Ferguson, A. B., Akahoshi, Y., Laing, P. G., and Hodge, E. S.,** Characteristics of the trace ions released from embedded metal implants in the rabbit, *J. Bone Jt. Surg. Am. Vol.,* 44, 343, 1962.
3. **Von Ludinghausen, M., Meister, P., Rossmann, E., and Guckel, W.,** Spektral Analytische und Radio Chemische Untersuchungen bei Mettalose, *Verh. Dtsch. Ges. Pathol.,* 54, 414, 1970.
4a. **Coleman, R. F., Herrington, J., and Scales, J. T.,** Concentrations of wear products in hair, blood, and urine after total hip replacement, *Br. Med. J.,* 1, 527, 1973.
4b. **Coleman, R. F.,** Concentration of implant corrosion products in the body, *J. Bone Jt. Surg. Br. Vol.,* 55, 422, 1973.
5. **Lux, F. and Zeisler, R.,** Instrumentelle multi-element-aktivierungs-analyse von Biologischem Gewebe und Ihre Anwendung zur Metallose Untersuchungen, *Z. Anal. Chem.,* 361, 314, 1972.
6. **Taylor, D. M.,** Trace metal patterns and disease, *J. Bone Jt. Surg. Br. Vol.,* 55, 422, 1973.
7. **Anon.** Timely topics in clinical chemistry Toxicology and trace metals. II, *Am. J. Med. Technol.,* 35, 652, 1969.
8. **Green, D. E.,** Enzymes and trace substances, *Adv. Enzymol.,* 1, 177, 1941.
9. **Underwood, E. J.,** *Trace Elements in Human and Animal Nutrition,* 3rd ed., Academic Press, New York, 1971.
10. **Reinhold, J. G.,** Trace elements — a selective survey, *Clin. Chem.,* 21, 476, 1975.
11. **Cotzias, G. C.,** Importance of trace substances in environmental health as exemplified by manganese, in *Proc. 1st Annu. Conf. Trace Substances Environmental Health,* Hemphill, D.D., Ed., University of Missouri Press, Columbia, 1967, 5.
12. **Liebscher, K. and Smith, H.,** Essential and non-essential trace elements. A method of determining whether an element is essential or non-essential in human tissue, *Arch. Environ. Health,* 17, 881, 1968.
13. **Harper, H. A.,** *Review of Physiological Chemistry,* 14th ed., Lange, Los Altos, Calif., 1973, 145.
14. **Weinberg, E. D.,** Iron and susceptibility to infectious disease, *Science,* 184, 952, 1974.
15. **Moore, C. U.,** in *Modern Nutrition in Health and Disease,* Goodhart, R. S. and Shils, M. E., Eds., Lea & Febiger, Philadelphia, 1973, 372.
16. **Masane, A. E. J., Muindi, J. M., and Saai, G. B. R.,** Infections in iron deficiency and other types of anemia in the tropics, *Lancet,* 2, 314, 1974.

17. **Kochan, I.**, The role of iron in bacterial infections with special consideration of host-tubercle bacillus interaction, *Curr. Top. Microbiol. Immunol.*, 60, 1, 1973.
18. **Smith, G. K.**, Transport and Distribution of Corrosion Products from SS316L Implants and Some Related Systemic Effects, Ph.D thesis, University of Pennsylvania, Philadelphia, 1979.
19. **Brettle, J.**, A Survey of the literature on metallic surgical implants, *Injury*, 2, 26, 1970.
20. **Dobson, J. L., Mathews, R. S., and Stelling, F. H.**, Implant acceptance in the musculo-skeletal system, *Clin. Orthop. Relat. Res.*, 72, 233, 1970.
21. **Samitz, M. H. and Katz, S. A.**, Nickel dermatitis hazards from prosthesis, *Br. J. Dermatol.*, 92, 287, 1975.
22. **Fouscreau, J. and Laugier, P.**, Allergic eczemas from metallic foreign bodies, *Trans. St. John's Hosp. Dermatol. Soc.*, 52, 220, 1966.
23. **McKenzie, A. W., Aitken, C. V., and Ridsdill-Smith, R.**, Urticaria after insertion of a Smith-Petersen nail, *Br. Med. J.*, 4, 36, 1967.
24. **Barranco, V. P. and Solomon, H.**, Eczematous dermatitis from nickel, *JAMA*, 220, 1244, 1972.
25. **Evans, E. M., Freeman, M. A. R., Miller, A. J., and Vernon-Roberts, B.**, Metal sensitivity as a cause of bone necrosis and loosening of the prosthesis in total joint replacement, *J. Bone Jt. Surg., Br. Vol.*, 56, 626, 1974.
26. **Symeonides, P. P., Paschalaglow, C., and Papageorgiou, S.**, An allergic reaction after internal fixation of a fracture using a vitallium plate, *Allergy Clin. Immunol.*, 51, 251, 1973.
27. **Benson, M. K. D., Goodwin, P. G., and Brostoff, J.**, Metal sensitivity in patients with joint replacement arthoplasties, *Br. Med. J.*, 4, 374, 1975.
28. **Munro-Ashman, D. and Miller, A. J.**, Rejection of metal to metal prosthesis and skin sensitivity to cobalt, *Contact Dermatatol.*, 2, 65, 1976.
29. **Elves, M. W., Wilson, J. N., Scales, T. T., and Kemp, H. B. S.**, Incidence of metal sensitivity in patients with total joint replacements, *Br. Med. J.*, 4, 376, 1975.
30. **Deutman, R., Mulder, T. J., Brian, R., and Nater, J. P.**, Metal sensitivity before and after total hip arthroplasty, *J. Bone Jt. Surg. Am. Vol.*, 59, 862, 1977.
31. **Hueper, W. C.**, Experimental studies in metal carcinogenesis; nickel cancer in rats, *Tex. Rep. Biol. Med.*, 10, 167, 1952.
32. **Hueper, W. C.**, Experimental studies in metal carcinogenesis; cancer produced by parenterally introduced metallic nickel, *J. Natl. Cancer Inst.*, 16, 55, 1955.
33. **Hueper, W. C.**, Experimental studies in metal carcinogenesis. IX. Pulmonary lesions in Guinea pigs and rats exposed to prolonge inhalation of powedered metallic nickel, *Arch. Pathol. (Chicago)*, 65, 600, 1958.
34. **Heath, J. C.**, The production of malignant tumors by cobalt in the rat, *Br. J. Cancer*, 10, 668, 1956.
35. **Heath, J. C. and Daniel, M. R.**, The production of malignant tumors by cadmium in the rat, *Br. J. Cancer*, 18, 124, 1964.
36. **Heath, J. C. and Daniel, M. R.**, The production of malignant tumors by nickel in the rat, *Br. J. Cancer*, 18, 261, 1964.
37. **Doll, R.**, Cancer of the lung and nose in nickel workers, *Br. J. Ind. Med.*, 15, 217, 1958.
38. **Greene, N. D.**, Biomaterials: improved implant alloys, paper presented at Brighton Implant Meeting, Brighton, Utah, 1972.
39. **Bech, A. D., Kipling, M. D., and Heather, J. G.**, Hard metal disease, *Br. J. Ind. Med.*, 19, 239, 1962.
40. **Grogan, C. H.**, Experiment studies in metal carcinogenesis. VIII. On the etiological factor in chromate cancer, *Cancer*, 10, 625, 1957.
41. **Oppenheimer, B. S., Oppenheimer, E. T., Danishefsky, I., and Stout, A. P.**, Carcinogenic effect of metals in rodents, *Cancer Res.*, 16, 439, 1956.
42. **Heath, J. C., Freeman, M. A. R., and Swanson, S. A.**, Carcinogenic properties of wear particles from prostheses made in cobalt-chromium alloy, *Lancet*, 1, 564, 1971.
43. **Gaechter, A., Alroy, J., Anderson, G. B. J., Galante, J., Rostoker, W., and Schajowic, F.**, Metal carcinogenesis, *J. Bone Jt. Surg.*, 59, 622, 1977.
44. **McDougall, A.**, Malignant tumor at site of bone plating, *J. Bone Jt. Surg.*, 38, 709, 1956.
45. **Dube, V. E. and Fisher, D. E.**, Hemangioendothelioma of the leg following metallic fixation of the tibia, *Cancer*, 30, 1260, 1972.
46. **Siddon, A. H. M. and MacArthur, A. M.**, Carcinomata developing at the site of foreign bodies in the lung, *Br. J. Surg.*, 39, 542, 1952.
47. **Delgado, E. R.**, Sarcoma following a surgically treated fractured tibia: case report, *Clin. Orthop.*, 12, 315, 1958.
48. **Brodey, R. S., McGrath, J. T., and Reynolds, H.**, A clinical and radiological study of canine bone neoplasms. I, *J. Am. Vet. Med. Assoc.*, 134, 53, 1959.

49. **Banks, W. C., Morris, E., Herron, M. R., and Green, R. W.**, Osteogenic sarcoma associated with internal fracture fixation in two dogs, *J. Am. Vet. Med. Assoc.*, 167, 166, 1975.
50. **Sinibaldi, R., Rosen, H., Liu, S., and DeAngelis, M.**, Tumors associated with metallic implants in animals, *Clin. Orthop. Relat. Res.*, 118, 257, 1976.
51. **Harrison, J. W., McLain, D. L., Hohn, R. B., Wilson, G. P., Chalman, J. A., and MacGowan, K. N.**, Osteosarcoma associated with metallic implants, *Clin. Orthop. Relat. Res.*, 116, 253, 1976.
52. **Nunamaker, D., Newton, C., and McNeill, B.**, personal communication, 1977.
53. **Jacobs, A. and Worword, M.**, *Iron in Biochemistry and Medicine*, Academic Press, New York, 1976.
54. **Moore, C. V.**, in *Modern Nutrition in Health and Disease*, 6th ed., Goodhart, R. and Shils, M., Eds., Lea & Febiger, Philadelphia, 1980.
55. **Gitlow, L. and Beyers, J.**, Metabolism of iron; intravenous iron tolerance tests in normal subjects and patients with hemochromatosis, *J. Clin. Lab. Med.*, 39, 337, 1952.
56. **Tan, A. T. and Woodworth, R. C.**, Ultraviolet difference spectralstudies of conalbumin complexes with transition metal ions, *Biochemistry*, 8, 3711, 1969.
57. Chromium, in Medical and Biological Effects of Environmental Pollutants, Division of Medical Sciences, National Research Council, National Academy of Sciences, Washington, D.C., 1974.
58. Nickel, in Medical and Biological Effects of Environmental Pollutants, Division of Medical Sciences, National Research Council, National Academy of Sciences, Washington, D.C., 1975.
59. **Williams, D. F., Ed.**, *Fundamental Aspects of Biocompatibility*, CRC Press, Boca Raton, Fla., in press.

Chapter 2

METAL ION TOXICOLOGY IN HEMODIALYSIS

Michael K. Ward

TABLE OF CONTENTS

I.	Introduction	24
II.	Hemodialysis	24
III.	Sodium, Potassium, Calcium, and Magnesium	25
	A. Sodium and Potassium	25
	B. Calcium and Magnesium	26
IV.	Aluminum	26
	A. Dialysis Encephalopathy	26
	B. Osteomalacic Dialysis Osteodystrophy	27
V.	Other Metals	28
	A. Copper	28
	B. Zinc	28
	C. Cobalt	29
	D. Lead	29
	E. Cadmium	29
	F. Nickel	29
VI.	Fluoride	29
VII.	Other Considerations	30
VIII.	Removal of Metal Ions From Dialysis Fluid	30
	References	30

I. INTRODUCTION

The introduction of hemodialysis as a method of treatment for patients with end stage renal failure has been one of the most successful applications of biomedical engineering in the last 20 years. There are now more than 23,000 patients under treatment in Europe (52,000 started since 1970),[1] more than 30,000 in the U.S., and a similar number in Japan. The 6-year accumulative survival is in excess of 50% for patients under treatment. This is more successful than therapy for carcinoma of the breast or lung and survival following myocardial infarction.[1]

Trace metal ingestion in the diet from cooking utensils and food continues, despite reduction in renal function, and in addition, many theraputic agents contain metal ions, including antacids.[2] No features of uremia prior to the development of end stage renal failure can be attributed to metal-ion accumulation (other than sodium, potassium, calcium, magnesium, and phosphorus) though this may change in the future. However, trace metal deficiency syndromes due to severe, prolonged dietary restriction do occur, for example, loss of sex function has been linked to zinc deficiency.[3]

When renal function deteriorates to such an extent that life is threatened, replacement of renal function by hemodialysis has led to the emergence of syndromes attributed to trace element accumulation. Acute toxicity syndromes are easier to identify than chronic toxicity where the slow accumulation of one or more metallic ions may only produce subtle pathological changes over many years.[4]

With the current integrated practice of hemodialysis followed by renal transplantation and return to hemodialysis and further renal transplantation following homograft rejection, long-term survival (more than 10 years) will become commonplace.

II. HEMODIALYSIS

The principle of hemodialysis is to place a semipermeable membrane between blood and a physiological solution known as the dialysate. This principle was described by Graham[5] in 1854 and applied to animals by Abel et al. in 1915,[6] and applied to man by Haas in 1924.[7] Uremic toxins transfer from blood across the membrane to the dialysate by diffusion and ultrafiltration. As this is a passive device, substances and elements can transfer from dialysate to blood, often against a concentration gradient, due to plasma protein binding.[8] Between 60 and 360ℓ of dialysate pass through the artificial kidney, called dialyser, at each treatment, lasting 2 to 10 hr for one to three treatment periods per week, depending on the regime adopted. Thus, many times the volume of fluid ingested by normal individuals is exposed to the blood stream, approximately 45,000ℓ of dialysate being exposed to the blood of a patient per year. Semipermable membrane used for hemodialysis are less selective than human gut, so trace elements can enter the blood and be distributed throughout the body. Dialysate is prepared from a domestic water source and a chemical concentrate by a proportionating unit or in batch production in a tank. Potential sources of contamination or production of concentrations of elements that are potentially toxic are listed in Table 1. There is still debate about the composition of dialysis fluid and the molar concentration of the four major metallic ions, sodium, potassium, calcium, and magnesium, is variable depending on the individual requirement. Manipulation of the concentrations of these four metallic ions is undertaken to achieve mass balance. For further details, see the review of the composition of dialysis fluids by Parsons and Davison.[9] In order to achieve the constant chemical composition required, water used to prepare dialysate should have a constant chemical composition of sodium, potassium, calcium, and magnesium. This should be below the physiological concentration so that adjustment

TABLE 1
POTENTIAL COURSES OF TOXIC CONCENTRATIONS OF ELEMENTS OR SUBSTANCES IN DIALYSATE

Dialysate
   Chemicals used to produce the concentrate for dialysis
   Malfunctioning proportionating unit
Domestic water
   Natural contamination of wells rivers, reservoirs (As, Ba, Cd, Cr, Cu, F, Pb, Al, Ca, K, Sn, Mn, Mg)
   Organisms — humic acids from peat and forests
   Added by man
     1. Sulfur dioxide
     2. Pesticides, algicide, weed killers
     3. Fertilizer (nitrates, nitrites)
     4. Sewage
     5. Industrial effluent (Cd, Hg)
Acquired in distribution
   Pipes — Sn, Pb, Al, Cu
   Tanks — Zn, Al, Fe
   Water softeners/deioniser Metal contamination from casing or pipes
Used for water treatment
   Flocculating agents
   Chlorine or chloramines
   Fluoride
   Pyrethreins

of the concentrate can produce the prescribed concentration. In practice, the solute content of a domestic water supply fluctuates from day to day, month to month, and year to year, and is dependent on pH and prevailing rain fall. So in practice, the majority of water supplies require some form of treatment to provide a constant chemical composition of the raw water used. The alternative of measuring the chemical composition of the raw water daily and making appropriate adjustments to the concentrate, although done in the early days of hemodialysis, is now impractical. The dialysis fluid should not contain any organic or inorganic element or substance known to cause acute or chronic pathological syndromes. This paper will discuss those metal ions shown to be toxic or thought likely (but not proven) to be harmful to patients maintained on hemodialysis in the short and long-term.

III. SODIUM, POTASSIUM, CALCIUM, AND MAGNESIUM

A. Sodium and Potassium

The fluctuating levels of sodium and potassium in the raw water supply have not, in practice, been a major cause of concern except in very rare supplies in some countries where high levels of these ions in the raw water cannot be adjusted by alteration of the concentrate. The most frequent cause of acute change with sodium is malfunction of a proportionating unit leading to dialysis against a sodium concentration either greater than 150 meq or less than 130 meq/ℓ, which produces acute hyper- or hypotension, confusion, and coma associated with nausea, vomiting, and obtundation. Extreme changes of sodium concentration, being dialysed against raw water from total failure of the proportionating unit, leads to intravascular hemolysis, fits, coma, and death. This accident with modern proportionating units is rare, though was more common when batch production of dialysate in bulk tanks was the rule and fail-safe devices on the units were in the development stage. The raw water concentration of potassium is usually of no significance in the majority of water supplies, although some do have an extremely high level of potassium, approaching that of blood. Unless the chemical composition of the concentrate is adjusted or the potassium removed from the water supply, acute hyperkalemia occurs with the risk of cardiac arrhythmias, still a major cause of death in patients treated by hemodialysis.[1]

B. Calcium and Magnesium

The fluctuating water content of calcium and magnesium often exceeds the ionized serum concentration leading to acute hypercalcemia and hypermagnesemia during hemodialysis. To prevent this, water softeners have been used to remove calcium and magnesium from the raw water. However, water softener failure or an unexpected rise in the raw water calcium and magnesium concentration can occur due to the fluctuation in the water supply from changes in reservoirs, rivers, and wells. Unless the calcium and magnesium content of the raw water is monitored regularly and the appropriate alteration to the calcium and magnesium concentrate made, the narrow range in which these two ions are kept can be exceeded. The hard water syndrome due to an acute rise in serum calcium and magnesium during hemodialysis has been shown to cause nausea, vomiting, extreme weakness and lethargy, a variable response in blood pressure, a burning sensation in the skin, and in extreme cases, prostration and coma.[10] The long-term effects of prolonged hypercalcemia due to this problem are not well documented, but soft tissue deposition of calcium in muscle, blood vessels, and myocardium is likely to be detrimental to long-term survival.

IV. ALUMINUM

Prior to 1976, aluminum was thought to be nontoxic and of little significance to patients with chronic renal failure. A review by Sorensen et al.[11] showed little evidence for toxicity in man. However, in 1970 Berlyne showed raised serum aluminum content in patients with chronic renal failure taking aluminum ion exchange resins and warned that aluminum may be toxic in the long-term in this group of patients.[12] Several studies have confirmed his findings.[13,14] The main source of aluminum in patients with chronic renal failure appears to be aluminum-containing phosphate-binding agents used in the management of secondary hyperparathyroidism.[15] However, the most convincing evidence that aluminum is toxic to patients with chronic renal failure comes from studies of patients on intermittent hemodialysis. It is now thought that aluminum intoxication is responsible for the development of dialysis encephalopathy,[16] and probably is one etiological factor in the development of osteomalacic dialysis osteodystrophy[17] and possibly one factor responsible for unexplained anemia, cardiac failure, and generally nonspecific ill health in patients on hemodialysis.[18] The evidence for this is reviewed below.

A. Dialysis Encephalopathy

Dialysis encephalopathy described by Alfrey, et al.[19] has been reported in other centers in the U.S.,[20] Britain,[18,21,22] Australia,[23] Netherlands,[24] France,[25] Ireland,[26] and individual cases reported for 17 of 22 European countries.[27] The syndrome is often preceded by subtle changes in personality with some evidence of intellectual impairment in the presymptomatic phase.[28] This is followed by dysarthria or dysphasia, usually towards the end of a hemodialysis treatment, later becoming permanent and appearing in the interdialysis period. Progression of the speech disorder often leads to a total inability to speak and articulate. Later, myoclonic spasms of the facial muscles and limbs occur, often extending to involve the whole trunk.[29] Grand Mal seizures are not infrequent and may be the presenting feature in a minority. Ataxia and fatigue ability of handwriting and speech may occur. Progressive intellectual deterioration leads to global dementia and death occurs from aspiration pneumonia due to disorganized oropharyngeal coordination.[29]

The most characteristic finding is on electroencephalography where episodic bursts of spiky high voltage activity occur on a background of slow wave activity.[30] At autopsy the brain looks normal, but there is considerable glyosis and cell loss on microscopic examination.[29,71] Diazepam has proved the most effective theraputic agent in controlling the myoclonic spasms and other features of this syndrome.[31]

Alfrey, et al.[16] demonstrated a strikingly significant elevation of aluminum in the grey matter of brain tissue, and also is other tissues, which has been confirmed by groups in Britain,[32] Holland,[24] and the U.S.[33] Normal brain grey matter aluminum content is less than 4 $\mu g/\ell$.[32] Grey matter aluminum content in patients with dialysis encephalopathy has been reported to be 24.9 ± 9 $\mu g/g$,[16] 20.4 ± 16 $\mu g/g$,[32] and 12.4 ± 9.7 $\mu g/l$).[33] In addition, epidemiological evidence shows a striking association between centers with a high incidence of dialysis encephalopathy and the aluminum content of the water supply used to prepare dialysate.[21,34] There is a negative association with different methods of water treatment, and the fact that the syndrome can be prevented in centers that have previously had outbreaks by effective removal of aluminum and other solutes from the water used to prepare dialysate, supports this. In addition, the accidental contamination of the water supply of a single dialysis center in Eindhoven, Holland[24] documented patients who developed dialysis encephalopathy and osteodystrophy when patients dialysed at another center on the same water supply did not develop these syndromes. This is probably the nearest to an accidental trial of the addition of aluminum to the dialysate that is ever likely to occur in man.

Oral ingestion of aluminum-containing compounds, phosphate binders, was thought to be the source of aluminum [16] but the water supply now appears to be the major contributor. Though oral aluminum compounds can raise serum aluminum content in patients with chronic renal failure prior to hemodialysis,[13,14] there is a striking rise in serum aluminum during hemodialysis, even with a dialysate concentration as low as 50$\mu g/\ell$. Reducing the raw water aluminum content to less than 5$\mu g/\ell$ fails to lower the elevated serum levels during dialysis in the short term,[35] but prevents elevation of serum aluminum in patients not previously exposed to aluminum.[35]

Treatment for the established encephalopathic syndrome has not been successful. Plasma exchange to remove aluminum failed because of contamination of replacement protein solutions with aluminum.[36] Removal of all sources of aluminum, i.e., adequate water treatment and discontinuing aluminum-containing compounds used as phosphate binders, has occasionally been reported to be successful.[37] Renal transplantation has prevented further deterioration of intellectual function in some patients while in others the encephalopathic syndrome has progressed despite restoration of renal function.[38] The most successful treatment appears to be prevention by removal of aluminum from the water supply and careful monitoring of prescriptions for aluminum-containing phosphate binders.[17]

B. Osteomalacic Dialysis Osteodystrophy

Osteomalacia is a histological diagnosis made after microscopic examination of bone, the etiology of which is multifactorial.[39] This has probably led to some of the confusion in the current literature. For many years the distinct geographical distribution of an osteomalacic syndrome seen in patients on hemodialysis has puzzled nephrologists.[40] This syndrome has a high morbidity and not infrequent mortality. Clinical features are bone pain, myopathy, pathological fractures, and progressive severe disability with many patients using walking sticks, wheelchairs, or being confined to bed. Death ensues from collapse of the thoracic cage and respiratory failure.[17] Histologically and biochemically, the distinguishing features from vitamin D deficiency osteomalacia are wide osteoid seams with numerous osteoid lamilla associated with few inactive osteoblasts and little or no evidence of secondary hyperparathyroidism.[41] Total serum alkaline phosphatase is normal or slightly elevated and the alkaline phosphatase isoenzyme patterns show that the serum alkaline phosphatase is mainly of liver and intestinal origin, and the bone isoenzyme is greatly reduced or absent.[42] The evidence that this syndrome is also caused by aluminum is as follows:

1. The close geographical association with aluminum-containing water supplies[17,21]
2. The association with dialysis enciphalopathy[17,21]
3. Raised bone[24] and serum aluminum content[13]
4. The prevention of the syndrome when adequate water treatment is used from the initiation of hemodialysis[17]
5. The reproduction in an animal model of a similar osteomalacic syndrome following aluminum intoxication[43]

It is possible that aluminum is only a marker for some other toxic metal ion or substance, though this seems unlikely. Raised brain (white matter) manganese content also occurs in patients with dialysis encephalopathy,[44] and in one geographical survey manganese was associated with aluminum in its distribution.[21] The clinical syndrome associated with manganese toxicity is that of Parkinson's Disease.[45] The features of Parkinson's Disease are not those described with dialysis encephalopathy.[19]

V. OTHER METALS

A. Copper

The most commonly used hemodialysis membrane is a cellulose membrane produced by the cupramonium process. Sufficient residual copper was thought to be contained within these membranes to raise the serum copper in patients on hemodialysis. This was true for early days, but modern cellulose acetate membranes contain little residual copper and the small elevation of serum copper seen across hemodialysis may have explanations other than leeching from the membrane. Only small elevations of copper in tissues have been found to which no pathological significance has yet been established.[46]

Acute copper intoxication is a well recognized syndrome with fever, chills, headache, generalized pain, nausea, and fatigue recurring at each dialysis when patients are subjected to copper contamination of the dialyzate. Death due to acute intravascular hemolysis associated with metabolic acidosis and leukocytosis has occurred with extreme exposure where copper heating coils or copper pipes were exposed to deionized water or dialysate at acid pH. This has probably been produced where the deionizer has been overrun and the resin bed exhausted.[47,48] Even a short length of copper pipe in contact with stationary water can leech significant quantities of copper to cause fever and chills and because of this, the practice of running to waste for a few minutes the water used to prepare dialysate can significantly reduce the exposure.

B. Zinc

Although zinc deficiency is probably the most common abnormality in patients with uremia,[49] acute zinc toxicity has also been recognized. Acute hemolytic anemia with fever and chills, vomiting and nausea have been reported in two outbreaks.[50,51]

The zinc sources were a galvanized iron rain water storage tank and the galvanized container of a water softener. Although water softeners are extensively used for water treatment, toxic syndromes attributed to failure of the lining material and exposure of the underlying metal container has been rarely reported!

C. Cobalt

The most common cause of death in patients on intermittent hemodialysis is cardiac in origin, though the EDTA statistics do not give further details. Unexplained cardiac failure with hypertrophied myocardium is not an uncommon pathological finding at autopsy. A grossly elevated myocardial cobalt content 3 to 80 times normal has been found in dialysis patients, one of whom had a cardiomyopathy.[52-54] One patient also had elevated serum cobalt concentrations and whose clinical features improved when serum cobalt levels fell following hemodialysis and renal transplantation. Further data are needed, but cobalt accumulation in myocardial tissue may be toxic. The source of cobalt is not only the water supply, but also cobalt-chloride given to raise the hemoglobin levels in patients with renal failure.[55] Therefore, it would seem sensible to abandon this form of therapy in patients with chronic renal failure until more is known about the long-term effects of cobalt accumulation. In addition, little information is available about the cobalt content in the majority of the domestic water supplies used to prepare dialysate.

D. Lead

Although no reports of lead intoxication in patients on intermittent hemodialysis have appeared, lead is a likely candidate. As with copper, much of the transport of domestic water has been through lead-contaminated pipes. Standing water can also leech toxic quantities, so in practice, the running of water to waste for several minutes prior to starting hemodialysis, especially after an overnight lull in water demand, would be a reasonable procedure. In centers where high lead water supplies have been found, elevated serum lead content has been documented in patients on hemodialysis and with chronic renal failure before requiring hemodialysis.[56,57]

E. Cadmium

Raised serum levels of cadmium have been found in patients on hemodialysis from contamination of the dialysate with zinc oxide plaster used around coil dialysers,[58] but more significantly, cadmium is elevated by smoking,[56] though no pathological significance has yet been attributed to this observation.

F. Nickel

Pipe work of dubious quality was often used in the early day of hemodialysis on heat exchanges. An outbreak of nausea, vomiting, headache, and dizziness was attributed to nickel toxicity from stainless steel heating coils.[46]

VI. FLUORIDE

Plasma fluoride levels rise in early renal failure with accumulation in bone.[59] Patients hemodialyzed on water in fluoridated areas have raised blood levels often rising during dialysis to equivalent concentration to that of the dialyzate supply and falling in the intradialysis period, probably due to sequestration in bone.[60] Previously, a strong association was found between fluoridated dialysate and renal bone disease of the osteomalacic type,[61] though in some dialysis centers on fluoridated water supplies a poor association was observed.[59]

In Toronto, where a prospective double blind trial of the addition of fluoride to dialysis fluid was undertaken, no excess of bone disease occurred.[62] The direct comparison between two dialysis centers, Newcastle upon Tyne with an 80% incidence of fracturing osteomalacic osteodystrophy, and Birmingham with an incidence of less than 15% were both on fluoridated water supplies.[60] It was later established that these two cities were on differing water supplies with regard to aluminum content.[17] It is possible that fluoride is synergistic with other elements; for example aluminum fluoride complexes may pass through softeners and deionizers with greater ease. Though it now seems likely that aluminum is the main culprit, fluoride has been shown to cause a mineralization defect and fluorosis[63] and it would seem sensible to remove this from water supplies used to prepare dialysate.

VII. OTHER CONSIDERATIONS

Blood and certain tissues (myocardial muscle, skeletal muscle, liver, skin, bone, and brain) have been found to have altered concentrations of other metals, including tin, molybdenum, cesium, selenium, rhubidium, arsenic, chromium, and uranium, though the change is small and no pathological significance has yet been attributed to these elements. Little is known about the interaction of several trace metals and their ensuing long-term pathological effects.[46]

VIII. REMOVAL OF METAL IONS FROM DIALYSIS FLUID

The major source of metal ion contamination in dialysis fluid is the raw water supply. Naturally high aluminum-containing waters are not uncommon. The major contributor to the aluminum load in the water supply is treatment with aluminum sulfate as a flocculant to clear the water of coloring material for domestic purposes. Although only a small number of water supplies have so far been studied extensively with regard to use for hemodialysis, there is considerable fluctuation in the metal ion content of domestic water supplies and deionization does not seem to be adequate for all supplies.[64,65] Only reverse osmosis reduces the aluminum content consistently to less than 5 $\mu g/\ell$.[65] It appears that each individual water supply used for hemodialysis should be studied and an appropriate water treatment prescribed depending on the solute content of that water supply.[66,67]

REFERENCES

1. Combined report on regular dialysis and transplantation in Europe, *Proc. Eur. Dial. Transplant Assoc.*, in press.
2. Data Sheet Compendium Association of the British Pharmceutical Industry Pharmid Publications Limited, 1979.
3. Antoniou, L. D., Salhoub, R. J., Sudhark, T., and Smith, J. D., Reversal of uremic impotence by zinc, *Lancet*, 2, 895, 1977.
4. Perry, H. M., Thind, G. S., and Perry, E. F., Biology of cadmium, *Med. Clin. North Am.*, 60 (4), 759, 1976.
5. Graham, T., Liquid diffusion applied to dialysis, *Philos. Trans. R. Soc. London*, 15, 185, 1861.
6. Abel, J. J., Rowntree, L. G., and Turner, B. B., Plasma with return of corpuscles, *J. Pharmacol. Exp. Trans.*, 5, 625, 1919.
7. Haas G., Dialysiem des strömenden Blutes am Lebenden, *Klin. Wochensc hr.*, 2, 1888, 1923.
8. Kaehny, W. D., Alfrey, A. C., Holman R. E., and Schorr, W. J., Aluminum transfer during haemodialysis *Kidney Int.*, 12, 361, 1977.
9. Parsons, F. M. and Davison, A. M., The composition of dialysis fluid, in *Replacement of Renal Function by Dialysis*, Drukker, W., Parsons, F. M., and Maher, J. F., Eds., Martinus Nijhoff, The Hague, Netherlands, 1979.
10. Freeman, M. R., Lawton, L. R., and Chamberlain, A. M., Hard-water syndrome, *N. Engl. J. Med.*, 276, 1113, 1967.

11. Sorenson, J. R. J., Campbell, K. R., Tepper, L. B., and Lingy, R. D., Aluminum in health, *Environ. Health Perspect.*, 8, 3, 1974.
12. Berlyne, G. M., Ben-Ari, J., Pert, D., Weinberger, J., Stern, M., Gilmore, G. N., and Levine, R., Hypoaluminemia from Aluminum. Resin in renal failure, *Lancet*, 2, 494, 1970.
13. Boukori, M., Rottenburg, J., Jardon, M. C., Cavel, J. P., Legrain, M., and Galli, A., Influence de la prise prolongée de gels d'alumin sur les taux sérique d'aluminium chez les patients atteints d'insuffisance rénale chronique, *Nouvre Presse Med.*, 7, 85, 1978.
14. Marsden, S. N. E., Parkinson, I. S., Ward, M. K., Ellis, H. A., and Kerr, D. N. S., Evidence for aluminum accumulation in rénal failure, *Proc. Eur. Dial. Transplant Assoc.*, in press.
15. Coburn, J. W. and Llaca, F., Renal osteodystrophy, in *Replacement of Renal Function by Dialysis* Drukker, W. Parsons, F. M. Maher, F. J., Eds., Martinus Nijhoff, The Hague, Netherlands, 1979, chap. 32.
16. Alfrey, A. C., Legrandre, G. R., and Kaehny, W. D., The dialysis encephalopathy syndrome, possible aluminium intoxication, *N. Engl. J. Med.*, 294, 184, 1976.
17. Ward, M. K., Feest, T. G., Ellis, H. A., Parkinson, I. S., Kerr, D. N. S., Herrington, J., and Goode, C. L., Osteomalacic dialysis osteodystrophy: evidence for a water borne aetiological agent, probably aluminium, *Lancet*, 1, 841, 1978.
18. Elliott, H. L., Dryburgh, F., Fell, G. S., Sabet, S., and MacDougall, A. I., Aluminium intoxication in dialysis patients, *Br. Med. J.*, 1, 1101, 1978.
19. Alfrey, A. C., Mischell, J. M., Burks, J., Contigiylia, S. R., Rudolph, M., Levin E., and Holmes, J. M., Syndrome of dyspraxia and multifocal seizures associated with chronic haemodialysis, *Trans. A.S.A.I.O.*, 18, 257, 1972.
20. Mahurker, S. D., Khar, S. K., Salta, R., Meyers, L., Smith, E. C., and Dunea, G., Dialysis dementia, *Lancet*, 1, 1412, 1973.
21. Platts, M. M., Goode, C., and Hislop, J. S., Composition of the domestic water supply and the incidence of fractures and encephalopathy in patients on home dialysis, *Br. J. Med.*, 2, 657, 1977.
22. Bone, I., Progressive dialysis encephalopathy, in *Dialysis Review*, Davison, A. M., Ed., Pitman Medical, London, 1978.
23. Barratt, L. J. and Lawrence, J. R., Dialysis associated dementia, *Aust. N. Z. J. Med.*, 1, 62, 1975.
24. Flendrig, J. A., Kruis, H., and Das, H. A., Aluminium intoxication: the cause of dialysis dementia, *Proc. Eur. Dial. Transplant. Assoc.*, 12, 355, 1976.
25. Cartier, F., Allain, P., and Gay, J., Dialysis encephalopathy, Abs. 7th Int. Cong. Nephrology, 5, 1978.
26. Silke, B., Fitzgerald, G. R., Hanson, S., Carmody, M., and O'Dwyger, O. F., Dialysis dementia and renal transplantation, *Dial. Transplant.*, 7, 486, 1978.
27. Jacobs, C., Branner, F. P., Chortler, C., Jonderwaldke, R. A., Sterlord, M. J., Hathway, N. A., Selwood, H., and Wing, A. S., Combined report on regular dialysis and transplantation in Europe, *Proc. Eur. Dial. Transplant Assoc.*, 14, 52, 1977.
28. English, A., Savage, R. D., Britton, P. G., Ward, M. K., and Kerr, D. N. S., Intellectual impairment in chronic renal failure, *Br. Med. J.*, 1, 888, 1978.
29. Ward, M. K., Pierides, A. M., Fawcett, P., Shaw, D. A., Perry, R. H., Tomlinson, B. E., and Kerr, D. N. S., Dialysis encephalopathy syndrome, *Proc. Eur. Dial. Transplant Assoc.*, 13, 348, 1976.
30. Chakroverty, S., Brutman, M. E., Berger, V., and Reyes, M. G., Progressive dialytic encephalopathy, *J. Neurol. Neurosurg. Psychiatry*, 30, 411, 1976.
31. Burks, J. S., Alfrey, A. C., Huddlestone, J., Norenberg, M. D., and Levin, E., A fatal encephalopathy in chronic haemodialysis patients, *Am. Int. Med.*, 90, 741, 1979.
32. McDermott, J. R., Smith, A. I., Ward, M. K., Parkinson, I. S., and Kerr, D. N. S., Brain aluminium concentration in dialysis encephalopathy, *Lancet*, 1, 901, 1978.
33. Arieff, I. A., Cooper. J. D., Armstrong, D., and Lazorowitz, V., Dementia renal failure and brain aluminium, *Am. Int. Med.*, 90, 741, 1979.
34. Parkinson, I. S., Ward, M. K., Feest, T. G., Fawcett R. W. P., and Kerr, D. N. S., Fracturing dialysis osteodystrophy and dialysis encephalopathy: an epidemiological survey, *Lancet*, 1, 406, 1979.
35. Parkinson, I. S., Ward, M. K., and Kerr, D N. S., unpublished observations.
36. Winney, R., personal communication.
37. Poisson, M., Mashaly, R., and Lebkiric, B., Dialysis encephalopathy: recovery after interruption of aluminum intake, *Br. Med. J.*, 2, 1610, 1978.
38. Davison, A. M., Renal transplantation in relation to dialysis encephalopathy in *Proceedings of the European Dialysis and Transplant Association*, North Holland, Amsterdam, in press.
39. Stanbury, S. W., Renal osteodystrophy, *Clin. Endocrinol. Metab.*, 1, 239, 1977.
40. Siddiqui, J., and Kerr, D. N. S., Complications of renal failure and their response to dialysis, *Br. Med. Bull.*, 27, 153, 1971.

41. Pieridies, A. M., Ellis, H. A., Feest, T. G., Ward, M. K., and Kerr, D. N. S., Variable response to long term 1-δ-Hydroxydole calcified in haemodialysis osteodystrophy, *Lancet*, 1, 1092, 1976.
42. Pierides, A. M., Skillen, A. W. and Ellis, H. A., Alkiline phosphatase in renal failure, in press.
43. McCarthy, J., Ellis, H. A., and Herrington, J., Bone aluminium in haemodialysis patients, *J. Clin. Pathol.*, 32, 832, 1979.
44. Allain, D., Thebaud, H. E., Dupouet, L., Caville, P., Disont, M., Spiesser, J., and Alquier, P., Study of blood levels of a number of metals (Al. Mn. Cd. Pb. Cu. Zn) in chronic haemodialysis patient before and after dialysis, *La Nourvelle Presse Medicale*, 3(7), 92, 1978.
45. Cooke, D. G., Fahn, S., and Braik, A, Chronic manganese intoxication, *Arch. Neurol. (Chicago)*, 30, 59, 1974.
46. Alfrey, A. C. and Smythe, R. W. in *Trace Metals and Regular Dialysis in Replacement of Renal Function by Dialysis*, Drukker, W., Parsons, F. H., and Mahes, J. F., Eds., Mactinus Nijhoff, The Hague, Netherlands, 1979.
47. Lyle, W. H., Payton, J. E., and Hui, M., Haemodialysis and copper fever, *Lancet*, 1, 1324, 1976.
48. Klein, W. J., Metz, E. M., and Price, A. R., Acute copper intoxication, Arch. Int. Med., *129*, *578*,
49. Leinderman, R. D., Baxter, D. J., and Yunice, A. A., Kraikitpanid S serum concentrations & urinary excretions of zinc in cirrhosis, nephrotic syndrome and renal insufficiency, *Am. J. Med. Sci.*, 275, 17, 1978.
50. Gallery, E. D. M., Blomfield, J., Dixon, S. R. Acute zinc toxicity in haemodialysis, *Br. Med. J.*, 4, 331, 1972.
51. Petrie, J. J. B. and Row, P. G., Dialysis anaemia caused by subacute zinc toxicity, *Lancet*, 1, 1178, 1977.
52. Manifold, I. M., Platts, M. M., and Kennedy, A., Cobalt cardiomyopathy in a patient on haemodialysis, *Br. Med. J.*, 2, 1609, 1978.
53. Curtis, J. R., Goode, G. C., and Herrington, J., Urdantenal, *Clin. Nephrol.*, 5, 61, 1976.
54. Duckholm, J. A. and Lee, H. A., Cobalt cardiomyopathy, *Lancet*, 1, 1350, 1976.
55. Linz, L. E. and Pehrsson, K., Cobalt intoxication in uramic cardiomyopathy, *Lancet*, 1, 1/92, 1976.
56. Kerr, D. N. S., Ward, M. K., Parkinson, I. S., Pinchin, M. J., and Newham, J., Trace elements in chronic renal failure, *Adv. med.*, 15, 290, 1979.
57. Campbell, B. C., Beattie, A. D., Moore, M. R., Goldberg, A., and Reid, A. G., Renal insufficiency associated with excessive lead exposure, *Br. Med. J.*, 1, 492, 1977.
58. Willden, E. G. and Hyne, B. E. D., Blood and urinary cadmium in chronic renal failure, *Nephron*, 13, 253, 1974.
59. Nielsen E., Solomon, M., Goodwin, M. J., Siddhiram, M., Galowsky, R., Taves, D., and Friedman, E. A., Fluoride metabolism in uraemia, *Trans. Am. Soc. Artif. Intern. Organs*, 19, 450, 1973.
60. Siddiqui, J. T., Simpson, S. W., Ellis, H. A., Kerr, D. N. S., Appleton, D. R., and Robinson, B. H., Fluoride and bone disease in patients on regular dialysis, *Proc. Europ. Dial. Transplant Assoc.*, 15, 586, 1978.
61. Comty, C., Laehmann, W., Wathen, R., and Shapiro, F., Prescription water for chronic haemodialysis, *Trans. Am. Soc. Artif. Intern. Organs*, 20, 189, 1974.
62. Comty, C. and Shapiro, F. L., Pretreatment and preparation of water for haemodialysis, in *Replacement of Renal Function by Dialysis*, Drukker, W., Parsons, F. M., and Maher, J. F., Eds., Mactinus Nijhoff, The Hague, Netherlands, 1979.

Chapter 3

HYPERSENSITIVITY TO METALLIC BIOMATERIALS*

Katharine Merritt and Stanley A. Brown

TABLE OF CONTENTS

I. Introduction 34

II. Nonspecific Defense Mechanisms 34

III. Immunologic Defense Mechanisms 34
    A. Antigens 34
    B. Pathways of the Immune Response 35
    C. Sensitivity Reactions 35
        1. Mechanisms of Cell Damage in Sensitivity Reactions 35
            a. Antibody 35
            b. Cell-Mediated Immunity 36
            c. Lymphokines 36

IV. Sensitivity to Metallic Biomaterials 37
    A. Immune Responses to Metal Salts 37
        1. Humans 37
            a. Testing for Sensitivity 37
        2. Sensitivity to Metal Salts in Animals 39
    B. Induction of Sensitivity by Metallic Implants 40
        1. Corrosion 40
         2. Generation of Sensitizing Complexes by Wear 40
    C. Sensitivity Reactions to Metallic Implants 40
        1. Human 40
            a. General Surgery 40
            b. Dental Surgery 41
            c. Orthopedic Surgery 42
        2. Sensitivity Reactions to Metallic Implants in Experimental Animals 45

V. Conclusions 45

References 46

---

* This work was funded in part by the Educational Foundation of America and USPHS grant AMAI 20271-01A1.

I. INTRODUCTION

The development of new materials or new devices requires the thorough testing for biocompatibility in the body. One of the major considerations is the reaction of the body's host defense mechanisms to the material itself, to the device when in use, or to any wear or dissolution products of the device. In this chapter we will discuss briefly the host defense mechanisms involved, both nonspecific and specific (immune responses reacting specifically with an antigen). We will then focus on the reaction of the immune system to metal salts and to metallic biomaterials.

II. NONSPECIFIC DEFENSE MECHANISMS

The response of tissue when confronted by substances recognized as foreign (nonself) to the body is to maintain the status quo and eliminate the material. This attempt at elimination is universal for all foreign materials. With inert materials, the reaction may simply be a walling off of the foreign object with a coating of fibrin, enabling the body and foreign object to coexist for years. However, with materials that are not inert, a more rigorous attempt at elimination may ensue with the accumulation of phagocytic cells such as polymorphonuclear leukocytes, monocytes, macrophages, and histiocytes which attempt to degrade the offending object. If the phagocytic system is capable of eliminating the material rapidly, there is no residual effect on the body. (However, unless the material was intended to function for only a short time, such as with an absorbable suture, it has failed as a biomaterial.) If the phagocytic system is not capable of eliminating the object rapidly, a more intense reaction may follow with accumulation of many phagocytic cells, some of which coalesce to become giant cells, and a large inflammatory reaction, the so-called foreign body reaction, forms. The enzymes released during the process of phagocytosis and the process of forming a foreign body reaction may cause severe tissue damage.

Concomitant with the effect of the body tissues on the implant is the effect of the implant on the body tissues. If the implant is toxic to cells, then the phagocytic system not only attempts to eliminate the foreign object, but also must clean up the damaged tissue. If the implant releases materials that are recognized as antigens by the body, then the body mounts an immune response against them and the result may be to further magnify the foreign body reaction.

III. IMMUNOLOGIC DEFENSE MECHANISMS

The fundamental aspects of immunity have been reviewed in another volume of this series,[1] and the reader is referred to that chapter or a general immunology textbook for more detail. A brief disucssion of the immune system will be given here for orientation.

A. Antigens

In order to stimulate the immune system, there must be an antigen. Whether or not a substance will be antigenic is not entirely predictable; however, there are certain features which are involved. First, the substance must be recognized as foreign by the body. Second, the substance must be a large molecule with a molecular weight of several thousand daltons. Third, the chemical composition is of importance with proteins being highly antigenic, carbohydrates less antigenic, and lipids very weakly antigenetic. Biomaterials being used in or on the body can become antigenic since they are foreign to the body. It is usually the abrasion or degradation products which activate the immune system, and these are generally of low molecular weight. However, they can meet the requirement for size and chemical composition by binding to body substances so that the body substance provides the necessary molecular size and chemical makeup while the product of the biomaterial provides the message of being foreign. The small molecular weight substances that serve as antigens in this manner are called haptens.

B. Pathways of the Immune Response

There are two pathways of the immune response that can be stimulated by antigens. One of these is the humoral response and leads to the production of a circulating protein called antibody. There are five classes of antibody — IgG, IgA, IgM, IgD, and IgE — produced in response to an antigen. Humoral responses occur following the stimulation by antigen of a particular class of small lymphocytes, called B cells. These stimulated B cells differentiate into plasma cells which make the antibody.

The other pathway is called cell-mediated immunity and involves the stimulation of a second class of small lymphocytes which develop in the thymus and are called T cells. Cell-mediated immunity results in the production of many T cells capable of reacting with the antigen.

The immune response is designed to detoxify or eliminate the harmful foreign substances. However, the immune response in the process of reacting with and eliminating the antigen may also cause harm to the body tissues, and this is referred to as allergy or sensitivity. Sensitivity reactions can occur either as a result of stimulating antibody production or as a result of stimulating cell-mediated immunity.

C. Sensitivity Reactions

The reactions involving the production of antibody to eliminate the foreign material generally result in the accumulation of antibody, plasma cells, and polymorphonuclear leucocytes in the tissue. (It should be noted here, however, that reactions involving antibody of the IgE class are associated with a particular type of sensitivity such as that to pollen and dust. This is called atopic or Type I sensitivity. Mast cells, basophils, and/or eosinophils accumulate at the reaction site of IgE antibody.) In contrast, the cell-mediated response usually involves the accumulation of small lymphocytes, monocytes, and macrophages at the site.

1. Mechanism of Cell Damage in Sensitivity Reactions
a. Antibody

The mechanism of tissue damage by antibody varies. When antigen and antibody react in vivo, another set of proteins [called complement (C')] becomes activated. Complement then functions to kill those cells to which antibody has bound. Thus, haptens which attached to host cells to become an immunogen can react with the antibody produced and secondarily lead to the destruction of the host cell to which they were bound. The activation of complement also causes the production of chemotactic factors (C'_{3a} and C'_{5a}) which attract the polymorphonuclear leukocytes and the enzymes from these phagocytic cells can cause tissue damage. Capillary permeability is also altered causing the accumulation of fluid leading to the final manifestation of the inflammatory response.

In the case of IgE mediated responses, histamine is released from mast cells, basophils, or eosinophils, and this results in contraction of smooth muscles, fluid accumulation, and the release of other substances which further alter the function of smooth muscle and also affect the clotting mechanism.

Antibody as a cause of tissue damage can be inferred from a biopsy where there is accumulation of plasma cells and polys at the site or detected with special stains. More often, plasma or serum is obtained from the patient and in vitro tests for antibody to the antigen are performed. Tests for antibody are quite specific, sensitive, and reliable.

b. Cell-Mediated Immunity

The mechanism by which cell-mediated immunity causes tissue damage is not as clear as with the antibody-mediated systems. Our knowledge of antibody and any associated damage far surpasses our knowledge of cell-mediated immunity. However, much progress has been made in the last 15 years in the understanding of the process of cell-mediated immunity. Careful in vitro studies of the lymphocytes have demonstrated that reaction of the lymphocytes with antigen for which they have receptors causes the production by the lymphocytes of various soluble factors which are termed lymphokines and are referred to as "in vitro correlates of cell-mediated immunity." Unlike antibody which is easy to measure and is present in detectable amounts in both in vivo and in vitro studies, the lymphokines are produced in low levels, and their actual presence in vivo has only rarely been demonstrated. Nevertheless, these lymphokines have various activities which help explain what is seen in the in vivo response and are, therefore, accepted as measures of cell-mediated immunity.

c. Lymphokines

There are many lymphokines which have been detected and some have been partially characterized. For more detail on the lymphokines, the reader is referred to the chapter by Elves[1] of this series, or to a monograph on cell-mediated immunity. Similarly, the details of the techniques for assaying for these lymphokines can be found in immunologic methods books.[2,3] There are five lymphokines which are of importance for our discussion.

Blastogenic factor — This is a protein, produced by lymphocytes, which causes other lymphocytes to differentiate, divide, and produce more lymphocytes. This accounts for the accumulation of lymphocytes at the reaction site and the increased number of lymphocytes with receptors for the stimulating antigen. The increased cell division can be measured by cell counts or by the uptake of radioactive precursors and provides an in vitro test for cell-mediated immunity.

Chemotactic factor — This protein, produced by lymphocytes reacting with antigen, causes the directional movement toward the reaction site and accumulation of monocytes, macrophages and polys.

Migration inhibition factor (MIF, LIF) — This glycoprotein, produced by lymphocytes reacting with an antigen, prevents the migration of macrophages or monocytes (MIF) or polymorphonuclear leukocytes (LIF) away from the site. Presumably its function is to retain the phagocytes to digest the foreign material. MIF and LIF can be distinguished by sophisticated biochemical analyses. For purposes of our general discussion, they have the same function and will be considered as the same lymphokine.

Lymphotoxin — This is a protein elaborated by lymphocytes which serves to kill other cells. This lymphokine is presumed to be important in eliminating foreign cells (such as tumor cells or tissue grafts), but also can be a major cause of host cell damage.

Osteoclastic activating factor (OAF) — This lymphokine has received only limited study and does not usually appear in the list of lymphokines. Nevertheless, its production appears to be analogous to that of other lymphokines,[4-6] and it would appear to be an important substance in generation of bone loss. Presumably, therefore, it is of special importance in complications following the use of orthopedic and dental implants in sensitive individuals, and further study is needed.

The generation of these lymphokines requires (1) the presence of lymphocytes which have cell surface receptors for the antigen, (2) the presence of macrophages, and (3) the presence of the antigen. Certain plant lectins such as phytohemagglutinin (PHA) are potent stimulators of lymphokines and substitute for the antigen since many lymphocytes have surface receptors for lectins. These lectins can serve as control stimulating substances for in vitro tests and as eliciting agents for production of these lymphokines for analysis and purification. These lectins are often referred to in articles on cell-mediated immunity and in vitro studies.

Cell-mediated immunity as a cause of tissue damage is inferred from the observation in the biopsy specimen of the accumulation of lymphocytes, monocytes, and macrophages. More often, sensitivity to the antigen is confirmed by skin testing which involves the intradermal injection of the antigen or the application of the antigen to the skin with observation for any subsequent reaction at 24, 48, and 72 hr. In vitro studies for lymphokines are becoming increasingly important in detecting cell-mediated reactions as a cause of tissue damage.

Immune responses of either humoral or cell-mediated type may occur to a given antigen. Thus, when considering the response of the body to a biomaterial, the possibility of either type of response must be considered and evaluated. The discussion will now focus on what is known about the immune response to metal salts and metallic biomaterials.

IV. SENSITIVITY TO METALLIC BIOMATERIALS

A. Immune Responses to Metal Salts
1. Humans

Immune responses to metal salts have been documented in humans with exposure to metal salts. Asthma (dyspnea due to bronchial constriction) has been reported as occurring in individuals working with nickel salts[7] and individuals working with chromium salts.[8]

Asthma is often caused by a sensitivity reaction mediated by IgE antibody. However, reactions of cell-mediated immunity and reactions involving antibodies of other classes may also cause asthmatic reactions. In these cases,[7,8] inhalation of the metal salt stimulated the immune response with resultant bronchial constriction and in some cases also a dermatitis.

The more common problem is that of contact dermatitis. In this situation, workers with skin contact with metal salts may develop sensitivity. The metal salts bind to skin proteins and this complex then stimulates the immune response. Development of sensitivity from occupational exposure in metal foundries, metal plating companies, paint manufacturers, shoe tanners, and cement workers has been well documented with a sensitivity to nickel, cobalt, and chromium. Similarly, metal sensitivity to nickel and chromium has developed from the wearing of stainless steel jewelry, including watches. Only a small percentage of people at risk to develop sensitivity do. The reasons for this remain obscure and are presumed to be due to the genetics of the immune response. Reference to sensitivity in the various metallic elements in biomaterials has been made in the preceding chapters.

a. Testing for Sensitivity

The classical test for sensitivity to metals is the skin test which detects cell-mediated immunity. In this procedure, the metal salt is placed on a gauze patch and applied to the skin. The site is then observed at 24, 48, and 72 hr for development of swelling and redness, indicative of an allergic reaction. It is difficult to distinguish a toxic reaction from an allergic reaction, and care must be taken to use the correct testing dose.[9] The use of patch testing or intradermal testing for sensitivity to haptens poses the problem that this is how the sensitivity is acquired. Thus, it is possible to cause sensitivity by testing, and conversion from skin test negative to skin test positive may be due to acquisition of sensitivity from contact from occupation, jewelry, an implant, or from the skin testing.

Because of the problems of discomfort of testing, observing the tests at 24, 48, and 72 hr, distinguishing toxicity from sensitivity and the hazard of repeated testing, investigators have pursued the use of in vitro tests for the production of lymphokines as indicators of cell-mediated responses to specific antigens (described in Section IV C).

The in vitro tests also have their problems and not every laboratory that has attempted to use tests for lymphokines has been successful. However, the fact that some laboratories have been successful is important, and only the successful studies are presented here. Further work on these techniques should define more carefully the culture conditions required. Care must be taken in each laboratory undertaking this type of testing to titrate the level of metal salts causing toxicity and the minimum concentration required to stimulate the lymphokine production from lymphocytes from patients known to be sensitive. The production of blastogenic factor[10-13] or of migration inhibition factor[14-17] has been detected following stimulation of lymphocytes with nickel, cobalt, and chromium salts, and this has correlated well with the sensitivity status of the patients. The reader is referred to these articles and the general methods books[2,3] for details of testing.

Tests for migration inhibition factor do not require the use of radioisotopes as do the tests for blastogenic factor, and the results are obtained 4 to 5 days sooner than tests for blastogenic factor. Our preference has been to use the technique of cell migration in agarose.[3,16] This technique has been found to be very simple to perform, inexpensive, and rapid. Results can be obtained in 24 hr if the culture plate is examined for cell migration directly with a microscope, or 48 hr if the plate is fixed with methanol, dried, and stained. Again, choice of culture conditions may be important; we have had success for several years with RPMI 1640, but recently have found it necessary to use Click's Medium.[18]

One additional advantage found with the use of the migration inhibition test is that it will detect the situation in which the patient is sensitive and reacting to the implant. When a sensitive individual has skin contact with the metal salt, a clearly visible reaction is seen in 24 to 48 hr. When a sensitive individual has deep tissue contact with an implant from which metal ions have been released, the reaction is not visible. It is not until the reaction has progressed sufficiently to cause pain, fluid accumulation, loosening of the implant, or become systemic so as to cause a dermatological disorder, that a reaction to the implant is suspected. We have found that the leukocytes from patients who are sensitive to a metal and are reacting to the presence of a metallic implant show no migration of the cells in the migration inhibition test in any of the test conditions.[16,17] Thus, the cells are already inhibited and require no further stimulus for the release of migration inhibition factor. In essence, "the migration control doesn't work." If the implant is removed from these patients, migration in the control well returns to normal in about 6 weeks with migration inhibition factor being produced following stimulation with one or more of the metal salts. Patients with infections at the implant site show normal cell migration in the test. We consider this situation of nonmigration to be diagnostic for sensitivity reactions to an implant.

The use of cobalt salts in in vitro testing has been difficult. These salts seem to be more toxic than the salts of nickel or chromium, and the longer the incubation time, the lower the concentration tolerated by the cells.[10,13,19] We have had success in testing for cobalt sensitivity using migration in agarose. The shorter incubation period for migration inhibition compared to blastogenesis may be important in the success of testing.

Tests for antibody-to-metal salts are rarely performed, even though techniques are available. The simplest technique for measuring antibody-to-metal salts is by the use of passive hemagglutination. The metal salts will adhere to human type O or sheep red blood cells,[7,20] with or without the use of tannic acid, and antibody of the IgG or IgM class will cause agglutination of the cells. The plasma or serum can be serially diluted so that the concentration can be determined as the highest dilution causing agglutination.

Another simple technique which we have used is the Farr technique.[21] Nickel or cobalt chlorides are attached to albumin, the complex reacted with serum, then all antibody and antigen-antibody complexes are precipitated with $(NH_4)_2SO_4$. The amount of nickel or cobalt in the precipitate can be measured by atomic absorption spectroscopy. This will detect antibodies of the IgG, IgA, and IgM classes. Increasing levels of the metal indicate increasing antibody. Antibody from humans can also be measured by the more expensive technique of passive cutaneous anaphylaxis in the guinea pig.[22]

The question of IgE antibody-to-metal salts poses problems since the above methods are not sensitive enough to detect IgE antibody which occurs in ng/ml quantities as opposed to mg/ml quantities for IgG, A, and M. IgE is the immunoglobulin which is important in atopic type of sensitivity and is a cause of asthma. Thus it may be important in some metal sensitivity cases. Antibody of the IgE class to nickel has been measured with a sensitivity radioassay in one study[23] and in those cases the presence of antibody and a history of sensitivity were not well correlated.

2. Sensitivity to Metal Salts in Animals

The guinea pig was found in studies on immunity to be the best model for human allergy. Sensitivities mediated by antibody and sensitivities of cell-mediated immunity have both been studied effectively. The most detailed and exhaustive work has been undertaken with chromium allergy and is beautifully reviewed by Polak et al.[22] The in vitro migration inhibition tests correlated well with the skin test. The question of valency and sensitivity was thoroughly studied and would indicate that hexavalent chromium is required for skin testing, but that the other valencies will cause sensitization if injected and will stimulate production of MIF in vitro.

Sensitivity to nickel has been caused in guinea pigs, although with more difficulty than with chromium.[24] One problem with nickel sensitivity in the guinea pig is the ease with which they become desensitized (skin test negative) which may be associated with the presence of detectable antibody. Studies with migration inhibition factor and nickel sensitivity are not extensive. Similarly, cobalt has been used to cause sensitivity in the guinea pig,[25] but the studies are not yet as extensive as those with chromium.

The rabbit is suitable as an experimental animal because it is large enough both to receive an orthopedic implant and large enough to permit drawing 10 ml of blood for LIF testing every few weeks without undue risk to the animal. We have been able to cause sensitivity to nickel by injection of $NiCl_2$ in complete Freund's adjuvant and have been able to show good correlation between skin test results and the migration inhibition test.[26] We have also studied sensitivity to cobalt chloride in the rabbit and found production of migration inhibition factor by 8/10 of the injected rabbits and skin test sensitivity in 4/10 of these rabbits (3 of these 4 produced LIF). Thus, the correlation between migration inhibition and skin reaction is not good in this case, but the results indicate that the LIF test is more sensitive and probably, therefore, better. None of the uninjected rabbits produced LIF or were skin test positive. We have studied only two rabbits with chromium chloride, and they were skin test and LIF positive for chromium. Thus, the rabbit is a good animal for studying nickel sensitivity and probably is also suitable for studying sensitivity to cobalt and chromium.

B. Induction of Sensitivity by Metallic Implants

In order for an implant to cause sensitization in an individual or elicit a sensitivity reaction in an already sensitive individual, there must be release of metal ions from the implant. It has been stated by a leading dermatologist[27] that "stainless steel sutures can be used safely in a sensitive individual since the nickel is bound tightly in the stainless steel. However, the nickel is not bound tightly in the stainless steel, but rather is in solid solution in the alloy and can be released.

1. Corrosion

Release of metal ions by corrosion or leaching has been demonstrated by several laboratories. Analyses of tissue surrounding implants in muscle in rabbits and guinea pigs[28,29] have revealed that nickel, chromium, and cobalt can be detected in significant levels. Similarly, analysis of human tissue from the site where implants are being removed has indicated a high level of metal ions.[30,31] Other body fluids such as blood and urine and other body tissues such as hair may also show elevated levels of metal ions in patients bearing metallic implants.[32] The suspension of stainless steel in biological fluids in vitro has also resulted in the release of metal ions into the fluid.[33] Analyses of implants reveals pits or holes also indicative of metal loss. These experiments prove what the metallurgist knows: the metal ions are not bound tightly in the metal, but are essentially a metal solution, and the ions can be leached out.

2. Generation of Sensitizing Complexes by Wear

The articulation of one metal component against another in the use of metallic implants such as screw-plate combinations or metal-metal total joints can lead to the generation of metal particles and metal ions in solution. This debris can cause tissue darkening and has led to the presence of high levels of metal detectable in the adjacent tissue as well as in distant tissues.[32,34,35] The wear products from a joint simulator have been shown to have deleterious effects on phagocytic cells.[19] These wear products do go into solution and certainly as ions can bind to protein and stimulate the immune response. The particles themselves may be immunogenic. Thus, they may elicit a reaction in a sensitive individual or cause sensitization in a previously nonsensitive individual.

C. Sensitivity Reactions to Metallic Implants
1. Human
a. General Surgery

Cases have been reported in the literature of dermatitis following the use of implants or instruments containing nickel in nickel sensitive individuals. For instance, a case of dermatitis following the use of nickel plated retractors in abdominal surgery has been reported.[36] The retractors were found to be corroded. Two patients have been described[37] who reacted following the use of cannulas for i.v. infusions. The cannulas caused a response when placed on the skin of another subject known to be sensitive to nickel. Cutaneous lesions after i.v. infusions were reported by Smeenk and Teunissen[38] in seven patients. They performed an in vitro test passing 500 mℓ of physiologic solution through the needles and obtained 7 μg Ni/ℓ in 24 hr and 80 μg/ℓ in 48 hr. Fisher[39] could not repeat the in vitro test using the dimethylglyoxime test. However, the dimethylglyoxime test has a minimum sensitivity of 10 to 100 μg/ℓ[40] and thus may not have been sensitive enough to detect the nickel. Fisher attributes the difference to the quality of the needles. Certainly it would appear that at least some stainless steels elute levels of nickel high enough to elicit sensitivity.

One question that remains is the quantity of nickel needed to elicit sensitivity in an already sensitive individual or to cause sensitivity in a previously nonsensitive individual. Malten and Spruit[41] carried out a study on the skin of patients known to be sensitive using $NiSO_4$ and determined that 1 $\mu M/\ell$ (58.7 $\mu g/\ell$) of nickel would not elicit a response but that 10 μM(587 μg) could. The nickel leaching reported by Smeenk and Teunissen[38] described above is just over the safe level for skin sensitization, but below that known to elicit reactions. The dimethylglyoxime test is probably not sensitive enough to detect the minimal sensitization level of nickel.

The case of a lady who was sensitive to nickel and reacted to a stainless steel aneurysm clip used for gallbladder surgery has also been reported.[42] She developed an excematous reaction around the incision site, frequent bouts of pain in the region of the clip, and angioedema 3 months after surgery. Removal of the clip at 6 months after the gallbladder surgery relieved the symptoms within 2 weeks. A case has also been reported[43] of a lady with nickel sensitivity who rejected a nickel-containing heart valve prosthesis; this was replaced with a second nickel-containing prosthesis, and this was also rejected. She is tolerating a nickel-free prosthesis.

b. Dental Surgery

There are reports in the literature of reactivity to metal dental appliances. In one report[44] there was oral sensitivity to stainless steel wire used for mandibular immobilization. This patient was found to be sensitive to nickel and a history of reaction to jewelry was obtained. In another report, dermatitis resulted from a sensitivity reaction to a chrome-cobalt denture.[45] A complicated case with stainless steel wire and Erlich's arch bars in a jaw fracture has been described.[46] Within 8 months there was bone resorption under the stainless steel cerclage wire, and there was an infiltrate. Within 1 month after removal of the wire the alveolar process was normal and the infiltrate gone, and by 2 months there was evidence of bony healing. There had been no previous history of nickel sensitivity, but the patient was skin test positive to nickel when tested before removal of the wire. Sensitivity to a chrome-alloy dental plate has been reported in a patient who had a history of dermatitis to a metal buckle.[47] The symptoms appeared in 4 weeks and were manifest as an acute generalized eczema. The patient was skin test positive to nickel and cobalt. The eczema flared up again in a year and the plate was removed in 1½ years. The problem was assumed to be due to cobalt sensitivity and the presence of the dental plate. The oral mucosa appeared normal but was sore and tingling while the skin was reactive.

This last case points out the questions associated with sensitivity in the oral mucosa and the problems of antigens entering by oral route. The continual washing, diluting, and buffering effects of saliva undoubtedly interfere with the binding of the metal ions to the tissue proteins and the developent of sensitivity. It is also possible that the oral mucosa does not have the protein that bind metals or that they bind preferentially to materials in saliva. For some reason, the sensitivity often appears to be manifest at sites other than the oral mucosa.

The other interesting aspect of sensitivity to dental materials is that of an unresolved question in immunology: immunity and route of administration. We use oral vaccines very successfully, but these are living organisms that replicate in the tissue. We also know that there are food allergies that are elicited by ingestion. However, it was also shown in early immunology studies [48] that if the nonliving antigen, especially a hapten, was fed to an animal, sensitivity would not develop (tolerance) or would develop with more difficulty following the application of the antigen onto or into the skin. It has also been shown in guinea pigs that when nickel had been ingested by intratracheal or gastric intubation[49] the development of the sensitivity response was delayed.

There is much active work on this subject, and it would appear that with some antigens the ingestion of the antigen will prevent or delay the development of sensitivity, but that once sensitivity has developed, oral ingestion will elicit a sensitivity reaction. Certainly for nickel sensitivity it is evident that sensitive individuals can react to nickel in food or in food prepared with stainless steel utensils.[50-52]

The question of hypersensitivity to dental materials also brings up the subject of sensitivity to other metals such as gold, mercury, and copper. These sensitivities are rare, but reported. When attributing a reaction to gold, mercury, or copper, one must take care to exclude the presence of nickel, cobalt, or chromium in the alloy.[53,54]

c. Orthopedic Surgery

The bulk of the literature on sensitivity reactions to metallic implants relates to those used in orthopedic surgery. The most significant problem has been associated with the metal-metal total joint replacements.[55-62] The wear products from the articulation of the two metal surfaces are sufficient to elicit a local sensitivity reaction associated with loosening the device. Loosening of the prosthesis was correlated with a positive skin test to nickel, cobalt, or chromium. In a few of these patients a history of previous sensitivity to metals was obtained. There is no evidence for or against the premise that the metal-metal total joints *cause* sensitivity, since the patients were not tested prior to insertion of the total joint. There is evidence that the debris will *elicit* the sensitivity response in some individuals.

In none of the articles on metal sensitivity and total joint replacement do the authors conclude that sensitivity is the only cause of loosening. Infection, trauma, and faulty insertion of the device are far more common modes of failure. However, the absence of skin sensitivity in a series of patients with loose total hips does not rule out sensitivity as a cause of loosening in other patients. A recent report on 20 patients with loose total hip replacements[63] should be carefully scrutinized before excluding sensitivity as a possible cause of loosening. The authors have the right to state that since their 20 patients were skin test negative, implant failure was not correlated with a positive skin test. However, because *their* 20 patients were negative, this does not imply, as does their abstract, that sensitivity reactions to implants do not occur. This article also poses other problems that need to be considered when evaluating sensitivity. The authors carried out many tests on these patients, and the results are fascinating. In all cases the metal ion applied to the skin gave no evidence of sensitivity; the details of testing are not available, but one should assume that the procedure was standard. In 15 to 17 patients, discoloration of the tissue was noted, and there was necrotic tissue. In all patients there was evidence of inflammation, and in half the patients plasma cells (which are antibody-producing cells) were found in the tissue. The metallic debris was not intimately associated with the chronic inflammation.

The authors[63] also looked for migration inhibition factor and blastogenic factor in the actual joint fluid in five patients. The correlation of these lymphokines and cell-mediated immunity has been based on the in vitro production of the lymphokines by stimulated lymphocytes. The in vivo demonstration has been limited. In this articule, two of the five patients tested had lymphokines in the joint fluid. The authors ruled out the MIF on the basis of molecular size, assuming in vivo to be the same as in vitro. The blastogenic factor was ruled out because it was present in middle dilutions only (a common biological effect of drugs and hormones) and because it had minimal stimulation. The detection of these lymphokines is fascinating. Despite the presence of plasma cells in the tissues, the possible role of antibody was never mentioned.

Since the skin test was negative, they concluded there was no sensitivity. The possibility of false negatives on the skin test was not considered. It is our contention that it is possible that the hapten-protein complex formed in the deep tissue may be different antigenically from the hapten-protein complex of skin proteins. The cell-mediated response recognizes both the hapten and the carrier, whereas antibody will react to the hapten regardless of the carrier. Thus negative skin tests cannot rule out the presence of a sensitivity reaction — even of the cell-mediated type. The cases reported were, therefore, better classified as loosening of total hips not related to *skin* sensitivity to metals.

It is important to note that the total joint failures associated with sensitivity reactions were all of the metal-metal appliances. The problem with the metal-metal total joints arises from the metallic debris from the articulation of the two surfaces. Hemi-arthroplasty of the hip has similar problems due to abrasion and corrosion from metal-bone contact. In general, replacement of the prosthesis with one articulating with plastic has been successful. We have seen cases of loosening and pain in hip hemi-arthroplasties associated with sensitivity[16,17,42,64] and replacement with a metal-plastic total hip has been well tolerated. The stems of total hips are fully embedded in poly (methylmethacrylate) in the femoral canal so there is no bone-metal contact, and the ball articulates with the soft plastic so there is no metallic debris. The only tissue contact is between the fluid and the highly polished neck. It has been our experience and that of others that these appliances seem to be well tolerated by metal-sensitive individuals. In fact, the first patient in our series was a lady with a total hip replacement who reacted to the stainless steel wire used to reattach the greater trochanter. The wires were removed, and dacron suture was used for reattachment. She has done well now for 4 years with the stainless steel total hip. It is probably important to have the cement fully line the femoral component to prevent implant-bone contact, and to avoid the use of wire if an osteotomy is performed.

There are reports in the literature on sensitivity reactions to other orthopedic implants. These involve metal-bone contact as with the isolated bone screws,[65,66] the intramedullary nail,[67] or bone-metal-metal contact as with the bone plates and screws.[68-72] The devices have been of stainless steel and of cobalt-chrome. In most cases the sensitivity has been associated with a dermatitis or muscle reaction which resolved upon removal of the device. In our own series we have evidence in 45 patients of sensitivity on the LIF test (nonmigrators) with only a few cases of dermatitis, while others are associated with pain and/or loosening. The conclusions we draw from our data to date and from our animal studies are that these appliances are tolerated well for several months by individuals with metal sensitivity. Then, symptoms of pain, effusion, or loosening of the device may appear. The presence of sensitivity and symptoms seems to be associated with abrasion or corrosion of the implant. Apparently the implant must be in place several months before metal ion release is sufficient to elicit sensitivity.

With the organization and continued support by Michael B. Mayor, M.D., and the cooperation of the rest of the orthopedic staff, we are undertaking a study on the orthopedic patients at Mary Hitchcock Memorial Hospital, Hanover, N.H. to assess the role of metal sensitivity in the response to metallic internal fixation devices and joint arthroplasties. All patients receiving metallic implants are being tested for sensitivity by the LIF test at the time of implantation. LIF tests are also being done at various follow-up visits after implantation. All patients having metallic implants removed are also being tested for sensitivity. This should enable us to correlate morbidity with sensitivity. Skin testing is not being done because of the risk of sensitization.

Individuals who are interested in the problem of metal sensitivity and metallic implants in patients would do well to review the critique by Fisher.[73] Fisher has maintained for a long time that sensitivity reactions to stainless steel cannot occur. His discussion of the literature is of value in describing what is required by the dermatologist to prove it is sensitive. It also points out the areas of frustration involved for the physician and patient. The final points he makes are worth noting and commenting on.

1. "The patients must be patch test positive to 2.5% aqueous $NiSO_4$." Certainly any patch testing must be done with care and with the correct solutions. However, we are of the belief that there are sensitivity reactions with negative skin tests because of antibody or a different metal-protein antigen.
2. "There should be a positive test for nickel on the removed metal alloy with the dimethylglyoxime test." We believe this test is not very sensitive and know it is difficult to interpret in the presence of tissue and blood. After thorough ultrasonic cleaning, the test can be read, but the results are meaningless since any ionic or protein-bound nickel will probably have been washed off. We have had positive DMG test reactions on retrieved implant as well as on some accelerated corrosion specimens. It is the metallurgical analysis of the implants that is the most useful and meaningful. Examination even with a hand lens can often reveal evidence of metal loss as polished areas or pitted areas due to wear and corrosion. If metal has been lost from the implant, the lost metal is in the patient's tissues surrounding the implant. The physiologic solutions we have used for accelerated corrosion tests have also been tested for nickel with DMG using ASTM D-1886.[40] These measurements have shown significant (3 to 400 $\mu g/m\ell$) concentrations of nickel in solution, especially when the solutions contained serum proteins. It is not the reactive state of the metal after removal that is of importance; it is the reactive state in vivo that determines the possible patient reaction.
3. "The patient should show a positive reaction to the corroded implant and not to a noncorroded implant." His point that the reaction may be due to irritation and pressure is well-founded, and care should be taken. However, the alloys used for surgical implants are all self-passivating, which means that after removal and cleaning, the surface has reestablished a passive oxide layer which makes a previously corroded implant not unlike a noncorroded implant. As with item two, the issue is not the condition of the implant after removal, but the dynamics of the metal-tissue interactions in vivo that are important. If the implant is corroded, then the tissues have been exposed to the corrosion products.
4. "The dermatitis should clear within a week after implant removal." The problem here arises from general problem of cause and effect. The dermatitis should clear when the source of sensitivity is removed. The level of metal ions and debris in some patients, especially those with metal-metal total joints, is quite high, and it may take longer than a week to reduce the level of the offending material sufficiently. The reaction is to the corrosion products, not to the implant, and these products may not be removed with implant removal.

Fisher makes a statement in another article[74] which needs some clarification, claiming that cobalt-chromium alloys and stainless steel do not yield chromium and thus do not cause sensitivity. It is true that these metals passivate by the formation of a chrome oxide film on the surface. However, analysis of tissues from patients and animals with metal implants certainly show that chromium is released into the tissues.[28,35] In our experience and in our survey of the literature and the survey of Polak et al.,[22] the bulk of the reactions have been to nickel and cobalt, with chromium sensitivities occurring only rarely. An implant sensitivity reaction due to a sensitivity to chromium alone is only occasionally observed, and its importance remains to be assessed. Clearly the major problem is with nickel and cobalt.

2. Sensitivity Reaction to Metallic Implants in Experimental Animals

There are many studies on tissue reactions to metallic implants for toxicity and biocompatibility and release of metal ions, and these have been reviewed elsewhere.[28,29,33]

The studies on tissue reactions to metallic implants in experimental animals made sensitive to metal salts are limited. One group[29] has been placing metal implants of steel (Cr Ni Mo 18.10 steel,) with different surface preparations under the muscles of guinea pigs, some of whom were sensitive to chromium or nickel. Tissue reaction to the implant and elemental analysis of the internal organs revealed the differences were associated with the corrosion resistance of the implant surface rather than to the immunologic status of the guinea pig. This would certainly agree with the clinical findings of the time course for development of sensitivity. There must be degradation of the implant to elicit a reaction of sensitivity. The tissue reaction at the site of the implant was not thoroughly described, so the small differences between the groups are difficult to interpret, but apparently there was not a dramatic difference.

We have been modeling the problem of metal sensitivity and orthopedic implants using the rabbit and the guinea pig.[16,17,26] We have been successful in causing metal sensitivity in rabbits by the injection of nickel or cobalt chloride. These animals were sensitive by skin test and by the LIF test. Surgical grade stainless steel screws were placed in the proximal humerus to model the use of isolated bone screws in the human. The tissue reaction to the screw was then studied at 3, 9, and 16 weeks. The response to the screw in the unsensitized animals was to form bone at the interface. The response of the sensitized animals was an inflammatory reaction at the interface.[26] The most severe reactions, which included areas of necrosis, were in animals showing no migration in the LIF test. Similar studies are under way in the guinea pig, and preliminary evidence shows a similar inflammatory response by sensitized animals. Since the sensitivity response of the guinea pig to chromium is so well studied,[22] the use of bone screws in chromium-sensitized guinea pig should provide information on the effects of chromium sensitivity on tissue response to metallic implants.

V. CONCLUSIONS

In conclusion, it is evident from the reports in the literature that sensitivity reactions to metallic biomaterials do occur. These reactions may take a variety of forms including dermatological reactions, pain, swelling, or loosening of the implant. Considering the hundreds of thousands of metallic implants that are used each year in surgery, the reaction rate is very small. The problem has arisen mostly with the use of metal-metal total joint prostheses or devices with bone-metal contact that lead to the generation of metallic debris or situations of fretting or crevice corrosion. The patients with the devices tolerate them for quite some time before problems develop. In order for the implant to elicit sensitivity, it must release metal ions, and thus the quality of the metal is of extreme importance.

REFERENCES

1. **Elves, M. W.**, Immunological aspects of biomaterials, in *Fundamental Aspects of Biocompatibility*, Vol. 2 Williams, D. F., Ed., CRC Press, Boca Raton, F., chap. 9, in press.
2. **Bloom, B. R. and David, J. R., Eds.** *In Vitro Methods in Cell-Mediated and Tumor Immunity*, Academic Press, New York, 1976.
3. **Rose, N. R. and Friedman, H., Eds.**, *Manual of Clinical Immunology*, American Society for Microbiology, Washington, D.C., 1976
4. **Horton, J. E., Oppenheim, J. J., Mergenhagen, S. E., and Raisz, L. G.**, Macrophage-lymphocyte synergy in the production of osteoclast activating factor, *J. Immunol.*, 113, 1278, 1974.
5. **Mundy, G. R., Raisz, L. G., Shapiro, J. L., Bandelin, J. G., and Turcotte, R. I.**, Big and little forms of osteoclast activating factor, *J. Clin. Invest.*, 60, 122, 1977.
6. **Horton, J. E., Ed.**, *Mechanisms of Localized Bone Loss*, Information Retrieval, Washington, 1978, WE 200 14486 1977.
7. **McConnell, L. H., Fink, J. N., Schlueter, D. P., and Schmidt, M. G.**, Asthma caused by nickel sensitivity, *Ann. Intern. Med.*, 78, 888, 1973.
8. **Williams, C. D., Jr.**, Asthma and metals, *Ann. Intern. Med.*, 79, 761, 1973.
9. **Fisher, A. A.**, *Contact Dermatitis*, Lea and Febiger, Philadelphia, 1973, 87.
10. **Elves, M. W.**, Transformation in the presence of metals of lymphocytes from patients with total joint prostheses, *J. Pathol.*, 122, 35, 1977.
11. **Hutchinson, F., MacLeod, T. M., and Raffle, E. J.**, Nickel hypersensitivity, *Br. J. Dermatol.*, 93, 557, 1975.
12. **Svejgaard, E., Morling, N., Svejgaard, A., and Veien, N. K.**, Lymphocyte transformation induced by nickel sulfate: an *in vitro* study of subjects with and without a positive nickel patch test, *Acta Dermatol. Venereol.*, 58, 245, 1978.
13. **Veien, N. K. and Svejgaard, E.**, Lymphocyte transformation in patients with cobalt dermatitis, *Br. J. Dermatol.*, 99(2), 191, 1978.
14. **Jordan, W. P., Jr. and Dvorak, J.**, Leukocyte migration inhibition assay (LIF) in nickel contact dermatitis, *Arc. Dermatol.*, 112, 1741, 1976.
15. **Mirza, A. M., Perera, M. G., Maccia, C. A., Dziubynsky, O. G., and Bernstein, I. L.**, Leukocyte migration inhibition in nickel dermatitis, *Int. Arch. Allergy Appl. Immunol.*, 49, 782, 1975.
16. **Merritt, K., Mayor, M. B., and Brown, S. A.**, Evaluation of sensitivity to metallic implants, in *Evaluation of Biomaterials*, Winter, G. D., Leray, J., and de Groot, K., Eds., John Wiley & Sons, London, 1980, 315.
17. **Merritt, K., Mayor, M. B., and Brown, S. A.**, Metal allergy and implants: concepts and clinical significance, in *Internal Fixation of Fractures*, Uhtoff, H., Ed., Springer-Verlag, Berlin, 1980.
18. **Click, R. E., Benck, L., and Alter, B. J.**, Immune responses *in vitro*. I. Culture conditions for antibody synthesis, *Cell Immunol.*, 3, 264, 1972.
19. **Rae, T.**, Comparative laboratory studies on the production of soluble and particulate metal by total joint prosthesis, in press.
20. **Cohen, H. A.**, Experimental production of circulating antibodies to chromium, *J. Invest. Dermatol.*, 38, 13, 1962.
21. **Farr, R. S.**, A quantitative immunochemical measure of the primary interaction between I*BSA and antibody, *J. Infect. Dis.*, 103, 239, 1958.
22. **Polak, L., Turk, J. L., and Frey, J. R.**, Studies on contact hypersensitivity to chromium compounds, *Prog. Allergy*, 17, 145, 1973.
23. **Wahlberg, J. E.**, Immunoglobulin E, atopy, and nickel allergy, *Cutis*, 18, 715, 1976.
24. **Turk, J. L. and Parker, D.**, Sensitization with Cr, Ni, and Zr salts and allergenic type granuloma formation in the guinea pig, *J. Invest. Dermatol.*, 68, 341, 1977.
25. **Wahlberg, J. C. and Boman, A.**, Sensitization and testing of guinea pigs with cobalt chloride, *Contact Dermatitis*, 4, 128, 1978.
26. **Merritt, K. and Brown, S. A.**, Tissue reaction and metal sensitivity: an animal study, *Acta Orthop. Scand.*, 51, 403, 1980.
27. **Fisher, A. A.**, Safety of stainless steel in nickel sensitivity, *JAMA*, 221, 1279, 1972.
28. **Laing, P. G., Ferguson, A. B., Jr. and Hodge, E. S.**, Tissue reaction in rabbit muscle exposed to metallic implants, *J. Biomed. Mater. Res.*, 1, 35, 1967.
29. **Von Höhndorf, H.**, Untersuchungen zur Verträglichkeit des x5 CrNiMo 18.10 Stahls (Königsee) im Organismus, *Z. Exp. Chir.*, 10, 120, 1977.
30. **Mears, D. C.**, Electron probe microanalysis of tissue and cells from implant areas, *J. Bone Jr. Surg. Br. Vol.*, 45, 567, 1966.

31. Williams, D. F. and Meachim, G., A combined metallurgical and histological study of tissue-prosthesis interactions in orthopaedic patients, *J. Biomed. Mater. Res.*, 5, 1, 1974.
32. Coleman, R. F., Herrington, J., and Scales, J. T., Concentration of wear products in hair, blood and urine after total hip replacement, *Br. Med. J.*, 1, 527, 1973.
33. Samitz, M. H. and Katz, S. A., Nickel dermatitis hazards from prostheses, *Br. J. Dermatol.*, 92, 287, 1975.
34. Charosky, C. B., Bullough, P. G., and Wilson, P. D., Jr., Total hip replacement failures, a histological evaluation, *J. Bone Jr. Surg. Am. Vol.*, 55, 49, 1973.
35. Smethurst, E. and Waterhouse, R. B., A physical examination of orthopaedic implants and adjacent tissue, *Acta Orthop. Scand.*, 49, 8, 1978.
36. Kvorning, S. A., Post-operative dermatitis following non-epidermal nickel contact, *Contact Dermatitis*, 1, 327, 1975.
37. Stoddard, J. C., Nickel sensitivity as a cause of infusion reactions, *Lancet*, 2, 741, 1960.
38. Smeenk, G. and Teunissen, P. C., Allergic reactions to nickel from infusion equipment, *Ned. Tijdschr. Geneeskd.*, 121, 4, 1977.
39. Fisher, A. A., Nickel - the ubiquitous contact allergen, *Cutis*, 22, 544, 1978.
40. Standard methods of test for nickel in water, *ASTM Annual Book of Standards*, American Society for testing & Materials, Philadelphia, 1979.
41. Malten, K. E. and Spruit, D., The relative importance of various environmental exposures to nickel in causing contact hypersensitivity, *Acta Derm. Venereol.*, 49, 14, 1969.
42. Brown, S. A., Shafer, J. W., Mayor, M. B., and Merritt, K., Leucocyte migration inhibition test for metal sensitivity, paper presented at Brunel University, Uxbridge, England, Materials for Use in Medicine and Biology. Problems of Biocompatibility, September 1976.
43. Lyell, A., Bain, W. H., and Thomson, R. M., Repeated failure of nickel-containing prosthetic heart valves in a patient allergic to nickel, *Lancet*, 2, 657, 1978.
44. Schriver, W. R., Shereff, R. H., Domnitz, J. M., Swintak, E. F., and Civjan, S., Allergic response to stainless steel wire, *Oral Surg.*, 42, 578, 1976.
45. Brendlinger, D. L. and Tarsitano, J. J., Generalized dermatitis due to sensitivity to a chrome-cobalt removable partial denture, *J. Am. Dent. Assoc.*, 81, 392, 1970.
46. Roed-Petersen, B., Roed-Petersen, J., and Jørgensen, K. D., Nickel allergy and osteomyelitis in a patient with metal osteosynthesis of a jaw fracture, *Contact Dermatitis*, 5, 108, 1979.
47. Levantine, A. V., Sensitivity to metal dental plate, *Proc. R. Soc. Med.*, 67, 1007, 1974.
48. Smith, R. T., Immunological tolerance of nonliving antigens, *Adv. Immunol.*, 1, 67, 1961.
49. Parker, D. and Turk, J. L., Delay in the development of the allergic response to metals following intratracheal instillation, *Int. Arch. Allergy Appl. Immunol.*, 57, 289, 1978.
50. Fisher, A. A., Dermatitis due to contaminated food, *Cutis*, 17, 229, 1976.
51. Kaaber, K., Veien, N. K., and Tjell, J. C., Low nickel diet in the treatment of patients with chronic nickel dermatitis, *Br. J. Dermatol.*, 98, 197, 1978.
52. Mennee, T. and Thorboe, A., Nickel dermatitis - nickel excretion, *Contact Dermatitis*, 2, 353, 1976.
53. Cohen, S. R., A review of the health hazards from copper exposure, *J. Occup. Med.*, 16, 621, 1974.
54. Dick, D., Local gold toxicity, *Br. Med. J.*, 1, 51, 1977.
55. Benson, M. K. D., Goodwin, P. G., and Boostoff, J., Metal sensitivity in patients with joint replacement arthroplasties, *Br. Med. J.*, 4, 374, 1975.
56. Duetman, R., Mulder, T. J., Brian, R., and Nater, J. P., Metal sensitivity before and after total hip arthroplasty, *J. Bone Jt. Surg.*, 59, 862, 1977.
57. Elves, M. W., Wilson, J. N., Scales, J. T., and Kemp, H. B. S., Incidence of metal sensitivity in patients with total joint replacements, *Br. Med. J.*, 4, 376, 1975.
58. Evans, E. M., Freeman, M. A. R., Miller, A. J., and Vernon-Roberts, B., Metal sensitivity as a cause of bone necrosis and loosening of the prosthesis in total joint replacement, *J. Bone Jt. Surg. Br. Vol.*, 56, 626, 1974.
59. Gschwend, N., Scherrer, H., Dybowski, R., Hohermuth, H., Razavi, R., Stabuli, A., Wüthrich, B., and Scherrer, A., Allergologische Probleme in der Orthopädie, *Orthopäde* 6, 197, 1977.
60. Jones, D. A. and Lucas, K., Metal sensitivity in patients with joint prosthesis, *Br. Med. J.*, 4, 647, 1975.
61. Munro-Ashman, D. and Miller, A. J., Rejection of metal to metal prosthesis and skin sensitivity to cobalt, *Contact Dermatitis.*, 2, 65, 1976.
62. Ridley, C. M., How relevant is cobalt sensitivity in a patient with an unsatisfactory total knee replacement, *Clin. Exp. Dermatol.*, 2, 401, 1977.
63. Brown, G. C., Lockshin, M. D., Salvati, E. A., and Bullough, P. G., Sensitivity to metal as a possible cause of sterile loosening after cobalt-chromium total hip-replacement arthroplasty, *J. Bone Jt. Surg. Am. Vol.*, V164, 1977.

64. Brown, S. A., Shafer, J. W., Mayor, M. B., and Merritt, K., Metal allergy and its role in the biocompatibility of orthopaedic implants, in *Proc. 5th N. E. Bioengineering Conf.*, Pergamon Press, Oxford, 1977, 6.
65. Barranco, V. P. and Soloman, H., Eczematous dermatitis from nickel, *JAMA*, 220, 1244, 1972.
66. Kubba, R. and Champion, R. H., Nickel sensitivity resembling bullous pemphigoid, *Br. J. Dermatol.*, 93 (Suppl. 11), 41, 1975.
67. McKenzie, A. W., Aitken, C. V. E., and Risdill-Smith, R., Urticaria after insertion of Smith-Petersen vitallium nail, *Br. Med. J.*, 4, 36, 1967.
68. Cramers, M. and Lucht, L., Metal sensitivity in patients treated for tibial fractures with plates of stainless steel, *Acta Orthop. Scand.*, 48, 245, 1977.
69. Foussereau, J. and Laungier, P., Allergic eczemas from metallic foreign bodies, *Trans. St. John's Hosp. Dermatol. Soc.*, 52, 220, 1966.
70. Halpin, D. S., An unusual reaction in muscle in association with a vitallium plate: a report of possible metal hypersensitivity, *J. Bone Jt. Surg. Am. Vol.*, 57, 451, 1975.
71. Pegum, J. S., Nickel allergy, *Lancet*, 1, 674, 1974.
72. Symeonides, P. P., Paschaloglou, C., and Papageorgiou, S., An allergic reaction after internal fixation of a fracture using a vitallium plate, *J. All. Clin. Immunol.*, 51, 251, 1973.
73. Fisher, A. A., Allergic dermatitis presumably due to metallic foreign bodies containing nickel or cobalt, *Cutis*, 19, 285, 1977.
74. Fisher, A. A. and Brancaccio, R. R., Spark source mass spectrographic study of metal allergenic substances on the skin, *J. Invest. Dermatol.*, 68, 394, 1977.

Polymer-Based Materials

Chapter 4

INTRODUCTION TO THE TOXICOLOGY OF POLYMER-BASED MATERIALS

David F. Williams

TABLE OF CONTENTS

I.	Introduction	50
II.	The Toxicology of Polymer Macromolecules	51
	A. Inert High Polymers	51
	B. Low Molecular Weight Components	51
	C. Polymer Degradation Products	52
	D. Particles Derived From Polymers	53
	E. Polymer Solutions	53
III.	Hypersensitivity to Polymers	53
IV.	Toxicological Aspects of Polymer Sterilization	54
References		55

I. INTRODUCTION

As noted elsewhere in this series, there are several differences between metallic and polymer-based materials from the toxicological point of view. These include the following:

1. Metals and alloys are usually well defined and readily characterized. Their toxicology is easily correlated to the major elements present and only marginally less readily associated with minor and impurity elements. Polymeric-based materials, on the other hand, are usually chemically and structurally complex. Even if it is a homopolymer that is involved, there will be a distribution of molecular weights which will influence any reaction with the environment. There may also be a significant amount of monomer still present. In addition, there are numerous additives which are present in commercially available polymers, and each of these poses quite different toxicological hazards.
2. Toxicity arising from an implanted metal must be consequent to release of metal ions into the tissue. These metal ions are no different from those in the metal itself and, generally, the toxicology of any metallic compounds produced will be similar to that associated with the parent metal. Molecules from a polymer or any organic substance used as an additive may be similarly released into the tissues. However, these molecules may break down into different structures, the toxicology of which may be quite different to that of the parent polymer.
3. Metallic elements occur naturally and their toxicology is associated with the effect that raised levels have on the body. Many metals are essential and most are found in measurable amounts in normal healthy humans. Toxicological effects only arise when the level of any metal exceeds the limit that can normally be dealt with by homeostatic mechanisms. On the other hand, most polymers and many of the organic additives present are totally synthetic. The presence of one molecule of such a material in the tissue can be considered an excess, and it is in this light that their toxicology must be viewed.
4. Metals are readily sterilized by steam autoclaving. Polymers, with much lower softening temperatures are more difficult to sterilize and normally either gamma ray or chemical sterilization must be performed. These processes, and especially the latter, which may result in polymers containing residues of the chemicals, can influence the toxicology.

These differences mean that it is not possible to treat the toxicology and systemic biocompatibility of polymeric-based materials in the same systematic way that was used for metals where relevant individual metals and specific types of metal-tissue interactions were discussed. Instead, a few selected examples of the toxicology of substances found in plastics have been chosen for review which will give the reader some idea of the extent of the effects observed. It will be apparent from the foregoing comments that the most significant effects will arise from residual or derived monomers and from additives. Therefore, selected monomers are discussed in great detail by Haley in Chapter 5. Perhaps the most widely used biomedical polymeric material is polymethyl methacrylate, and it is this substance that provides one of the biggest problems with respect to residual monomer. The toxicology and biocompatibility problems associated with methylmethacrylate monomer are therefore discussed by Borchard in Chapter 6.

The subject of additives in plastics is very difficult to treat in this context as there are so many commercially available additives which can be used with a variety of polymers, all of which have their own toxicological characteristics.

Moreover, the vast majority of the data available refers to toxicity arising from the oral administration routes and may not be particularly relevant to the usage of biomedical polymers. In order to draw attention to the potential problems, the author presents the toxicology of some common additives, in tabular form, in Chapter 7 and reviews in some detail the clinical effects of one important plasticizer, di-2-ethylhexyl phthalate.

The remainder of this chapter is devoted to a discussion of a few issues which are relevant to the biocompatibility and toxicology of polymeric materials. First, there is the question of the toxicity of the high polymers themselves. Second, it is necessary to review the subject of hypersensitivity to biomedical polymeric materials and third, the toxicity of chemical sterilizing solutions in the context of medical polymers is briefly discussed.

The final three chapters in this volume address a couple of related subjects that have attracted much interest in recent years. These are the pharmacological behavior of polymers and the possibility of their therapeutic use, which is discussed by Kopeček in Chapter 8, and the use of polymers for the delivery of drugs. Chang discusses the technique of microencapsulation for this purpose in Chapter 9, while Bagnall presents a philosophical view of controlled drug release in Chapter 10.

II. THE TOXICOLOGY OF POLYMER MACROMOLECULES

A. Inert High Polymers

Most current applications of polymers in medical or dental devices involve the use of high molecular weight polymers that are chemically very inert and insoluble in physiological fluids. It has always been assumed, therefore, that such polymers should have minimal toxicity and this certainly appears to be the case. Several experiments have been reported in which large quantities of such polymers have been included in the diets of experimental animals and usually no adverse effects have been found. Generally the polymers, by virtue of their high molecular weight and, where appropriate, cross-linking between molecules, are not absorbed in the gastrointestinal tract and are excreted in the feces unchanged. Lefaux[1] reports experiments in which polypropylene, vinyl chloride-vinylidene chloride copolymers, polycaprolactam, polycarbonate, polystyrene, polyethylene, silicone rubber, and polytetrafluoroethylene have all failed to produce signs of toxicity after oral administration. For example, an attempt to evaluate the oral LD_{50} for polyethylene in rats failed because it was impractical to deliver doses more than 7.95 g/kg at which level no lesions were demonstrated histopathologically nor were the rats in anyway affected. Similarly, nothing anomalous was reported in rats fed a diet containing 25% polytetrafluoroethylene.

Inert insoluble pure high molecular weight polymers appear to have no dermatological effects, and there is abundant evidence to show that such materials produce minimal response upon s.c. or i.m. implantation, in terms of the tissue responses described in *Fundamental Aspects of Biocompatibility*[17] of this series.

The only deviations that can arise from this benign behavior are when very low molecular weight components are present within the molecular weight distribution, when the polymer chemically degrades within the physiological environment, the reaction products being of low molecular weight, when the polymer degrades mechanically to produce small fragments or droplets which can produce tissue irritation purely because of their physical form, or when the polymer is used in solution and injected into the tissues. These possibilities are now considered briefly.

B. Low Molecular Weight Components

The process of polymerization results in a spectrum of molecular sizes in the result-

Table 1
LD$_{50}$ VALUES FOR ACRYLATE MONOMERS[2]

Monomer	LD$_{50}$ (mol/10^{6}g)
Methyl acrylate	2.9
Ethyl acrylate	6.0
Butyl acrylate	6.7
Isobutyl acrylate	5.9
2-Ethylhexyl acrylate	7.2

ing material. Typically there is a normal distribution of molecular weights which implies that there will be some very small molecules and some very large molecules in addition to the majority of intermediate sizes. This arises in addition polymers because of the random way in which monomer molecules take part in the propagation reactions and because, with a dense polymer, diffusion of small molecules becomes more and more difficult as the degree of polymerization increases. With most high polymers, the polymerization conditions are such that even the low molecular weight components still have a high molecular weight in comparison to the monomer. With a polyolefin, for example, that has a weight average molecular weight in excess of 10^6, there will be few molecules of less than 10^5. This molecular weight distribution should not, therefore, pose any toxicological hazards.

However, some polymerization processes take place under conditions which allow a much wider distribution and may even result in a significant quantity of residual monomer. Polymethylmethacrylate is the obvious example. The toxicology of monomers is discussed in the next two chapters, but it is of interest here to note the general dependence of toxicity on the molecular weight of series of organic substances. It would appear from this that with molecular chains of structure $(R)_n$, the greater the value of n the less toxic will be the molecule. Certainly with many series of organic compounds of similar structure, the higher the molecular weight the lower the toxicity. For example, Table 1 shows the oral LD$_{50}$ of a series of acrylate monomers where this pattern is clearly in evidence.[2] Because of the tremendous variation in the toxic effect exerted by different molecular species and groups, and the influence of variables such as chain branching, it is not possible to suggest any lower limit of molecular weight which avoids any toxicological hazard. This would clearly have to be identified in each individual case. It should be noted, however, that the body is unlikely to be able to catabolize polymers of large molecular chain length. Michael and Coots, for example, have studied the metabolism of glycerol and polyglycerol and found that the latter was absorbed and excreted rapidly by rats without being catabolized.[3]

C. Polymer Degradation Products

Mechanisms of polymer degradation have been reviewed by Gilding in *Fundamental Aspects of Biocompatibility*.[20] It is generally agreed that high molecular weight, hydrophic, nonhydrolyzable polymers do not normally degrade in physiological environments. It is conceivable, as discussed by the author,[17] that enzymes could cause the degradation of some of these polymers, but this has yet to be demonstrated unequivocally. The immunity of this type of polymer to degradation arises from the fact that it is usually physical agencies such as heat, light, and ionizing radiations which cause polymer degradation and these are clearly inoperative in the body. On the other hand, any polymer which is susceptible to hydrolysis is a clear candidate for degradation in the aqueous physiological environment. The nature of the degradation products in these cases will clearly depend on the chemistry of the polymer in question, but it is likely that very small molecules, including monomers and dimers will be produced. This subject is discussed by Gilding in *Biocompatibility of Clinical Implant Materials*,[8] but it is relevant to note here that if a polymer is selected for an implant application which relies on its intentional degradation, the polymer must be chosen which has readily and harmlessly metabolized degradation products. Polyglycolic and polylactic acids provide good examples here where the degradation products are glycolic and lactic acid which are themselves featured in normal metabolic pathways. It is also important, of course, that the rate of formation of these low molecular weight products is slow enough for them to be readily absorbed without causing their accumulation in the tissue around the device.

D. Particles Derived From Polymers

The above discussion has been solely concerned with the chemical interactions between polymers (and their derivatives) and tissues. In the context of biomaterials it is also necessary to consider physical or mechanical interactions. It is beyond the scope of this volume on the systemic aspects of biocompatibility to review the nature of these interactions and how they influence the local biocompatibility of implanted polymers, but it is relevant to discuss briefly the possibility of adverse effects in the tissues arising from the mechanical breakdown of polymeric devices.

The most important phenomenon here is that of the wear of implanted polymers, especially those used in orthopedic joint prostheses. It has been recognized for a long time that particulate wear products can have a most significant effect on tissues, even if the polymer itself is chemically inert with little or no effect on tissues when in bulk form. The classic example here is that of polytetrafluoroethylene, one of the most inert polymers, but it is highly irritant to tissues in particulate form.[4] Many other polymers have been shown to be equally irritant when presented to the tissue in the same form and it is clear that this has little to do with the chemistry of the material, but is more related to the physical interaction between cells and particles of equivalent dimensions.

In a similar manner, polydimethylsiloxane, when cured as solid silicone rubber, elicites minimal response from tissues. If low viscosity silicone fluids are injected into tissue, as described by Donahue et al., for example,[5] small droplets can form which invoke a fibrocystic and granuloma-like inflammatory response.

E. Polymer Solutions

There has been increasing interest in the use of soluble polymers in medicine and this subject is reviewed thoroughly by Kopeček in Chapter 8. Generally the effect of injecting polymer solutions into tissues depends on the molecular weight of the polymer and the chemical structure of the molecule. A very high molecular weight polymer may find elimination from the body difficult and be retained in the liver, spleen, or kidneys where it can have adverse effects. Low molecular weight polymers are usually rapidly eliminated.

III. HYPERSENSITIVITY TO POLYMERS

There has been much discussion and controversy in the past on the subject of hypersensitivity to polymeric materials used in surgery, and especially those used in dental surgery. It has frequently been argued, for example, that a major factor in the etiology of denture stomatitis is hypersensitivity to acrylic resin. It is now clear, however, that as with the case of general toxicology, the cured, high molecular weight polymer is inert and only rarely could they be responsible for hypersensitivity reactions. Such effects will only normally arise when residual monomer, degradation products, or additives are present.[6]

The situation with acrylic denture bases is discussed by both Dahl and Bastiaan in *Biocompatibility of Dental Materials*[19] of this series. It is clear that methyl methacrylate is a potent sensitizer[7] and there certainly are cases where technicians handling denture base materials and patients wearing dentures show hypersensitivity responses.[8] In the latter case it is usually a cold curing acrylic that has been used, which has a greater level of residual monomer. There is also a certain amount of evidence that allergies may be associated with formaldehyde which is present in and released from acrylic denture base resins.[10]

Among the polymer systems well known for their tendency to produce allergies are the epoxy, polyester, and polyurethane resins. As pointed out by Fisher,[6] individuals may acquire dermatitis from one or more of the following chemicals used in an epoxy resin system: uncured epoxy resin, catalysts (e.g., amines, anydrides), diluents, plasticizers, solvents, fillers, and pigments. Reactants used to produce the resin are typically epichlorhydrin and bisphenol A, the latter being a known contact sensitizer. Aliphatic amine catalysts such as diethylenetriamine cause severe allergic contact dermatitis. Similarly, diluents such as butyl glycidyl ether and plasticizers such as dibutyl phthalate can cause sensitivity. Thus in the precured state, epoxy resins offer a considerable risk of inducing hypersensitivity, and Fisher estimates that 10% of all people who come into contact with uncured or curing epoxy resins develop true allergic contact dermatitis. Rather similar situations exist with the polyester and polyurethane resins. In all these cases, contact with the fully cured resin is unlikely to result in any hypersensitivity response.

The influence of catalysts is clearly seen with the two dental materials Scutan® and Impregum.® The former is used as a temporary crown and bridge material and the latter as an impression material, hypersensitivity reactions being reported at an early stage in their use.[10] The offending component was found to be the catalyst, in both cases there being sufficient in the curing material to affect both patients and dental personnel.

IV. TOXICOLOGICAL ASPECTS OF POLYMER STERILIZATION

Because most plastics soften or degrade at elevated temperatures, autoclaving cannot be used for their sterilization. The two alternative procedures are gamma radiation and chemical sterilization. With some polymers, gamma irradiation may cause small structural changes, but these will not generally affect the toxicology within the range of doses that are used. Chemical sterilization, however, can result in changes to the toxicological status and this has been the subject of much controversy. The problem arises because the substance universally used in this process is ethylene oxide. This is a gas which is absorbed by many plastics and may be subsequently released into the tissue after implantation.

Early experimental and clinical evidence indicated that ethylene oxide was retained for a finite length of time in some medical plastics, especially polyvinyl chloride, after sterilization and that toxic responses may be observed if ethylene oxide sterilized materials are placed in contact with tissues shortly after the sterilization process.[11] It is now clear that in the majority of cases complete desorption occurs within 24 hr and that, provided aeration of the plastic device is carried out for a minimum of this period, no problems should arise. McGunnigle[12] et al., for example, showed that immediately after sterilization, polyvinyl chloride contained 17,200 ppm ethylene oxide and a sample placed in a culture of L cells produced a toxic zone of more than 25 mm. Two hours later the content was 5340 ppm and the zone was 8 mm. At 6 hr the content was 1510 ppm and the zone 2.8 mm, while only 34 ppm were detected at 24 hr with no toxic zone at all. Rather similar results were obtained with a polyether polyurethane, while silicone rubber absorbed much less and the total was desorbed in 2 hr. Anderson[13] showed similar desorption characteristics for polyvinyl chloride and found that polyethylene and PTFE absorbed very little (about 0.6 mg/g), all of which was desorbed within 5 hr.

One clinically important problem is the effect of residual ethylene oxide on blood that comes into contact with sterilized plastic devices, and particularly the ability to produce hemolysis. This has been studied by O'Leary and Guess[14] and more recently by Jones.[15] Jones found that at least 2 mg/g was necessary before any cell lysis was observed.

One other aspect associated with ethylene oxide sterilization is the reaction that may occur between ethylene oxide and certain materials containing chlorides, when ethylene chlorohydrin is produced.[16] This is a substance of high acute toxicity although the full significance of this is not clear.

REFERENCES

1. **Lefaux, R.**, *Practical Toxicology of Plastics*, Iliffe, London, 1968, 48.
2. **Lawrence, W. H., Bass, G. E., Purcell, W. P., and Autian, J.**, Use of mathematical models in the study of structure-toxicity relationships of dental compounds, *J. Dent. Res.*, 51, 526, 1972.
3. **Michael, W. R. and Coots, R. H.**, Metabolism of polyglycerol and polyglycerol esters, *Toxicol. Appl. Pharmacol.*, 20, 334, 1971.
4. **Charnley, J.**, The wear of plastics materials in the hip, *Plastics in Medicine and Surgery*, Proc. Conf. at Univ. of Strathclyde, Glasgow, Plastics and Rubber Institute, London, 1975.
5. **Donahue, W. C., Nosanchuk, J. S., and Kaufer, H.**, Effect and fate of intraarticular silicone fluid, *Clin. Orthop. Rel. Res.*, 77, 307, 1971.
6. **Fisher, A. A.**, *Contact Dermatitis*, Kimpton, London, 1967, Ch. 8.
7. **Fisher, A. A.**, Allergic sensitization of the skin and oral mucosa to acrylic resin denture material, *J. Prosthet. Dent.*, 6, 593, 1956.
8. **Stungis, T. E. and Fink, J. N.**, Hypersensitivity to acrylic resin, *J. Prosthet. Dent.*, 22, 425, 1969.
9. **Wennstrom, A.**, Formaldehyde content and cytotoxicity of some denture base materials, *Odontol. Tidskr.*, 74, 212, 1966.
10. **Hartmann, K. and Vollrath, C.**, Ein Fall von Kontaktallergie bei Scutan, *Dtsch. Zahnderztebl.*, 25, 489, 1971.
11. **Rendell-Baker, L.**, Medical users' views of ethylene oxide sterilization problems. Sterile disposable devices and sterilization, Proc. HIA Tech. Symp., Health Industries Association, 1969.
12. **McGunnigle, R. G., Renner, J. A., Romano, S. J., and Abodealy, R. A.**, Residual ethylene oxide: results in medical grade tubing and effects on an in vitro biologic system, *J. Biomed. Mater. Res.*, 9, 273, 1975.
13. **Anderson, S. R.**, Ethylene oxide toxicity, *J. Lab. Clin. Med.*, 77, 346, 1971.
14. **O'Leary, R. K. and Guess, W. L.**, Toxicological studies on certain medical grade plastics sterilized by ethylene oxide, *J. Pharm. Sci.*, 57, 12, 1968.
15. **Jones, A. B.**, In vitro evaluation of haemolytic and cell culture toxicity potential of residual ethylene oxide in selected medical materials, *J. Biomed. Mater. Res.*, 13, 207, 1979.
16. **Lawrence, W. H., Turner, J. E., and Autian, J.**, Toxicity of ethylene chlorphydrin, *J. Pharm. Sci.*, 60, 568, 1971.
17. **Williams, D. F., Ed.**, *Fundamental Aspects of Biocompatibility*, CRC Press, Boca Raton, Fla., in press.
18. **Gilding, D. K.**, Biodegradable polymers, in *Biocompatibility of Clinical Implant Materials*, Vol. 2, Williams, D. F., Ed., CRC Press, Boca Raton, Fla., in press.
19. **Bastiaan, R. J.**, Denture stomatitis, in *Biocompatibility of Dental Materials*, Williams, D. F., Ed., CRC Press, Boca Raton, Fla., in press.
20. **Gilding, D. K.**, Degradation of polymers: mechanisms and implications for biomedical applications, in *Fundamental Aspects of Biocompatibility*, Williams, D. F., Ed., CRC Press, Boca Raton, Fla., in press.

Chapter 5

BIOCOMPATIBILITY OF MONOMERS

Thomas J. Haley

TABLE OF CONTENTS

I. Introduction 58

II. Vinyl Chloride..59
    A. Synthesis and Physical Properties................................59
    B. Analysis 60
    C. Toxicology...61
        1. Animal..61
        2. Human..62
            a. Allergic Manifestations 63
            b. Acroosteolysis......................................63
    D. Teratogenesis 64
        1. Animal..64
        2. Human..65
    E. Mutagenesis...65
        1. Animal..65
        2. Human..65
    F. Pharmacokinetics and Biotransformation.......................65
        1. Animal..65
            a. Absorption 65
            b. Distribution...66
            c. Metabolism 66
        2. Human..67
    G. Carcinogenesis 67
        1. Animal..67
        2. Human..71
    H. Comment..73

III. Silicones..73
    A. Synthesis and Physical Properties................................73
    B. Analysis 76
    C. Toxicology...77
        1. Animal..77
        2. Human..83
    D. Teratogenesis 86
        1. Animal..86
        2. Human..86
    E. Mutagenesis...87
        1. Animal..87
        2. Human..87
    F. Pharmacokinetics and Biotransformation.......................87
        1. Animal..87
            a. Absorption 87

| | | | b. | Distribution.................................87 |
| | | | c. | Metabolism..................................87 |

        2.   Human..88
   G.   Carcinogenesis 88
        1.   Animal...88
        2.   Human..88
   H.   Comment..88

IV.   Cyanoacrylic Adhesives...88
   A.   Synthesis and Physical Properties...............................88
   B.   Analysis...88
   C.   Toxicology...88
        1.   Animal...88
        2.   Human..91
   D.   Teratogenesis 92
        1.   Animal...92
        2.   Human..92
   E.   Mutagenesis..92
        1.   Animal...92
        2.   Human..92
   F.   Pharmacokinetics and Biotransformation........................92
        1.   Animal...92
        2.   Human..92
   G.   Carcinogenesis 92
        1.   Animal...92
        2.   Human..92
   H.   Comment..92

References..93

I. INTRODUCTION

With the scarcity and increased cost of essential metals and other materials during World War II, it became necessary to obtain substitutes which would serve in their place. The development of polymers and elastomers, therefore, became a reality. Table 1 gives a partial list of such products now available and the year of their development.[1] A whole new area concerning the toxicology of these materials has now been opened up. The monomers are the most significant in this respect, but essential to the understanding of the toxicity and metabolic fate of these materials is the knowledge, not only of the monomers themselves, but of the various additives which are used to stabilize them, the amounts of unreacted monomer in a given polymer, the additives added to stabilize the polymer, and the degree and amount of leaching of these materials into the environment with which they come in contact. This can become a serious problem where food and drugs are concerned and has resulted in restrictions on the use of polymers for packaging materials. In this chapter the toxicity of monomers used

Table 1
PLASTICS AND DATES OF
THEIR INTRODUCTION

Date	Material
1868	Cellulose nitrate
1909	Phenol-formaldehyde
1909	Cold molded
1919	Casein
1926	Alkyds
1926	Aniline-formaldehyde
1927	Cellulose acetate
1928	Polyvinyl chloride
1929	Urea-formaldehyde
1935	Ethyl cellulose
1936	Acrylic
1936	Polyvinyl acetate
1938	Cellulose acetate butyrate
1938	Polystyrene
1938	Nylon
1939	Polyvinyl acetals
1939	Melamine-formaldehyde
1939	Polyvinylidiene chloride
1942	Polyester
1942	Polyethylene
1943	Silicones
1943	Fluorocarbons
1945	Cellulose propionate
1947	Epoxy
1948	Acrylonitrile-butadiene-styrene
1956	Acetal resin
1957	Polypropylene
1957	Polycarbonate resin
1959	Chlorinated polyether

in a selected number of these polymers is discussed. The materials to be covered are vinyl chloride, silicones, and cyanoacrylic adhesives. In each case the discussion of the toxicity is preceded by a description of the synthesis techniques and the physical properties of the monomer.

II. VINYL CHLORIDE

A. Synthesis and Physical Properties

There are numerous procedures available for synthesis of vinyl chloride and all have problems associated with obtaining a good yield and reducing the possible side reactions and undesirable products. Vinyl chloride has been synthesized by reacting acetylene with hydrogen chloride in the presence or absence of $HgCl_2$ or dichloroethane with alcoholic potash. Fairhall[2] passed gaseous ethylene dichloride over alumina, activated charcoal, or pumice at elevated temperatures to obtain vinyl chloride. Deacon catalyst, a eutectic mixture of 20% or more of Cu with either KCl, NaCl, $CaCl_2$, or $PbCl_2$ on a support of alumina, silica, or other porous materials, is used in the oxychlorination of ethylene to make vinyl chloride. The ratio of Cu:K ions must be greater than 0.5:1 to allow the reaction to proceed at a temperature of 250°C. The granular catalyst is diluted with graphite, silicon carbide, or nickel and the size and porosity must be controlled; temperature control of the exothermic reaction is also critical for good yields. Rare earths are used in other catalysts. Increased HCl in the final product

Table 2
PHYSICAL AND CHEMICAL PROPERTIES OF VINYL CHLORIDE

Physical state	Gas (usually handled as a liquid under pressure)
Molecular weight	62.5
Specific gravity	0.9121 (20°/4°C)
Melting point	−159.71°C
Boiling point	−13.8°C
Vapor density	2.15
Vapor pressure	2580 mm Hg (20°C)
Solubility	sl. sol. H_2O, sol. ETOH, $(ET)_2O$, CCl_4, C_6H_6
Flash point	−78°C (open cup)
Explosive limits	lower 4% and upper 22% (vol. in air)
Autoignition temperature	472.22°C
Viscosity, liquid	0.281 centipoise at 20°C
Specific heat, liquid	0.27 cal/g C° at 20°C
Specific heat, gas	0.206 cal/g C° at 25°C, 1 atm
Critical temperature	158.4°C
Critical pressure	774.7 psia at 52.7 atm
Refractive index	$n^1_D l$ 1.4066

Note: Technical vinyl chloride contains the following impurities in mg/kg: unsaturated hydrocarbons-10; acetaldehyde-2; dichloro compounds-16; H_2O-15; HCL-2; nonvolatiles-200; Fe-0.4; and phenol as a stabilizer-25-50.

occurs when the ratios of ethylene, HCl, and O_2 are not maintained at 1:2:0.5 and the temperature is lower than 250°C. Pressures may vary from 1 to 10 atm. Packed-bed, fluidized-bed, or liquid-phase reactors may be used. Albright[3] has pyrolyzed 1,2-dichloroethane to vinyl chloride with catalytic or noncataylic processes at temperatures of 450 to 650°C and pressures of 20 to 35 atm. Vinyl chloride can be polymerized by emulsion, bulk, and solution processes.[4,5] Albright[6] has also discussed the factors involved in the suspension process for polymerizing vinyl chloride. Although Shelley and Sills[7] have discussed the procedure for storage and safe handling of vinyl chloride, the duration of storage is critical because increased times result in raised unstable polyperoxide content, increasing the explosive capability.[8] Table 2 lists the physical and chemical properties of vinyl chloride.[9] The list of impurities and the presence of phenol as a stabilizer should be noted.

B. Analysis

In any evaluation of the biological effects of a compound, its contaminants and decomposition, or metabolic products, it is essential that an accurate and sensitive analytic method be available. Baretta et al.[10] showed that vinyl chloride in air and breath samples could be determined by infrared spectrophotometry at a sensitivity of 0.05 ppm. This is below the limit of 1 ppm set by OSHA for occupation exposure. Marine pollution from vinyl chloride has been monitored by Jensen et al.[12] using gas chromatography with an electron capture detector. Sensitivity of the method was 0.1 to 0.6 mg/g. Boettner and Weiss[13] determined the combustion products of polyvinyl chloride by gas chromatography, infrared absorption, ultraviolet absorption, and mass spectroscopy. O'Mara et al.[14] showed that pyrolysis of vinyl chloride produced 27,000 ppm HCl, 58,100 ppm CO_2, 9500 ppm CO, 40 ppm phosgene, and a trace of vinyl chloride. Thermal decomposition of PVC gave rise to CH_4, C_2H_6, C_2H_4, benzene, and toluene as well as the above gases.[15] Popov and Yablochkin[16] showed that PVC film releases untreated vinyl chloride and dibutyl phthalate along with the above products. Standardized gas chromatographic procedures for vinyl chloride and other halogenated aliphatic compounds have been developed, and it has been shown that microcoulometric detectors are better than electron capture detectors.[17,18] Arena[19] showed that gas chromatographic analysis of vinyl chloride was 100 times more sensitive than infrared analysis. Meshkova et al.,[20] using a polarographic method, found unreacted vinyl chloride, polymerization inhibitors, and metal chlorides in waste water from a PVC plant. Automatic gas chromatographs, infrared spectrophotometers, chlorine-sensitive strip chart recorders, and organic vapor analyzers have been developed for atmospheric monitoring of vinyl chloride.[21]

C. Toxicology
1. Animal

Patty et al.[22] reported that guinea pigs exposed to 10% vinyl chloride were narcotized and died within 30 to 60 min. At necropsy, the findings were congestion and edema of the lungs and hyperemia of the kidneys and liver. Lower concentrations produced ataxia and narcosis. Vinyl chloride produces local irritation of the skin and mucous membranes. Low concentrations have no effect on heart rate or blood pressure, but high concentrations cause hypertension. At a concentration of 20 vol % the dog has respiratory paralysis, salivation, and emesis, but no liver or kidney pathology. Mice, rats, guinea pigs, and rabbits give similar responses.[23] Vinyl chloride anesthesia of dogs causes muscle incoordination, serious cardiac arrhythmias, and cardiac sensitization to catecholamines.[25,26] Similar results were previously reported by Peoples and Leake[26] and have been confirmed by Lester et al.,[27] Mastromatteo et al.,[28] and Prodan et al.[29] Vinyl chloride perfusion of the frog leg causes no effect.[30] In animal studies, the LD_{50}/2 hr ranged from 117,500 ppm for mice to 230,800 ppm for rabbits. Smyth and Weil[31] reported that chronic feeding with diets containing 1.5 to 15% vinyl chloride-vinyl acetate copolymer for 2 years had no effect on rats.

The implantation of plastics in guinea pig and rabbit muscle has caused severe fibroblastic effects, but it is not known whether the reaction is caused by monomer polymer, plasticizer, or other additives.[32,33] Furthermore, Autian et al.[34] have shown that some PVC products do not cause such reactions and related the effect to the types of additives used. However, an extremely low quantity of monomer could also be involved. Guess et al.,[35] using in vitro techniques of tissue culture and antigen-antibody reactions, showed the toxic potential of materials migrating from polyvinyl chloride. Haberman et al.[36] showed that these migrating materials had an anticomplement effect, hemolyzed erythrocytes, interfered with the agglutinating capabilities of blood grouping antibodies, and produced abnormalities in chicks hatched from eggs injected with these materials. Polyvinyl chloride was not involved itself, but the plasticizers, dibutyl sebacate, dimethyl sebacate, dimethyl phthalate, di-n-propyl phthalate, and butyl oleate were implicated. Such alterations in cell surfaces and serum proteins must be avoided. Chick embryo heart cells in culture grown in media exposed to polyvinyl chloride tubing showed blistering and granularity and stopped beating. DeHaan[37] did not identify the toxic agent.

Acute inhalation of concentrations of 0.5 to 40 vol % of vinyl chloride by guinea pigs resulted in death at concentrations of 2.5 to 40 vol %. The compound caused unsteadiness and ataxia, rapid jerky respiration followed by shallow respiration, surgical anesthesia, and respiratory paralysis. Necropsy findings included congestion and edema of the lungs and hyperemia of the liver and kidneys.[22] Von Oettingen[38] reported tachycardia, bradycardia, inversion of the R-wave, abnormalities in the QRS interval, sinus arrhythmia, ventricular multiform extrasystoles, transitory left axis deviation, A-V block, ventricular tachycardia, and inversion of the T-wave with an elevated ST segment in dogs exposed to 10 vol % of vinyl chloride. Mastromatteo et al.[28] reported that acute exposure of mice, rats, and guinea pigs to 10, 20, or 30 vol % of vinyl chloride for 1 to 30 min produced pulmonary edema and hemorrhages, congestion of the liver and kidneys, and hypocoagulability of the blood. Repeated exposures to 200 to 500 ppm 7 hr daily for 4.5 months produced histological changes in the liver and kidneys of rats and rabbits but not in guinea pigs and dogs. All species tolerated 50 ppm for 6 months.[39] Rats exposed to 2% vinyl chloride 8 hr/day for 3 months showed changes in liver and spleen weights, a decrease in leukocytes, and an increase in erythrocytes, but no effect on growth rate, hemoglobin, hematocrit, or prothrombin time.[37] Schotter[40] also reported liver and kidney changes. Vazin and Plokhova[41] reported cardiac arrhythmias and bradycardia and changes in the phonocardiogram in rats exposed to 0.03 to 0.05 mg/ℓ of vinyl chloride for 5 months.

Vinyl chloride exposures caused increased secretion of catecholamines in rabbits and changes in the biopotential in the posterior hypothalamus.[42] It has been suggested that the cardiovascular changes in rabbits induced by vinyl chloride would be helpful in studying the angioneurotic syndrome.[43]

No compound-related symptoms or pathology were observed in rats, rabbits, and monkeys exposed to 250 or 500 ppm of vinyl bromide for 6 months. However, 20 periods of 7-hr exposure to 10,000 ppm of the chemical caused a decrease in body weight and activity in male rats.[44] Basalaev et al.[45] reported central nervous system and cardiovascular dysfunction, bone resorption, and osteoporosis in rats and rabbits exposed to 0.03 to 0.04 mg/ℓ of vinyl chloride for 6 months. These bone changes were similar to the human acroosteolysis produced by vinyl chloride. Exposure to vinyl chloride for 4 months induced a significantly slower formation of defensive conditional reflexes in rats.[46] These same investigators showed that vinyl chloride affected the bioelectric activity of the cortex and anterior and posterior hypothalamic nuclei of rabbits.[47] Viola[48] showed that exposure of rats to 30,000 ppm of vinyl chloride 4 hr/day for 12 months resulted in degeneration of the brain, liver, and kidneys. There were histopathological changes in the skeleton and connective tissue similar to human acroosteolysis. Bones undergo processes of intense periosteal growth and diffuse chondroid metaplasia. Connective tissue dissociates into collagen bundles with reduced numbers of cells, the elastic reticulum is markedly reduced and fragmentary, and the dermal vessels lumen is hypertrophied. Fibrous tissue surrounds and infiltrates the nerve endings.

2. Human

Danziger[49] has reported on accidental deaths from vinyl chloride exposure, with the following findings at cyanosis: local burns of the conjunctiva and cornea, congestion of the lungs and kidneys, and failure of the blood to clot. Blood levels of vinyl chloride could not be obtained. Vinyl chloride has anesthetic properties and causes respiratory tract irritation with dryness and mucous membrane atrophy, followed by chronic bronchitis. Tribukh et al.[50] reported that the chemical produces hepatitis. Intoxication and euphoria from vinyl chloride have been reported and hypersomnia persists even after leaving the contaminated area. The chemical produces the skin sensations of formication and heat and repeated exposures lead to neurological asthenia. There are other toxic effects including dyspeptic disturbances, epigastric pain with swelling of the hypochondrium, anorexia, hepatomegaly, splenomegaly, hepatitis without jaundice, ulcers, Raynaud's syndrome, allergic dermatitis, and scleroderma.[51] Antonyuzhenko[52] reported that this poisoning persists even in the absence of further exposure. Smirnova and Granik[53] stated that the long-term side effects of vinyl chloride intoxication caused functional disturbances of the CNS with adrenergic polyneuritis. ECG recordings showed changes in rhythm, conductance and repolarization processes, and an increased systolic index.[54] Juhe et al.[55] reported skin disease, Raynaud's syndrome, osteolytic syndrome, and thrombocytopenia in German workers. Stalova[56] found changes in thermoregulation in chronic vinyl chloride poisoning with body temperature increasing 1 to 2°C as a result of peripheral vasoconstriction. Maricq et al.[57] examined PVC workers with a wide-field capillary microscope and observed capillary abnormalities. They suggested that such examinations might be useful in the early detection and prevention of vinyl chloride associated diseases. Soviet workers had changes in the visual, auditory, taste, and olfactory analyzers referable to the effects of vinyl chloride on the brain stem reticular formation.[58] Chronic vinyl chloride exposure for 1.75 to 18 years produced scleroderma with thickening and homogenization of collagen bundles and fragmentation and rarefaction of elastic fibers. Finger clubbing, osteolysis of the distal phalanges, and a Raynaud's syndrome were also observed. Thrombocytopenia, splenomegaly, liver malfunction with marked fibrosis in the portal areas, and pulmonary insufficiency with restrictive changes were also reported.[59] Kramer and Mutchler[60] made correlations between clinical and environmental measurements for workers exposed to vinyl chloride. Clinical parameters including blood pressure, bromsulphalein, icterus index, hemoglobin, and beta-protein were collected along with ambient air concentrations of vinyl chloride. The results indicated some impairment of liver function.

Clinical symptoms of chronically poisoned vinyl chloride workers include: headache, dizziness, increased fatigue, sleep disorders, decreased memory capacity, diaphoresis, paresthesia, pain in the extremities, paling of fingers and toes, edema of these extremities, and pain and unpleasant sensations in the region of the heart. An ACTH test related these changes with modification in adrenal cortical function.[61] Grigorescu and Tiba[62] have correlated the increased urinary excretion of monochloroacetic acid with atmospheric vinyl chloride and the duration of such exposures. Gabor et al.[63] reported a decrease in blood catalyase activity and an increase in peroxidase, indophenoloxidase, and glutathione content after 1 year of vinyl chloride exposure. Suciu et al.[64] have observed thyroid impairment and skin and muscle collagenosis. Gabor et al.[65] found decreases in blood albumin, serum cholinesterase, and pseudoestrase. They found no effect on blood catalase of serum or pyruvic acid activity. The beta and gamma globulins increase but there is a decrease in the beta-to-alpha lipoprotein ratio. Makk et al.[66] reported increases in serum levels of gamma-glutamic transpeptidase, alkaline phosphatase, serum glutamic pyruvic transaminase, serum glutamic oxaloacetic transaminase, lactic dehydrogenase, and bilirubin.

a. Allergic Manifestations

Morris[67] reported sensitization dermatitis in vinyl chloride PVC, and vinyl acetate workers. Key[68] suggested it was caused by the plasticizers used. Hypersensitivity has been seen in consumers using polyvinyl chloride products containing epoxy resin plasticizers and stabilizers.[69] Pegum[70] reported that PVC containing tricresylphosphate as a plasticizer-stabilizer caused contact dermatitis. Kalmanovich[71] observed eczematous dermatitis in children exposed to vinyl chloride and dibutylphthalate liberated from floor coverings, and volatile materials from PVC floor tiles caused mucosal irritation of the eyes and respiratory tract of infants and adults.[72] In contrast to vinyl chloride, vinyl acetate does not appear to cause contact dermatitis.[73]

Ward et al.[74] provided data suggesting that vinyl chloride induced an immune complex disorder. Immunological and immuno-chemical examination of vinyl chloride workers showed the presence of circulating immune complexes in 19 of 28 workers. Even with the lack of clinical signs, abnormalities were found in some vinyl chloride workers.

b. Acroosteolysis

Because of the special nature of this phase of vinyl chloride disease, it will be discussed separately and in detail. Cordier et al.[75] were the first to describe acroosteolysis of the distal phalanges accompanied by a Raynaud's syndrome. Wilson et al.[76] reported 31 cases from one American company working with vinyl chloride, but were unable to distinguish between physical insult, chemical insult, or personal idiosyncrasy as the specific cause of the disease. Dissolution of the bones, acroosteolysis of the terminal phalanges of the fingers and sacroiliac joints occurs in PVC workers. The patella and phalanges of the feet may also be involved and there are skin lesions and Raynaud's phenomena.[77] Chatelain and Montillon[78] found the same lesions in French PVC workers. The progression of the bone lesion can lead to fragmentation of the distal phalanx.[79] Nitti et al.[80] examined 20 Italian PVC workers and attributed the syndrome to sympathetic neurocirculatory dystonia associated with an unidentified pathogenic cofactor. Basalaev[81] suggested the higher centers in the hypothalamus were involved in the acroosteolysis. The latent period for the development of the disorder can be only 5 days or up to 30 days. Moreover, the portal of entry, lungs, or skin, is not known precisely.[82] Dinman et al.[83] performed an epidemiological study of 5011 workers in 32 vinyl chloride and PVC plants in the U.S. and Canada and discovered 25 cases of acroosteolysis. The condition was associated with hand cleaning of polymerizers and appeared to be a systemic rather than a local disease. Cook et al.[84] reported that there was a strong association between the development of the disease and manual cleaning of the reactors used in production of PVC. Clinical evaluation of four cases of the disease revealed that Raynaud's disease antedated the osteolytic lesions. Negative Ca and PO_4 balance could be present and ^{18}F scintiscans revealed variable fluoride uptakes correlated with the radiographic lesions. Other clinical parameters were normal.[85] Gitsios[86] suggested a wide variety of industrial hygiene precautions to reduce exposure and prevent the disease, including a water wash and reduction of ambient air vinyl chloride content to 50 ppm, the use of gloves, a yearly physical examination, and X-ray examination of the hands.

German experience with long-term vinyl chloride exposure (1 to 3.5 years) not only showed circulatory, skin, and bone disorders, but also deafness, vision failure, giddiness, and liver dysfunction.[87] Markowitz et al.[88] have suggested that 3% or less of the workers in the PVC industry are affected by the disease. The occurrence of idiopathic acroosteolysis with papular skin lesions makes it essential that vinyl chloride or PVC exposure be documented in all cases diagnosed as acroosteolysis.[89]

Misgeld et al.[90] have pointed out that the bone lesions can progress even after vinyl chloride exposure has ceased before recovery occurs. Rapid diagnosis to prevent vinyl chloride induced scleroderma, Raynaud's syndrome, and acroosteolysis is essential if mortality, now at least 10% is to be reduced. Periodic platelet counts have been suggested as the diagnostic tool because thrombocytopenia occurs much earlier than any other sign of vinyl chloride intoxication. However, other disease conditions and chemicals can affect the thrombocyte count.[55]

D. Teratogenesis
1. Animal
Studies using mice, rats, and rabbits showed that inhalation of vinyl chloride did not induce gross abnormalities in offspring of dams exposed to concentrations ranging from 50 to 2500 ppm for 7 hr daily. The mice and rats were exposed on days 6 and 15 and the rabbits on days 6 to 18 of gestation. An excess of minor skeletal abnormalities were observed and the high concentrations caused increased fetal death in all three species.[91] Radike et al.[92] also did not find gross abnormalities in the pups of rats exposed for 4 hr daily by inhalation of 600 or 6000 ppm of vinyl chloride on days 9 through 21 of gestation. There were an excess of minor skeletal abnormalities.

2. Human

The data on human teratogenesis from environmental exposure to vinyl chloride in the vicinity of PVC plants lacks information on air concentrations and duration of exposure. However, it has been reported that there are high rates of malformations in three small communities where PVC plants are located. Significantly greater numbers of malformations of the central nervous system, upper alimentary tract, genital organs, and club foot were reported.[93,94] These effects require further study to document the degree of exposure to vinyl chloride, its duration, and the possibility that other drugs and chemicals are or are not involved.

E. Mutagenesis
1. Animal

Vinyl chloride does not appear to be mutagenic per se, but when converted to chloroethylene oxide or 2-chloro acetaldehyde by a liver activating system, does produce mutants in the Salmonella typhimurium TA1535, TA1536, TA1537, and TA1538 strains.[96] This effect has been confirmed by Bartsch et al.,[97] McCann et al.,[98] Elmore et al.,[99] and Garro et al.[100] Greim et al.[101] have obtained similar results with *Escherichia coli* K12 bioauxotrophic strain and Loprieno et al.[102,103] showed that vinyl chloride was mutagenic in several species of yeast. Mutagenicity of vinyl chloride has also been demonstrated in the germ cell of Drosophila[104] and Chinese hamster V79 cells.[105]

The mutagenicity of inhaled vinyl chloride at concentrations of 3000, 10,000, or 30,000 ppm for 6 hr/day for 5 days was determined in fertile male CD-1 mice with the dominant lethal assay. Based upon postimplantation deaths, preimplantation egg losses, and reduction in fertility, vinyl chloride was not found to be mutagenic at these concentrations. Positive controls indicated that these CD-1 mice expressed a dominant lethal effect.[106]

2. Human

Ducatman et al.[107] and Purchase et al.[108] reported an increased incidence of chromosomal breakage in workers exposed to vinyl chloride. Heath et al.[109] observed cytogenetic effects in three groups of workers: PVC polymerizers (presumed high exposure), PVC processors (presumed low exposure), and rubber and tire manufacture workers (presumed negligible exposure). All three groups had a significantly greater chromosome breakage than nonexposed controls and overall breakage levels were similar in the three exposed groups, indicating that other chemicals could have been involved.

Waxweiler et al.[110] reported cytogenetic studies of vinyl chloride workers, plastics workers, and rubber workers. Chromosomal breakage was highest in plastic workers followed by vinyl chloride workers then rubber workers. All were higher than nonindustrial controls. Wives of vinyl chloride workers had a significant increase in fetal loss rate.

Picciano et al.[111] reported no significant differences in chromatid and chromosome aberrations or the proporton of abnormal cells in a group of 209 vinyl chloride workers who had been exposed for 1 to 332 months to concentrations of vinyl chloride ranging from 0.3 to 15.2 ppm. Killian et al.[112] confirmed these findings.

F. Pharmacokinetics and Biotransformation
1. Animal
a. Absorption

Duprat[113] exposed rats to 20,000 ppm (^{14}C) vinyl chloride for 5 min and found the chemical in the liver, bile duct, digestive lumen, and kidneys 10 min after start of the exposure. The amount and distribution of vinyl chloride and its metabolites increased

Table 3
PERCENTAGE OF ^{14}C ACTIVITY PER GRAM OF TISSUE 72 HR AFTER ORAL ADMINISTRATION OF ^{14}C-VINYL CHLORIDE

Tissue	Dose (mg/kg)		
	0.05	1.0	100
Liver	0.172 ± 0.025[a]	0.182 ± 0.005	0.029 ± 0.002
Skin	0.070 ± 0.023	0.076 ± 0.010	0.010 ± 0.002
Carcass	0.027 ± 0.007	0.046 ± 0.002	0.007 ± 0.001
Plasma	0.041 ± 0.004	0.053 ± 0.007	ND[b]
Muscle	0.028 ± 0.003	0.031 ± 0.003	0.006 ± 0.001
Lung	0.050 ± 0.003	0.061 ± 0.003	0.011 ± 0.001
Fat	0.030 ± 0.004	0.045 ± 0.008	0.006 ± 0.001

[a] Mean ± SE, five rats per dose.
[b] Not detectable.

From Watanabe, P. G., McGowan, G. R., and Gehring, P. J., *Toxicol. Appl. Pharmacol.*, 36, 339, 1976. With permission.

up to 3 hr postexposure and ^{14}C activity was found in the urinary tract, salivary, and lacrimal glands, skin, and thymus. Watanabe et al.[114] and Bolt et al.[115] also reported rapid uptake and equilibration of atmospheric vinyl chloride in rats via this route.

Watanabe et al.[116] showed that oral vinyl chloride followed a pattern of absorption similar to that seen in the inhalation studies. When repeated exposures were compared to single exposures, the fate and rates of excretion were the same. Similar results were obtained with microsomal enzyme activity, but covalent binding to hepatic micromolecules was greater with repeated exposures. Also hepatic nonprotein sulfhydryl concentration was 79% in the repeated exposure rats and only 37% in single exposure animals as compared to the controls.[117] Aqueous vinyl chloride (22.6 to 28.2 mg per animal) or oil solutions (12.55 to 25.1 mg per animal) given per os rapidly pass from the gastrointestinal tract to the blood. The vehicle has no influence on the rates of uptake and elimination. After inhalation exposure to 7000 ppm of vinyl chloride for 5 hr, the chemical is rapidly eliminated from the blood.[118]

b. Distribution

Watanabe et al.[116] showed that the liver retains the largest percentage of vinyl chloride and/or its metabolites 72 hr after oral administration of the chemical. Table 3 gives the organ distribution of the chemicals at 72 hr. Bolt et al.[115] reported the percent of ^{14}C activity per gram of tissue was kidney, 2.13; liver, 1.86; and spleen, 0.73, immediately after exposure of rats to 50 ppm of vinyl chloride for 5 hr. Activity was still detectable 48 hr after exposure. No information is available on distribution of vinyl chloride in humans after exposure.

c. Metabolism

Vinyl chloride exposure of a cell-free hepatic microsomal preparation results in its conversion to ethylene chlorohydrin with further conversion to chloral and chloroacetic acid. Gothe et al.[119] used 3,4-dichlorobenethiol to trap the metabolites and identify them by gas chromatography-mass spectrometry. Figure 1 gives the overall scheme. In vivo, the reactions may involve the glutathione S-transferases which catalyze the reaction of glutathione with α,β-unsaturated compounds.[120] Hefner et al.[121] reported that vinyl chloride is converted to polar compounds which are conjugated with glutathione

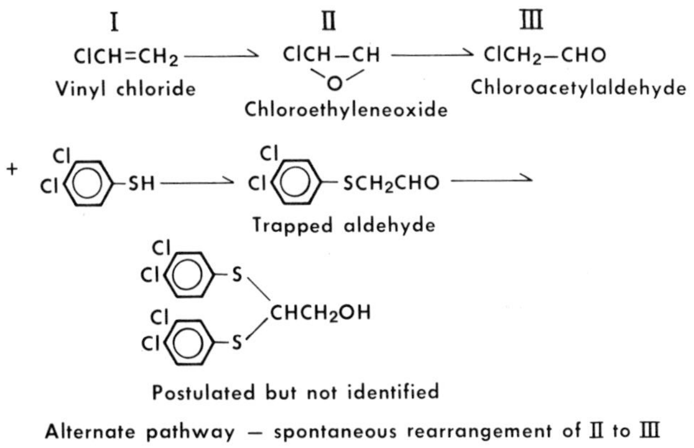

FIGURE 1. Biotransformation of vinyl chloride by mixed function oxidases.

and/or cysteine and excreted in the urine. Watanabe et al.[114,116] and Bolt et al.[115] reported that the in vivo vinyl chloride metabolite in vivo appears to be saturable. Hefner et al.[121] postulated a primary metabolic pathway involving alcohol dehydrogenase because ethanol or pyrazole inhibits vinyl chloride uptake. This pathway is saturated by exposure to concentration above 220 to 250 ppm. A secondary pathway involving epoxidation and/or peroxidations appears to be active at higher concentrations. It seems that vinyl chloride is metabolized to an activated carcinogenic electrophile which covalently reacts with nucleophilic groups or cellular macromolecules.[122-124]

Mixed function oxidases metabolize vinyl chloride and the enzyme inducer, phenobarbital, enhances vinyl chloride liver toxicity.[125] Rat liver microsomes catalyze the covalent binding of vinyl chloride metabolites to protein and nucleic acids.[124,126] Laib and Bolt[127] suggest that chlorethylene oxide is the metabolite involved. Hathway[128] reported in vitro depurination of calf thymus DNA by chloroacetaldehyde identical to that observed in hepatocyte DNA following administration of vinyl chloride to rats in vivo. Chloroacetaldehyde increases pentobarbital sleeping time in mice, decreases protein synthesis in mouse fibroblasts in culture, but did not sensitize guinea pigs.[129] Vinylidene chloride causes hepatic injury, giving an increase in serum alanine α-ketoglutarate transaminase. This effect can be prevented by simultaneous exposure to vinyl chloride.[130]

2. Human

Human exposure to vinyl chloride has been shown to result in the urinary excretion of chloroacetic acid indicating that the metabolic pathway is similar to that reported in rats.[62]

G. Carcinogenesis
1. Animal

The first report of vinyl chloride carcinogenesis from inhalation exposure appeared in 1971. Viola et al.[131] showed that rats exposed to 30,000 ppm of vinyl chloride 4 hr per day, 5 days a week for 12 months developed epidermoid carcinomas, papillomas and mucoepidermoid carcinomas of the skin, adenocarcinoma of the lungs, and osteochondroma of the metacarpal and metatarsal regions of all four limbs. Using larger groups of male and female rats, Caputo et al.[132] reported all concentrations of vinyl chloride except 50 ppm produced carcinomas and sarcomas (Table 4). Another study using vinyl chloride concentrations from 50 to 10,000 ppm produced Zymbal gland

Table 4
TUMORS INDUCED IN RATS AND RABBITS BY INHALATION OF VINYL CHLORIDE

Conc. VC (ppm) 4 hr/day, 5 days/wk 12 months	Number of animals	Liver angiosarcomas cholangiomas	Lung adenoalveolar carcinomas	Skin squamous cell carcinoma acanthoma	Other
Rats					
20,000	150	31	21	67	7
10,000	200	16	16	34	8
5,000	200	12	4	20	2
2,000	200	10	8	6	6
500	150	4	—	3	—
50	200	—	—	—	—
No treatment	200	—	—	—	—
15 months Rabbits					
10,000	40	—	6	12	—
No treatment	20	—	—	—	—

From Caputo, A., Viola, P. L., and Bigotti, A., *IRCS*, 2, 1582, 1974. With permission.

carcinomas, nephroblastomas, and hepatic and extrahepatic angiosarcomas in rats, and pulmonary tumors, mammary carcinomas, and liver angiosarcoma in mice. No carcinoma was produced at 50 ppm.[133] Maltoni and Lefemine[134] established a direct dose-time relationship and neoplastic response from exposure to vinyl chloride. They also exposed rats, mice, and hamsters to concentrations of vinyl chloride varying from 50 to 10,000 ppm for varying time intervals, then observed the animals for their lifetime.[135] All three species had hepatic angiosarcomas as well as tumors at other sites (Tables 5 to 7). Maltoni[136] obtained similar results in mice and hamsters with shorter exposure times (Tables 8 and 9). He also observed four s.c. angiosarcomas, four Zymbal gland carcinomas, and one nephroblastoma in 66 pups of 60 Sprague-Dawley rats exposed to 6000 to 10,000 ppm of vinyl chloride 4 hr/day from the 12th to the 18th day of gestation.[136]

Lee et al.[137] have confirmed the carcinogenicity of vinyl chloride in CD-1 mice and CD rats. The animals were exposed to 50, 250, or 1000 ppm of vinyl chloride 6 hr/day, 5 days/week. Serial sacrifices were performed at 1, 2, 3, 6, 9, and 12 months. After 12 months, the mice developed bronchoalveolar adenomas, mammary gland tumors, and hepatic angiosarcomas at all levels of exposure. Rats exposed to 250 or 1000 ppm of vinyl chloride developed angiosarcoma in the liver, lungs, and other sites.[138] Prior ingestion of 5% ethanol for 4 weeks followed by inhalation of 600 ppm of vinyl chloride 4 hr/day, 5 days/week for 12 months reduced the induction time for hepatic angiosarcoma to 39 weeks from 53 weeks.[139]

Oppenheimer et al.[140,141] produced 31.8% sarcoma in rats by embedding PVC film in their anterior abdominal wall. It was suggested that the slow rate of plastic breakdown and the release of degradation products was related to the long latent period for cancer development.[142] Russell et al.[143] confirmed these observations. Kogan and Tugarinova[144] implanted PVC film in the rat kidney and produced cancer. The stages were granulation, hyperplasia, presarcoma, and sarcoma.[145] A significant proliferation of fibroblastic cellular elements surrounds the plastic implant.[146,147] Raikhlin and Kogan,[148] using histochemical methods, showed that abnormal collagen was formed in the capsule surrounding the plastic insert prior to tumor development. Brand et al.[149] suggested that a single specific premalignant cell clone resides dormantly on the plastic, and it detaches and produces a tumor 4 weeks later.

Table 5
INCIDENCE OF TUMORS IN SPRAGUE-DAWLEY RATS EXPOSED 4 HR/DAY 5 DAYS/WK 52 WEEKS BY INHALATION TO VARIOUS CONCENTRATIONS OF VINYL CHLORIDE: RESULTS AFTER 135 WEEKS

Conc. VC (ppm)	Number of animals		Liver		Kidney		Zymbal gland		Other	Total number of rats with one or more tumors
	Total	Corrected	Angiosarcomas	Average latency (weeks)	Nephroblastomas	Average latency (weeks)	Carcinomas	Average latency (weeks)		
10,000	69	61	9	64	5	59	16	50	25	38
6,000	72	60	13	70	4	65	7	62	19	31
2,500	74	59	13	78	6	74	2	33	18	32
500	67	59	7	81	4	83	4	79	11	22
250	67	59	4	79	6	80	—	—	9	16
50	64	59	1	135	1	135	—	—	12	10
No treatment	68	58	—	—	—	—	—	—	10	6

From Maltoni, C. and Lefemine, G., *Ann. N.Y. Acad. Sci.*, 246, 195, 1975. With permission.

Table 6
INCIDENCE OF TUMORS IN SWISS MICE INHALING VINYL CHLORIDE 4 HR/DAY, 5 DAYS/WK, 30 WEEKS: RESULTS AFTER 41 WEEKS

Conc. VC	Number of animals		Liver angiosarcomas	Pulmonary tumors		Mammary carcinomas		Other	Total number of mice with one or more tumors
	Total	Corrected		No.	Average (weeks)	No.	Average (weeks)		
10,000	60	50	4	27	34	9	28	9	28
6,000	60	54	2	22	33	8	33	5	27
2,500	60	53	4	12	35	4	32	3	13
500	60	58	4	16	34	2	33	2	17
250	60	58	3	11	34	6	30	1	15
50	60	57	—	—	—	7	35	4	8
No treatment	150	141	—	1	39	—	—	—	1

From Maltoni, C. and Lefemine, G., *Ann. N.Y. Acad. Sci.*, 246, 195, 1975. With permission.

Table 7
INCIDENCE OF TUMORS IN GOLDEN HAMSTERS
INHALING VINYL CHLORIDE 4 HR/DAY, 5 DAYS/WK
FOR 30 WEEKS: RESULTS AFTER 48 WEEKS

Conc. VC (ppm)	Number of animals		Liver angiosarcomas	Other	Total number of animals with one or more tumors
	Total	Survivors			
10,000	35	19	—	3	3
6,000	32	21	—	8	5
2,500	33	19	—	5	4
500	33	23	1[a]	4	4
250	32	18	—	2	2
50	33	23	—	5	5
No treatment	70	49	—	2	2

[a] More than 18 weeks postexposure.

From Maltoni, C. and Lefemine, G., *Ann. N.Y. Acad. Sci.*, 246, 195, 1975. With permission.

Table 8
INCIDENCE OF TUMORS IN SWISS MICE INHALING VINYL
CHLORIDE 4 HR/DAY, 5 DAYS/WK, 30 WEEKS: RESULTS
AFTER 81 WEEKS

Conc. VC (ppm)	Number of animals	Number of animals with tumors		
		Liver angiosarcomas	Pulmonary tumors	Other
10,000	60	8	35	26
6,000	60	5	38	24
2,500	60	11	30	25
500	60	11	38	26
250	60	11	33	35
50	60	1	2	25
No treatment	150	—	8	1

From Maltoni, C., *Ann. N.Y. Acad. Sci.*, 271, 431, 1976. With permission.

2. Human

Although a Soviet survey of PVC workers covering the period 1957 to 1960 showed 15% of them had chronic epithelial hepatitis, no cancers were recorded.[150] Masteller et al.[151] reported liver dysfunction in German PVC workers but no cancer.

The first hepatic angiosarcoma cases in PVC workers were reported in the U.S. in 1974. However, retrospective studies discovered one case in 1968, one in 1971, and one in 1973. Regardless of the treatment instituted, death occurred within 14 months of diagnosis.[152] Manucuso[153] pointed out the difficulties associated with the diagnosis of liver cirrhosis and angiosarcoma and suggested both clinical and experimental studies to improve the situation. Popper and Thomas[154] pointed out the importance of proper diagnosis because characteristic hepatic fibrosis was present in all cases of angiosarcoma. The transition from the fibrotic stage to angiosarcoma is suggested by the focal proliferation of the sinusoidal lining cells and as the hepatocytes, seen in the fibrotic stage, become more pronounced in the initial stages of angiosarcoma devel-

Table 9
INCIDENCE OF TUMORS IN GOLDEN
HAMSTERS INHALING VINYL CHLORIDE 4
HR/DAY, 5 DAYS/WK, FOR 30 WEEKS:
RESULTS AFTER 76 WEEKS

Conc. VC (ppm)	Number of animals		Liver tumors angiosarcoma angiomas hepatomas	Other
	Total	Survivors		
10,000	35	1	—	13
6,000	32	3	3	10
2,500	33	4	4	10
500	33	4	2[a]	8
250	32	4	—	4
50	33	5	—	10
No treatment	70	14	—	4

[a] Angiosarcomas

From Maltoni, C., *Ann. N.Y. Acad. Sci.*, 271, 431, 1976. With permission.

opment. These observations suggest that the fibrotic lesions without angiosarcoma seen in vinyl chloride workers may be only the prestage of the developing neoplastic lesions.[155] Berk[156] pointed out that the series of hepatic changes appear to represent a multicentric development of angiosarcoma and are similar to the changes induced by Thorotrast® and inorganic arsenic. Diagnostic difficulties in vinyl chloride-induced angiosarcoma are illustrated by three cases in which the primary diagnosis was gastric ulcers. It was recommended that serum alkaline phosphatase and SGOT determinations be coupled with photoscans using ^{131}I rose bengal and ^{198}Au for a more precise diagnosis. Needle biopsy should be avoided because the vascular nature of the tumor could lead to an exsanguinating hemorrhage.[157] Makk et al.[158] suggested α-glutamic transpeptidase determinations are the most useful in detecting abnormalities and reflecting the extent of liver damage.

The first epidemiologic study of the vinyl chloride problem occurred in 1974 and covered 7129 workers. Exposures of 20 years or longer occurred in 854 workers and 15 years or more in 1640 workers. Specific causes of death were no greater than expected and no deaths were attributable to angiosarcoma. Increasing exposure level and duration showed an increase in standardized mortality ratios for malignant neoplasms, particularly for the liver, brain, and respiratory system, but they were not statistically significant.[159] Holder[160] studied mortality of 594 vinyl chloride workers and found only 13 cases of neoplasms, none of which involved the liver. Monson et al.[161] conducted a proportional mortality study of vinyl chloride workers and found cancer deaths to be 50% higher than expected and of these, five angiosarcomas were noted. Nicholson et al.[162] studied 257 vinyl chloride workers and found 24 cases of angiosarcoma. Ott et al.[163] reexamined the mortality data of Tabershaw and Gaffey[159] using more clearly defined vinyl chloride concentrations and reached the same conclusions on malignant neoplasms. Wagoner[164] studied 950 workers with at least 5-year exposure to vinyl chloride and found a 57% increase in cancer deaths with lesions in the respiratory system, blood forming tissues, brain, and central nervous system. Liver cancer was 12-fold greater than expected. Chiazze et al.[165] performed a mortality study on 4341 employees of 17 PVC plants and found no angiosarcoma deaths. However, there was an increase in breast and urinary tract cancers in white females.

Fox and Collier[166] studied 7000 British vinyl chloride workers and found four cases of malignant liver cancer of which two were angiosarcomas. Byren et al.[167] traced 21 of 771 Swedish vinyl chloride workers and found a four- to fivefold increase in pancreas and liver tumors and of the latter two were angiosarcomas. Makk et al.[166] found ten cases of angiosarcoma in a small Quebec PVC plant. Delorme and Theriault[168] in a restudy of these workers suggested these angiosarcomas were associated with high vinyl chloride exposure levels and overtime work. Alcohol and cigarettes were not involved.

Spritas and Kaminski[169] have documented 64 cases of angiosarcoma worldwide, and considering the number of workers exposed to vinyl chloride, this appears to be a very small number. However, when cancer latency is considered, this number may increase with time. Table 10 lists these cases by country. Baxter et al.[170] have suggested that people living near PVC processing plants may also be at risk. Kuzmak and McGaughy[171] estimated that the 4.6 million people residing within 5 miles of PVC or vinyl chloride production plants in the U.S. were exposed to 17 ppm of vinyl chloride. New regulations and equipment have reduced this exposure. Brady et al.[172] examined annual rates of angiosarcoma in New York State, except New York City, caused by arsenic, Thorotrast®, and vinyl chloride. Five cases of hepatic angiosarcoma lived in the vicinity of a vinyl chloride plant, suggesting that this chemical was the cause of their disease.

H. Comment

The difficulties encountered with vinyl chloride and PVC appear to be related to the presence of the monomer in the air of the work place, but the amount to which the public is exposed and the duration of that exposure are not exact. It is necessary for continuous epidemiological surveys to determine the overall extent of the problem, bearing in mind the long latency for the development of angiosarcoma and the degree of exposure prior to the imposition of controls at the manufacturing site. The appearance of liver malfunction, Raynaud's syndrome, scleroderma, and acroosteolysis in excess of what would usually be found in the population should be considered as indications of vinyl chloride exposure until other causes can be identified. Greater emphasis should be given to the development of better, more precise and rapid diagnostic methods.

III. SILICONES

A. Synthesis and Physical Properties

The silicones have alternating silicone-oxygen linkages where the common pendant groups attached to the silicone atom are methyl, phenyl, vinyl, longer-chain alkyl, polyoxyalkylene ethers, and fluoralkyl. Chlorosilane intermediates use methyl chloride or chlorobenzene to produce a variety of mono-, di-, and trifunctional groups. These pendent groups contribute significantly to such properties as high and low temperature characteristics, lubricity, solvent resistance, reactivity and general compatibility with other silicones, and organic polymers.[173] Figure 2 shows various silicone intermediates.

Silicone elastomers come in three grades for use in medical applications. The chemical and physical properties of these silicones have been discussed by Braley.[174] The structural formula and the chemical and physical properties of three medical grades silicones are given in Table 11.[175] The physical properties of medical grade elastomer are given in Table 12.[176] Although ethylene oxide can be used to sterilize Silastic® 382, the absence of this sterilizant must be assured prior to use of the plastic. The physical properties of tubing prepared from this elastomer are given in Table 13.[177]

Liquid silicones are also used for medical purposes and the physical properties of Dow Corning 360 Medical Fluid® are given in Table 14.[178] Several of these liquid

Table 10
ANGIOSARCOMA OF THE LIVER IN VINYL CHLORIDE/PVC WORKERS

Country	Case no.	Birth date	1st VC or PVC exposure	Diagnosis of angiosarcoma	Age at diagnosis	Years from 1st exposure to diagnosis	Total years of exposure	Date of death
Belgium	01	00-00-00	00-00-00	00-00-00	00	00	00	06-29-76
Canada	01[a]	12-15-13	00-00-44	00-00-55	41	11	11	09-02-55
	02[a]	03-06-14	00-00-43	00-00-57	43	14	14	12-21-57
	03[a]	08-26-19	00-00-41	00-00-62	42	21	20	03-22-62
	04[a]	04-05-19	00-00-45	00-00-67	48	22	22	01-21-68
	05[a]	05-07-11	00-00-44	00-00-68	57	24	05	07-05-68
	06[a]	12-15-19	00-00-47	00-00-71	51	24	23	04-10-71
	07[a]	11-09-19	00-00-46	00-00-72	53	26	25	12-24-72
	08	05-13-20	00-00-61	00-00-73	53	12	05	06-12-73
	09	07-19-21	00-00-46	00-00-74	53	28	26	09-04-74
	10	05-16-15	00-00-53	00-00-76	61	23	14	04-00-77
Czechoslovakia	01[a]	00-00-28	00-00-57	00-00-73	46	16	16	00-00-74
	02[a]	00-00-26	00-00-51	00-00-66	40	15	15	00-00-66
Germany	01[a]	06-04-30	10-01-56	09-19-68	38	12	12	01-25-69
	02[a]	07-26-31	10-14-57	09-25-70	39	13	12	12-41-71
	04	09-04-30	04-16-57	00-00-74	44	17	17	11-25-74
	05[a]	01-01-32	12-16-62	00-00-75	43	13	12	01-09-75
	07[a]	09-29-26	04-15-54	00-00-75	49	21	12	11-13-75
	08[a]	10-19-17	04-19-54	00-00-75	58	22	21	12-25-75
	09[a]	12-13-34	12-02-59	06-16-76	42	17	15	Alive
	10[a]	07-25-29	10-10-55	06-28-77	47	22	22	06-28-77
	11[a]	12-29-36	01-02-61	00-00-77	41	16	10	03-07-77
France	01[a]	04-15-24	01-00-46	02-18-67	43	21	19	02-19-67
	02	06-03-11	07-06-59	01-08-75	63	15	12	01-24-75
	03[a]	00-00-19	00-00-46	01-00-75	55	29	29	06-29-75
	04[a]	01-27-27	10-19-49	01-04-76	49	26	26	01-04-76
	05[a]	01-29-38	00-00-65	04-00-76	38	11	10	05-13-76
	06[a]	04-14-34	00-00-58	09-00-76	42	18	17	09-12-76
	07	00-00-27	07-01-50	07-00-76	49	26	23	07-02-76
	08[a]	04-01-34	05-23-57	12-03-76	42	19	19	01-30-77

Country	Case	Date	Date			Date
Great Britain	01[a]	04-20-01	00-00-44	71	28	22	12-00-72
	03	06-02-37	02-00-66	37	09	04	12-24-74
Italy	02[a]	11-13-29	00-00-57	43	15	06	12-00-72
	03[a]	03-14-20	00-00-53	55	22	21	07-10-75
Japan	01	08-01-22	04-00-53	52	22	22	10-24-75
Norway	01[a]	12-23-15	03-00-50	56	22	21	01-04-72
Sweden	01[a]	06-23-27	08-14-51	43	19	18	10-20-70
	03[a]	06-10-10	05-00-47	65	29	21	03-19-76
	04	11-16-14	00-00-46	62	31	31	05-12-77
U.S.A.	01[a]	10-17-23	12-09-48	49	24	21	03-03-73
	02[a]	08-19-33	11-15-55	37	14	13	09-28-71
	03[a]	05-25-15	11-28-45	58	28	28	12-19-73
	04[a]	01-15-24	07-06-52	43	15	15	01-07-68
	05[a]	01-25-12	06-19-44	52	20	20	04-09-64
	06[a]	11-23-28	01-17-62	46	12	12	07-24-75
	07[a]	05-03-22	08-27-44	45	24	17	03-23-68
	08[a]	05-06-20	10-07-46	41	15	15	08-29-61
	09[a]	11-08-31	05-28-45	43	29	24	03-00-75
	10[a]	08-16-13	06-12-51	55	17	17	05-10-68
	11[a]	05-27-09	10-14-46	61	23	23	03-16-70
	12[a]	11-17-18	09-13-49	50	20	19	05-02-69
	13[a]	12-01-21	12-11-42	52	32	26	07-04-74
	16[a]	11-04-27	05-08-50	41	19	04	03-27-69
	17[a]	05-06-31	06-23-55	43	19	19	Alive
	18[a]	04-22-28	09-15-54	46	21	11	11-02-75
	19[a]	00-00-15	00-00-43	60	32	22	04-06-76
	20[a]	08-31-17	00-00-55	58	21	18	01-30-77
	21[a]	09-02-09	12-00-46	67	30	21	01-02-77
	22[a]	10-02-23	07-11-47	52	29	28	12-04-76
	23[a]	00-00-23	09-00-58	50	15	14	04-06-73
	24[a]	05-07-17	00-00-39	60	38	26	05-27-77
	25[a]	08-07-10	02-00-47	67	30	20	03-10-77
Yugoslavia	01[a]	04-05-14	00-00-53	59	20	20	04-08-73
	02[a]	11-15-31	00-00-50	42	23	18	07-12-73

Note: Total reported cases: 00 indicates unknown data.
[a] Diagnosis was microscopically confirmed.

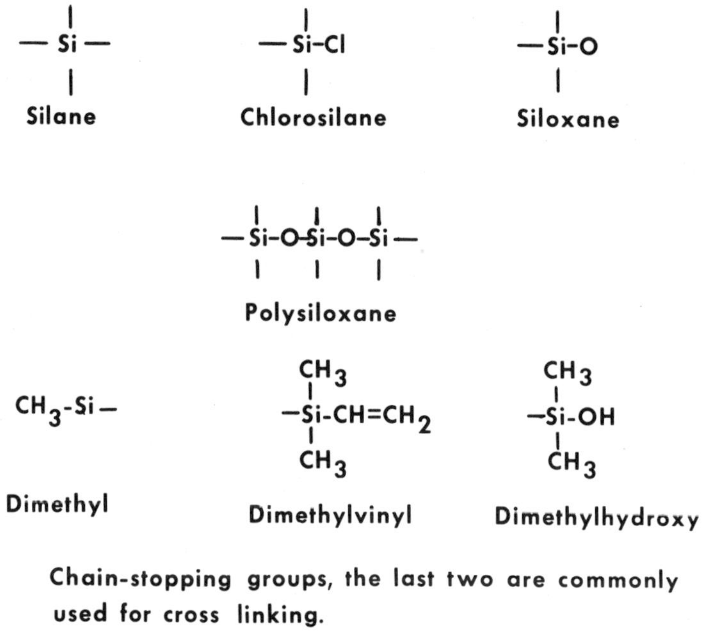

```
    —Si—              —Si—Cl              —Si—O
     |                  |                   |
   Silane            Chlorosilane         Siloxane
```

```
              —Si—O—Si—O—Si—
                |   |   |
                Polysiloxane
```

```
                    CH₃                CH₃
                     |                  |
   CH₃—Si—         —Si—CH=CH₂         —Si—OH
                     |                  |
                    CH₃                CH₃

   Dimethyl        Dimethylvinyl      Dimethylhydroxy
```

Chain-stopping groups, the last two are commonly used for cross linking.

FIGURE 2. Structural formulae of silicone derivatives.

Table 11
STRUCTURAL FORMULA, CHEMICAL AND PHYSICAL PROPERTIES OF SILICONES

	MDX4-4514	MDX4-4515	MDX4-4516
Siloxane polymer	Dimethyl, phenylmethyl, and methylvinyl	Dimethyl and methylvinyl	Dimethyl and methylvinyl
Filler	Fumed silica	Fumed silica	Fumed silica
Vulcanizing agent	0.71% 2,4-Dichlorobenzoyl peroxide	0.54% 2,4-Dichlorobenzoyl peroxide	0.54% 2,4-Dichlorobenzoyl peroxide
Color	Clear Translucent	Clear Translucent	Clear Translucent
Specific gravity	1.12	1.14	1.23
Durometer hardness	25	50	75
Tensile strength (psi)	850	1200	1000
Elongation %	600	450	350
Die B tear strength (pi)	70	75	75
Brittle point (°F)	−175	−100	−100
Stiffening temperature (°F)	−166	—	—

silicones have been used as antifoam agents and typical properties of one are given in Table 15.[179] The properties and medical uses of silicone rubber are reviewed in *Biocompatibility of Clinical Implant Materials* of this series, by Van Noort and Black.[343]

B. Analysis

Silicones have been analyzed using optical emission spectroscopy. Samples were excited by AC spark and determinations were made continuously during the excitation period. This method is applicable to silicone determination covering the range 5 to 100 µg/mℓ with a detectability of 5 µg/mℓ and a precision of ± 10%.[180,181]

Table 12
PHYSICAL PROPERTIES OF SILASTIC®
382 MEDICAL GRADE ELASTOMER

Property	Unvulcanized	Vulcanized
Color	Gray	—
Viscosity	50,000 centipoises	—
Filler content	23% by wt.	—
Catalyst	Pure stannous octoate	—
Durometer hardness	—	43
Tensile strength	—	400 psi
Elongation	—	160%
Specific gravity	—	1.13

Table 13
PHYSICAL PROPERTIES OF SILASTIC®
MEDICAL-GRADE TUBING

Test	Value
Durometer hardness	45—49
Tensile strength	1100
Elongation %	500—700
Specific gravity at 77°F	1.14—1.23
Brittle point (°F)	−100
Maximal workable temperature range	−65— +500°F
Die B tear strength (lb/in.)	120—160
Dielectric constant at 10^2 cps	2.8
Dissipation factor at 10^2 cps	0.0008
Tension set (%)	5—15%
Compression set, 22 hr at 77°F (%)	4—5
Odor	None
Taste	None
Color	Translucent

A spectrophotometric method for the determination of silicones in food, human lung tissue, blood, and animals employs combustion of the sample and assay of the silicone as heteropoly blue at 800 nm. An infrared spectroscopic method determines the silicone after solvent extraction. The colorimetric method is sensitive to 1 µg/mℓ and the infrared one to 2 ppm.[182] Phosphorus in digested biological samples is eliminated by converting to phosphomolybdate and bleached with oxalate.[183] Silicones have been removed from fatty foods by freezing, then determined by atomic absorption, visible, or ultraviolet spectroscopy.[184] Infrared spectroscopy has been used to determine silicones in beer and yeast. It is sensitive to 0.2 ppm.[185]

C. Toxicology
1. Animal
Single oral doses of various fluid silicones, detailed in Table 16, given to guinea pigs produced a minimal acute toxicity. The most pronounced effect was their laxative action. This effect was not seen with hexamethyldisiloxane, which caused a slight inebriation and central nervous system depression at the higher doses. Dodecamethylpentasiloxane did not have this effect, but was slightly laxative. When Dow® 200 fluid was fed to young adult female rats at doses of 1, 3, 5, 10, and 20 g/kg for 28 days, there was no effect on growth, hematology, bone marrow morphology, or organ weights.

Table 14
PHYSICAL PROPERTIES OF DOW CORNING 360 MEDICAL FLUID

Viscosity in cs at 25°C	20	50	100	200	350	500	1000	12500
Specific gravity	0.945—0.955	0.955—0.965	0.965—0.960			0.968—0.975		
Refractive index	1.398—1.402	1.40—1.404	1.401			−1.405		
Pour point °F	−76	−67	−67	−63	−58	−58	−58	−58
Surface tension dynes/cm	20.6	20.8	20.9	21.0	21.1	21.1	21.2	21.2
Volume resistivity ohm-cm	1×10^{14}	1×10^{14}	2×10^{14}		4×10^{14}	6×10^{14}	1×10^{15}	2×10^{15}

Table 15
PHYSICAL PROPERTIES OF DOW CORNING MEDICAL ANTIFOAM A COMPOUND

Percent active defoamer	100%
Color	Translucent gray
Consistency	Viscous liquid
Volatile content	12%
Filler content	4.25%
Viscosity[a]	450 centistokes
Specific gravity[a]	0.967
Refractive index[a]	1.402

[a] With filler removed by centrifugation.

Table 16
MORTALITY AND RESPONSE RESULTING FROM THE ADMINISTRATION OF SILICONE FLUIDS IN SINGLE ORAL DOSE - GUINEA PIGS

Silicone	Viscosity in Cstks at 25°C	Dose ml/kg	Mortality ratio	Observations on the laxative effects at various periods of time after administration			
				2 1/2 hr	8 hr	24 hr	48 hr
DC 200 fluid (hexamethyl-disiloxane)	0.65	3.0	0/7				
		10.0	0/7	−	−	−	−
		30.0	0/7	−	−	−	−
		50.0	1/10				
DC 200 fluid (dodecamethyl-pentasiloxane)	2.0	10.0	0/3	−	+	−	
		30.0	0/6	−	+ +	−	−
		50.0	3/3				
DC 200 fluid	50	10.0	0/2	+ + +	+ + +	+ + +	+
		30.0	0/6	+ + +	+ + +	+ + +	+ + +
		50.0	0/3				
DC 550 fluid	75	3.0	0/3				
		10.0	0/3	+	+	+	
		30.0	0/6		+ +	+ + +	+ + +
DC 702 fluid	35	3.0	0/3				
		10.0	0/3	+	+ +	+ + +	
		30.0	0/6		+ +	+ + +	+ +
DC 200 fluid	350	5.0	0/2	−	−	+	
		10.0	0/5		+	+	−
		30.0	0/6	−	+	+	−
		50.0	0/3	−	−	+ +	+ +
DC fluid	12,500		Could not be fed satisfactorily				
Mineral oil		10.0	0/2	+ +	+ +	+ + +	+
		30.0	0/3	+ + +	+ + +	+ + +	+

From Rowe, V. K., Spencer, H. C., and Bass, S. L., *J. Ind. Hyg. Toxicol.*, 30, 332, 1948. With permission.

Histopathological examination of the heart, spleen, liver, kidney, adrenals, pancreas, bone marrow, stomach, and intestine showed no pathological changes. Single i.p. injections of 0.1, 0.3, 1.0, 3.0, and 10.0 mℓ/kg of silicones to male rats produced only a nonirritating foreign body reaction. However, the three high doses of hexamethyldisiloxane caused death and all the animals had extensive local irritation reactions with adhesions throughout the viscera. A repeat experiment produced the same results.

Table 17
SUMMARY OF EYE IRRITATION STUDIES

Material	Viscosity in Cs at 25°C	Occurrence and persistence of irritation					
		Immediately	1 hr	4 hr	8 hr	24 hr	48 hr
DC 200 fluid	0.65	+	0	0	0	0	0
DC 200 fluid	2.0	0	0	0	0	0	0
DC 200 fluid	50.0	0	+	+	+	0	0
DC 550 fluid	75.0	0	0	+	+	0	0
DC 702 fluid	35.0	0	0	+	+	0	0
DC 200 fluid	350.0	0	+	+	+	+	0
DC 200 fluid	12,500	0	+	+	+	+	0

Intradermal injection of these silicones in rabbits caused no local irritation with the exception of hexamethyldisiloxane which caused inflammation, edema, and necrosis at the injection site. Subcutaneous injection of these liquid silicones in rabbits produced no reaction except with hexamethyldisiloxane which produced a marked irritation and necrosis. Twenty applications of liquid silicones to the ears and shaved abdomens caused no skin irritation. Eye instillation in rabbits indicated that the liquid silicones caused transitory irritation (Table 17), although fluorscein staining indicated no corneal damage.

Inhalation of 25,000 ppm of hexamethyldisiloxane for 30 min by guinea pigs caused no ill effects, but 40,000 ppm caused death from respiratory failure in 15 to 20 min. The animals survived if removed from the chamber. Rats exposed to 4400 ppm of hexamethyldisiloxane for 7 hr in 15 days and guinea pigs given the exposure for 20 days showed no signs of discomfort, but the growth curve of the latter species was slightly depressed. At necropsy the rats had slight increases in liver and kidney weights. Histopathological examination of the lungs, hearts, livers, kidneys, spleens, and adrenals revealed no pathological changes.

Feeding DC antifoam A silicone at 10 ppm for 3 months to female rats had no effect on body weight, organ weights, blood urea nitrogen, and caused no pathological lesions in the liver, spleen, kidney, adrenal, heart, lung, pancreas, and stomach. Feeding 3000 ppm of the chemical to male and female rats for 18 months gave similar results. DC antifoam A caused transient irritation to the eyes of rabbits. Continuous skin contact in rabbits for 28 days produced no skin irritation.[186]

The static 48-hr TL_{50} for Dow Corning 200 fluid 100 cs is 44.5 ppm and the dynamic 48-hr TL_{50} is 73.4 ppm. Bluegill sunfish and rainbow trout have 96-hr TL_{50} of Dow Corning antifoam C emulsion of >10,000 ppm. The TL_{50} values for Dow Corning 200 fluid 100 cs in cockles, shore crabs, mummichogs, and brown shrimp is >1000 ppm. In mallard ducks and bobwhite quail the LC_{50} is >5000 ppm. The chemical had no adverse effects on the body weight or food consumption of white leghorn chickens.[187]

No gross or histopathological effect were observed in rabbits exposed to 5% aerosolized polydimethylsiloxane 12,500 cs for 7.5 hr or exposed to a 0.2% aerosol for 90 days. Similarly, dogs, rats, and guinea pigs exposed to polydimethylsiloxane 300 cs at a concentration of 2.12 mg/ℓ for 6 hr showed no ill effects. Cynomolgus monkeys exposed to a 10% aerosol for 20 min twice a day for 90 days had no changes in neurophysiological functions, respiratory function tests, and electrocardiograms.[188]

The use of silicone fluid 20, 350, and 1000 cs did not prevent intestinal adhesions or granuloma formation in rats.[189] These lesions in rats were thicker in the silicone treated animals than in the controls.[190] When dimethylpolysiloxane in doses of 2, 3, or 16 mℓ was injected intraperitoneally into rats with peritoneal adhesion, only the largest dose suppressed adhesion formation.[191] In rabbits, dimethylpolysiloxane effectively prevented peritoneal adhesion when 50 ml were injected into the peritoneal cavity.[192] Silicone film prevented adhesions in animals with intestinal anastomoses and promoted healing.[193] Although silicone fluid, 350 and 1000 cs, was unable to prevent pleural and peritoneal adhesions in rats, it significantly reduced adhesions in dogs.[194]

Liquid silicones injected subcutaneously into the buccal pouch of hamsters showed the materials to be nontoxic and to induce only moderate tissue reactions. In rabbits, injection into the palate and alveolar ridges resulted in encapsulation by the proliferation of connective tissue and collagenous fibers.[195] I.p. injection of dimethylpolysiloxane in two strains of mice resulted in deposition of the compound in the kidneys, spleen, liver, adrenals, ovaries, lymph nodes, pancreas, and adipose tissue. It was suggested that the material is collected and stored in the reticuloendothelial system.[196] S.c. injection of 3 ml of dimethylpolysiloxane (20, 100, and 1000 cs) into mice resulted in deposition in the s.c. tissue as large and small cysts. Tissue concentrations decreased with time.[197] Subcutaneous or intraperitoneal injection of 1 ml of dimethylpolysiloxane in mice caused conglomerations of histiocytes filled with silicone at the adrenomedullary junction.[198] Subcutaneous injection of massive doses of dimethylpolysiloxane into guinea pigs, mice, and rats produced multiloculated cysts but no granulomas or tumors.[199]

Dimethylpolysiloxane, 50 cs, was injected into the right knee joint of rabbits and the animals followed for 14 to 168 days. Joint tolerance was acceptable and histologically satisfactory. There was no inflammatory or sclerogenic reaction or hyperplasia of the synoviocytic lining, but there was an intense macrophage reaction. The silicone was deposited in the deep layers of the synovial membrane making the viscosity material unsatisfactory as a joint lubricant.[200] A similar experiment using rabbits and a silicone with a viscosity of 200 cs demonstrated that the material left the joint within 1 week by leakage and phagocytosis by the cells of the synovial layer. A transitory inflammatory reaction was followed by fixation of droplets in the synovial layer. Electronmicroscopy showed vacuoles within the cytoplasm of swollen cells, indicating that this silicone was unsatisfactory as a joint lubricant.[201]

Irregularities or scratches in silicone rubber cardiac prostheses causes thrombi formation when implanted in dogs. Thrombus development is promoted by stasis, turbulence, and an active clotting system.[202] Vascular grafts of silicone rubber tubing placed in the carotid and femoral arteries of dogs rapidly thrombosed. The source of the thrombosis was in the area of intimal proliferation and fibrin deposition near the proximal vessel-silicone junction.[203] Bonding heparin to anion exchange resin attached to silicone rubber produces a nonthrombogenic surface. Such tubes implanted in the canine vena cava showed no clotting after 1 hr.[204] Molding cavity exudates, 2,4-dichlorobenzoic acid, the presence of traces of Nasteavate®, and exposed silica particles were shown by electronmicroscopy to determine the blood compatibility of silicone rubber implants.[205]

Injection of silicone under the anterior uvea in rabbits produces a decrease in ocular tension lasting 2 to 5 months when one fourth to one half of the ciliary body was detached. Detachment of an equal amount of anterior choroid results in hypotony for 1 month. Histological studies indicate that these implants cause hypotony by trauma and not by the detachment space they create.[206] Retinal degeneration may follow successful retinal detachment surgery in which intraviterous silicone injection is used. Histological examination of animal eyes revealed degeneration of ganglion cells and nerve fiber layers and alteration of the rod and cone layer of the retina. The ERG is also suppressed.[207] A few hours after liquid silicone injection, silicone particles appeared in the retinal tissue of monkeys. Electronmicroscopy showed that the intercellular gaps between Muller cells widened and there were openings in the internal limiting membrane allowing silicone particles to pass from the vitrous cavity into the retina.[208] It has been reported that liquid silicone injected into the rabbit eye to replace part of the vitreous causes no histological damage.[209] Silicone hydrophobic or hydrophilic contact lenses slow oxygen uptake by the rabbit cornea.[210] Silicone contact lenses had no effect on corneal epithelial glycogen, glucose, ATP, and lactate levels in rabbits after 8 hr, 16 hr, and 1 week.[211]

Silicone rubber has been used to reconstruct the trachea in 31 rabbits and 2 dogs. No histological pulmonary changes were seen, and the tracheal defect healed by the formation of connective tissue on which normal epithelium developed. Complete reconstruction with silicone rubber resulted in granulation tissue formation on the anastomosis and respiratory obstruction. The nonreactivity of silicone rubber is an advantage, but it does not get covered by epithelium and there is no restoration of ciliated epithelium.[212] Similar results were obtained when silicone rubber tubing was used in replacement of the ureter in 17 dogs. The graft was always rejected into the ureteral lumen and in some cases obstructed the flow of urine.[213] Reconstruction of the esophagus in dogs with dimethylpolysiloxane rubber caused a fibrous sheath to form, but 1 or 2 months later a constrictive stricture formed and the passage was impaired.[214]

Permanent silicone rubber urinary bladder implants in 14 dogs were ineffective and infection killed the animals because antibiotic treatment was impaired.[215] Silicone rubber was used to repair the anterior wall of the bladder in 16 rabbits and the defects were healed in 16 weeks with normal transitional epithelium and a firm layer of collagenous fibers. However, the patches were rejected toward the bladder wall.[216] In cystectomized sheep an artificial bladder composed of silicone rubber lined with Dacron® velour became covered with thick connective tissue followed by a muscle layer at 9 months. At 18 months the natural bladder regenerated and the implant remained in the lumen as a hard foreign body. The artificial bladder is effective as long as it remains functional.[217] Silicone rubber-covered Dacron® ribbons were sutured under tendons of rectus muscles in 24 dogs to produce ressions of 5 to 15 mm while preserving the original point of insertion and arc of contact. A fibrous capsule enveloped the implant and bound it to the sclera. Breakdown of the conjunctiva and Tenon's capsule occurred in 11 dogs from 2 weeks to 9 months.[218] Silicone rubber tubing was satisfactory as a replacement of the common bile duct in 13 of 19 dogs.[219] Muscle implants of silicone rubber in 12 dogs caused only slight foreign body reaction and slight degenerative changes after 1 year.[220] When silicone rubber reinforced with Dacron® was used in eight dogs to repair the femoral artery, all vessels were patent at 3.5 to 50 weeks and aortagraphy showed a normal vascular pattern.[221] Silicone rubber tubes are the best cuffing material for the repair of peripheral nerves. They were well tolerated in chimpanzees after 3 years.[222]

The intermediates, alkylchlorosilanes, used to produce the polysiloxanes, are highly irritating compounds. The mortality caused by four of them given by i.p. injection to rats are shown in Table 18. Their oral toxicity in rats is given in Table 19. Severe ocular damage was observed when these compounds were instilled in the rabbit eye and there were burns of the cornea and lids. When the compounds were placed on the skin of rabbits, denaturation occurred within 1 to 3 min. Exposure of rats to these compounds by inhalation resulted in lachrymation, breath holding, salivation, nasal discharge, and shallow and difficulty in breathing. There was marked damage to the eyes and burning of the edges of the ears.[186]

The mortality in rats given a single oral dose of ethoxysilanes is given in Table 20. These chemicals produced immediate transitory irritation of the rabbit's conjunctiva, but no corneal damage and the eyes were normal in 24 hr. Furthermore, they were not irritating to the rabbit's skin. Single exposure of rats to the vapors of the ethoxysilanes gave the mortality shown in Table 21. Tetraethoxysilane was the most toxic and produced eye and nose irritation, unsteadiness, tremors, salivation, respiratory difficul-

Table 18
MORTALITY RESULTING FROM THE INTRAPERITONEAL ADMINISTRATION TO RATS OF ALKYLCHLOROSILANES

Material	Dose (g/kg)	Mortality ratio
Dichlorodiethylsilane	0.01	0/2
	0.03	2/2
	0.10	2/2
Ethyltrichlorosilane	0.01	0/2
	0.03	2/2
	0.10	2/2
Dichlorodimethylsilane	0.01	1/2
	0.03	1/2
	0.10	2/2
Methyltrichlorosilane	0.01	0/2
	0.03	1/2
	0.10	1/2
	0.30	2/2

From Rowe, V. K., Spencer, H. C., and Bass, S. L., *J. Ind. Hyg. Toxicol.*, 30, 332, 1948. With permission.

Table 19
MORTALITY RESULTING FROM THE ORAL ADMINISTRATION TO RATS OF SINGLE DOSES OF ALKYLCHLOROSILANES

Material	Dose (g/kg)	Mortality ratio
Dichlorodimethylsilane	0.1	0/2
	0.3	0/3
	1.0	2/2
Methyltrichlorosilane	0.1	0/2
	0.3	0/3
	1.0	2/2
Dichlorodiethylsilane	0.1	0.2
	0.3	0.3
	1.0	2/2
Ethyltrichlorosilane	0.1	0.2
	0.3	0.3
	1.0	1/2

From Rowe, V. K., Spencer H. C., and Bass, S. L., *J. Ind. Hyg. Toxicol.*, 30, 332, 1948. With permission.

Table 20
MORTALITY RESULTING FROM THE ADMINISTRATION OF ETHOXYSILANES IN A SINGLE ORAL DOSE — RATS

Dose (g/kg)	Mortality ratio			
	Tetraethoxysilane	Methyltriethoxysilane	Diethoxydimethylsilane	Ethoxytrimethylsilane
0.6	0/5	0/1	0/1	0/3
1.0	1/5	0/1	0/1	0/5
1.4	1/5	—	—	4/5
2.0	2/3	0/1	0/5	5/5
3.0	4/5	0/10	3/5	1/1
5.0	5/5	3/5	4/5	1/1
7.0	1/1	4/5	4/5	1/1
10/0	1/1	5/5	5/5	1/1

From Rowe, V. K., Spencer, H. C., and Bass, S. L., *J. Ind. Hyg. Toxicol.*, 30, 332, 1948. With permission.

ties, and unconsciousness. Repeated inhalations of this chemical in concentrations of 125, 250, 500, and 1000 ppm caused kidney damage consisting of tubular degeneration and necrosis and pulmonary damage consisting of interstitial edema, leucocyte infiltration, and hemorrhages into the alveoli and smaller bronchioles. The effect of tetraethoxysilane on body and organ weights is given in Table 22. Repeated exposure at 125 ppm showed no effect on hematological parameters. Repeated exposures of female rats to 1000, 2000, or 4000 ppm of diethoxydimethylsilane had no effect on various organs or the blood and only caused slight pulmonary irritation.[186]

2. Human

When Dow Corning silicone 200 fluid®, 100 cs, was applied to the backs of five

Table 21
MORTALITY RESULTING FROM SINGLE EXPOSURES OF RATS TO THE VAPORS OF THE ETHOXYSILANES

Calculated concentration (ppm)	Length of exposure (hr)	Mortality ratio			
		Tetraethoxy-silane	Methyltriethoxy-silane	Diethoxydim-ethylsilane	Ethoxytrim-ethylsilane
4000 (some mist)	4	5/5			0/5
	6		0/5		
	8		4/5	0/5	4/5
2000	1	0/5			
	2	0/5			
	3	1/5		0/5	
	4	3/5			
	6	9/10		0/5	0/5
	7		0/5		
	8	5/5			0/5
1000	8	4/5			

From Rowe, V. K., Spencer, H. C., and Bass, S. L., *J. Ind. Hyg. Toxicol.*, 30, 332, 1948. With permission.

Table 22
AVERAGE BODY WEIGHTS AND ORGAN WEIGHTS OF MALE RATS THAT HAD RECEIVED REPEATED EXPOSURES TO THE VAPORS OF TETRAETHOXYSILANE

Vapor concentration (ppm)	No. of exposures	No. of rats	Body weights (g)		Organ weights (G./100 G.)				
			Initial	Final	Heart	Kidney	Liver	Spleen	Testes
Controls	0	4	258	268		0.70	3.31	0.22	1.01
1000	1	2	236	225		0.88	3.08	0.27	1.16
	2	2	306	290		1.08	3.32	0.22	0.92
	3	2	294	270		1.35	3.44	0.20	1.04
500	3	2	185	171	0.47	1.50	4.16	0.40	1.34
	5	5	264	197	0.46	2.16	3.70	0.32	1.21
250	4	2	254	244	0.34	0.86	2.81	0.22	1.11
	8	2	307	299	0.42	0.80	2.97	0.23	0.98
	10	2	282	267	0.40	1.05	2.92	0.26	1.01
	10[a]	2	226	236	0.36	0.74	3.57	0.25	0.99
	10[b]	2	226	244	0.30	0.66	3.07	0.22	0.92
Controls	0	8	232	282	0.34	0.66	3.33	0.17	0.93
125	5	2	178	192	0.41	0.83	3.91	0.26	1.16
	10	2	195	221	0.41	0.89	4.12	0.26	1.22
	15	2	174	204	0.36	0.79	3.45	0.28	1.20
	25	4	219	261	0.34	0.72	3.43	0.18	0.90
	30	10	206	250	0.35	0.71	3.30	0.22	1.04

[a] Killed 1 week after last exposure.
[b] Killed 2 weeks after last exposure.

From Rowe, V. K., Spencer, H. C., and Bass, S. L., *J. Ind. Hyg. Toxicol.*, 30, 332, 1948. With permission.

humans at a daily dose of 50 mg/kg for 10 days, only one showed an increased urinary silicone. There was no irritation.[180] Dow Corning 200 fluids, viscosities ranging from 0.65 to 12,500 cs, instilled into the human eye caused erythema of the conjunctiva and some eyelid edema which cleared up in 24 hr. Similar results were obtained with DC antifoam A.[186]

Severe degeneration of the silicone rubber ball in Starr Edwards prosthetic aortic valves occurred in 11 of 271 patients within 24 months of implantation. In six patients the ball had atrophied, fractured, and become dislodged from the cage and in five it had swollen and become impacted within the cage. All these patients died.[223] Silicone rubber balls caused less hemolysis than steel ones, as measured by serum lactic dehydrogenase activity and the half-life of ^{51}Cr labeled red cells.[224] Thin-walled silicone tubes have been used successfully in neural anastomosis.[225]

Silicone rubber catheters for long-term bilateral nephrostomy drainage have been used for 9 months without becoming encrusted.[226] Siliconizing of Foley catheters renders them less irritating to the urethra[227] and organic impotence has been treated successfully by implantation of a silicone rubber penile prosthesis.[228] A molded silicone tube has been used successfully to connect the lacrimal lake to the nose for tear drainage.[229] Injection of dimethylpolysiloxane beneath the eyelid skin results in formation of palpebral bags from a localized tissue reaction and the thinness of the skin. The bulk of the silicone must be removed surgically.[230] Silicone fluid has been used for the treatment of retinal detachment.[231] This procedure is successful when the silicone remains posterior to the transocular fibrous membrane and flattens the retina without residual folds.[232] Silicone buckling procedures for retinal detachment has resulted in eight cases of scleral abscesses with pain, proptosis, and a cloudy vitreous requiring removal of the implant if useful vision is to be maintained.[233] Although silicone fluid is inert, complications can arise if it comes in contact with the lens capsule, cornea, ciliary body, or iris.[234] Silicone rubber has been used to repair conjunctival defects. New tissue grows over the implant which is later removed.[235] Silicone rubber is very useful in the management of orbital fractures and the reconstruction of the orbital floor.[236] It has also been used successfully for restoration of defects of facial contour.[237]

Injections of silicone fluid into stiff, drying, grating joints of rheumatoid arthritis or osteoarthritis patients results in more mobile joints with painless movement. This treatment was shown to be free of complications over a period.[238] In patients undergoing revision arthroplasty, silicone rubber implants were removed and examined for breakage and cracks and analyzed for lipid content. No relationship was found between implant breakage and lipid content.[239]

Silicone fluid, 360 cs, was used as a vehicle for the prevention of postoperative pelvic adhesions. This high viscosity fluid did not decrease or prevent and even appeared to augment adhesion formation.[240] Three cases of siliconomas, two in the breast and one in the face, developed after injections of silicone fluid. Histological evidence showed these reactions to be foreign body granulomas.[241] Silicone implants have been suggested for women who have had submammary mastectomy, and while not giving a fine cosmetic result, was of great psychological benefit.[242] Silicone enema is definitely contraindicated in the diagnosis of colon injuries because it might perforate the colon.[243] Dimethylpolysiloxane implants impregnated with megestrol have been used as a contraceptive device. These s.c. implants were well tolerated, but after 2 years fibrotic tissue appeared containing granulomas and foreign body giant cells.[244]

From the results obtained in seven patients with burns, it was suggested that silicone immersion was of value in burn therapy.[245] Such treatment leads to early mobilization, early debridement, early removal of the eschar, control of infection, and a remarkable epithelialization of burned hands.[246] A silicone rubber otoplasta prosthesis is useful in

Table 23
SKELETAL FINDINGS AMONG FETUSES OF RATS TREATED WITH DOW CORNING 360 FLUID

	DC 360 (mg/kg)			
	0[a]	20	200	1000
Total no. fetuses	104	120	130	97
Pups with sternebrae not fully developed				
number	28	21	40	39
percent	27	17.5	31	40
Pups with less than 13 pairs of ribs				
number	3	1	0	0
percent	2.9	0.8	0	0
Wide separation between cranial bones	6	0	6	14
Scrambled ribs	3	1	1	5
Scrambled vertebrae	3	2	0	0
Bipartite sternebrae	8	26	16	37
Fused sternebrae	0	0	0	1
Concave frontal and/or parietal bones	6	1	0	11
Displaced femurs	0	0	3	0
Absence of vertebrae column below T_1	0	0	4	0
Bilateral shortening of humerus and radius	0	0	0	1

[a] 1,000 mg/kg sesame oil.

From Kennedy, G. L., Jr., Keplinger, M. L., Calandra, J. C., and Hobbs, E. J., *J. Toxicol. Environ. Health,* 1, 909, 1976. With permission.

the reconstruction of total or partial traumatic losses of the ear.[247] Dimethylpolysiloxane has been used in treatment of dyspepsia and was well tolerated.[248] Exposure of humans to tetrachlorosilane causes burning of the ocular and respiratory membranes and destruction of erythrocytes.[186]

D. Teratogenesis
1. Animal

Subcutaneous injection of dimethylpolysiloxane, 360 cs, to pregnant rats caused the teratogenic effects seen in Table 23. The two higher doses caused incomplete developed sternebrae with the effect being greatest at the 1000-mg/kg dose. At this dose there was a greater frequency of concave frontal and parietal bones. Similar results were obtained with rabbits (Table 24). A study of reproduction in rats showed that dimethylpolysiloxane at 20, 200, and 1000 mg/kg had no effect on gestation, viability, or lactation. Similar results were again obtained in rabbits.[249] A reproduction study in White Leghorn chickens showed that dimethylpolysiloxane at concentrations of 200, 1000, and 5000 ppm had no effect on egg production and quality, hatchability, and body weights and viability of chicks.[188]

2. Human

No teratogenic effects in humans have been reported.

Table 24
SKELETAL FINDINGS AMONG FETUSES OF
RABBITS TREATED WITH DOW CORNING 360
FLUID

	DC 360 (mg/kg)			
	0[a]	20	200	100
Total no. fetuses				
Alive	63	48	33	46
Dead	1	6	0	6
Pups with sternebrae not fully developed				
Number				
Alive	4	4	1	7
Dead	0	5	0	6
Percent				
Alive	6	8	3	15
Dead	0	83	0	100
Pups with less than 13 pairs of ribs				
Number				
Alive	49	31	12	29
Dead	1	6	0	6
Percent				
Alive	78	65	36	63
Dead	100	100	0	100
Bipartite sternebrae	0	0	1	1
Percent with bipartite sternebrae	0	0	3	2

[a] 1,000 mg/kg sesame oil.

E. Mutagenesis
1. Animal
Dimethylpolysiloxanes were not mutagenic in mice.[249]

2. Human
No mutagenic studies in humans have been reported.

F. Pharmacokinetics and Biotransformation
1. Animal
a. Absorption
Dimethylpolysiloxane-^{14}C residues were examined in bluegill sunfish exposed to 1 and 10 ppm of the chemical and it was shown that very little was absorbed. Ingestion of the compound by White Leghorn chickens indicated very little absorption and no accumulation in their eggs.[187] Sewage bacteria exposed to dimethylpolysiloxane were not able to biodegrade the chemical.[187]

b. Distribution
Dimethylpolysiloxane is so poorly absorbed that no tissue distribution could be detected in chickens.[187] When an oral dose of 18 mg/kg of Antifoam A silicone fluid was given to Rhesus monkeys 2.1 to 2.5% of the dose was recovered in the expired air, bile, and urine. Blood levels were below detectability at 1 ppm.[188]

c. Metabolism
These compounds are apparently not metabolized by bacteria, chickens, or monkeys.[188]

2. Human

Humans receiving a single oral dose of 100 mg/kg of Antifoam A silicone fluid excreted 0.5% of the dose in expired air. The chemical is metabolized to some extent because unidentified metabolites were found in the urine.[188]

G. Carcinogenesis
1. Animal

Subcutaneous implantation of silicone in Sprague-Dawley rats produced fibroadenoma and fibrosarcoma. These cancers contained hyaluronic acid, chondroitinsulfate A, and chondroitinsulfate B.[250]

2. Human

The s.c. injection of dimethylpolysiloxane for increasing volume and altering regional contour of tissue resulted in uptake and storage by histiocytes and a 35% reduction in the original volume.[251] Silicone injection of the breast produced foreign body granulomas.[252] Breast cancer developed in two patients after s.c. simple mastectomy and reconstruction with silicone prostheses.[253]

H. Comment

Various silicones have been shown to be inert and in most cases do not cause any adverse effects in animals or humans. There is limited data to show silicone fluid can, under some circumstances, be carcinogenic. The chemical intermediates from which the silicones have been synthetized are corrosive and highly irritating to both animal and human tissues.

IV. CYANOACRYLIC ADHESIVES

A. Synthesis and Physical Properties

Pure cyanoacrylic monomers have been prepared by condensing formaldehyde with 2-cyanoacetates in the presence of a catalyst comprising an acid and the salt of a primary or secondary amine with the same or stronger acid.[254] Chloroacetic acid was reacted with a cyanide salt to form cyanoacetic acid which was esterified with methanol or ethanol to cyanoacetic acid esters. Transesterification of this compound gave the propyl and butyl esters. The cyanoacrylic compound was prepared by condensing the cyanoacetate with formaldehyde in the presence of a basic catalyst.[255] High purity α-cyanoacrylate monomers have been prepared by polymerization of formaldehyde with alkyl cyanoacetates and depolymerization of the product under reduced pressure and anhydrous conditions.[256] Cyanoacetic acid in the form of hydroscopic crystals has the following physical properties: MP 66°C decomposition; BP 108°C; soluble in water, ethanol, and ether; slightly soluble in benzene and chloroform.[257]

B. Analysis

Alkyl 2-cyanoacrylates and alkyl cyanoacetates have been identified and determined by gas chromatography on a 10% XE-40® silicone gum nitrile/Chromosorb® P column at 170°C with helium carrier gas and a hydrogen flame ionization detector.[258]

C. Toxicology
1. Animal

The methyl, ethyl, N-propyl, N-butyl, N-amyl, N-hexyl, N-heptyl, N-octyl, and isobutyl 2-cyanoacrylates inhibited growth of *E. coli* and *S. aureus* on blood agar and nutrient broth. This inhibition was related to the length of the alkyl side chain and it was suggested that the bacteriotoxicity was the result of polymer degradation.[259] 2-

Cyanoacrylic glue used to suture wounds in mice exhibited no histological evidence of tissue damage. There was normal formation of granulation tissue and epithelial regeneration and no giant cell formation.[260]

Implants of sponges impregnated with methyl-, hexyl-, and decyl-2-cyanoacrylates in rats showed that the methyl compound was necrotizing to the tissue and inhibited collagen formation, while the other compounds did not.[261] Methyl 2-cyanoacrylate causes a marked histotoxicity by continuous interaction with surrounding tissues. This reaction is much reduced with higher molecular weight homologues.[262] Hexyl, heptyl, and octyl cyanoacrylates applied to the rat's tongue caused a local inflammatory reaction, and the material remaining after 30 days was surrounded by histiocytes and foreign body giant cells.[263] Isoamyl 2-cyanoacrylate is irritating to the skin.[264] First molars were extracted from rats and the wound sprayed with butyl cyanoacrylate. The area showed less inflammatory reaction than the control wound and better and faster collagenization and epithelization.[265] Butyl 2-cyanoacrylate induced a foreign body reaction in rat tissue.[266] N-Butyl and isobutyl cyanoacrylates produce satisfactory hemostasis and sealing of experimental wounds, but they elicit inflammatory tissue reactions.[267] It has been shown that butyl 2-cyanoacrylate degrades more slowly in rat tissue than the methyl derivative.[268] Feeding butyl 2-cyanoacrylate to rats for 10 days then sacrificing them at 90 days showed no gross or histopathological changes from ingestion of the compound.[269]

A severe inflammatory response was observed from intrastromal injection and surface application of cyanoacrylic adhesives to the central cornea of conventional and germ free guinea pigs.[270]

A comparison of tissue compatibility of methyl-, ethyl-, and butyl cyanoacrylates on experimental wounds in parenchymatous organs of rabbit showed that the butyl compound decomposed more rapidly and scar tissue formed without a marked inflammatory reaction.[271] No deaths or obvious toxicity was observed when the entire liver of rabbits was covered with methyl cyanoacrylate. However, a severe inflammatory reaction occurred involving the liver and diaphragm, and at 4.5 months there was marked fibrosis and residual adhesive.[272] Based upon the extent of leukocyte infiltration, necrosis of the media, and calcification of the rabbit abdominal aorta, butyl cyanoacrylate was less toxic than the methyl and ethyl cyanoacrylates.[273] Cyanoacrylate adhesives have been used successfully for anastomoses of small veins and the vena cava in the rabbit. Thrombus formation and an inflammatory reaction occurred in 20% of the animals.[274]

Butyl 2-cyanoacrylate was applied to perforated keratomies in rabbit eyes, and after 3 weeks the corneal scar showed good compatability to the adhesive. Histopathological examination at 5 months indicated no vascularization and no chronic symptoms of inflammation.[275] However, this adhesive caused necrosis when applied to conjunctival flaps.[276] Rabbits and monkeys tolerated a methyl methacylate lens attached to the corneal stroma with octyl cyanoacrylate.[277] Comparison of butyl 2-cyanoacrylate and isobutyl 2-cyanoacrylate adhesives as sutures for perforating cuts in the rabbit cornea showed no histological reaction and normal healing of the wounds.[278] Isobutyl 2-cyanoacrylate was well tolerated by the rabbits cornea and there was no evidence of vascularization, cellular infiltration, or infection.[279] Isobutyl 2-cyanoacrylate proved to be nontoxic when used to attach homologous sclera in scleral buckling procedures on rabbit eyes.[280] Experiments with $\beta^{14}C$ isobutyl cyanoacrylate on rabbit eyes indicated that the adhesive does not penetrate the cornea and no radioactive degradation products were found in the inner eye tissues.[281] Butyl 2-cyanoacrylate tightly closes lacerations and fistulating injuries of the rabbit's cornea.[282] Methyl 2-cyanoacrylate is less well tolerated by the rabbit cornea than the n-decyl, n-octyl, n-heptyl, n-hexyl, n-butyl, or isobutyl derivatives.[283] Methyl 2-cyanoacrylate adhesive was better than sutures when compared on corneal wounds.[284] Cyanoacrylate adhesives are well tolerated when under the retina.[285] Although octyl cyanoacrylate usually does not produce neovascularization or diffuse scarring of the rabbit cornea, such effects have been observed in a small number of experiments.[286]

When methyl 2-cyanoacrylate was used in blood vessel surgery in cats, the adhesive caused necrosis of the wall of the femoral vein.[287] This same adhesive caused a higher infiltration of leucocytes and lymphocytes in wounds in cats and was unsatisfactory mechanically as a suture material.[288]

Methyl 2-cyanoacrylate has been used to implant teeth in dogs, but it inhibits normal fibrosis and the implant falls out.[289] This adhesive was used to cement a fitted acrylic plug into the dental alveoli of dogs. The plugs remained an adequate length of time and no hemorrhage occurred at the time of their exfoliation.[290] Methyl 2-cyanoacrylate has been used in cementing bone grafts in dogs and it caused irritation of the overlying soft tissues and displacement and death of the graft.[291] In cases of acute peritonitis in dogs, isobutyl cyanoacrylate is capable of anastomosing the intestine. Methyl 2-cyanoacrylate is somewhat more toxic under these conditions.[292] Cyanoacrylate anastomosis of the small intestine of 37 dogs resulted in ten deaths from leakage and peritonitis, while moderate constriction of the lumen and adhesions were observed in the successfully treated dogs.[293] Vesicovaginal, vesicocolonic, and rectovaginal fistulae in dogs were successfully repaired with cyanoacrylate adhesive.[294] Cyanoacrylic adhesive was successful in the repair of liver wounds in dogs.[295] The longer chain alkyl 2-cyanoacrylate are more effective homeostatic aids in liver wounds in dogs than methyl 2-cyanoacrylate.[296] The use cyanoacrylate adhesives for the repair of liver injury in dogs did not produce lethality or necrosis.[297] Application of ethyl 2-cyanoacrylate to the cut section of the dog's liver, pancreas, kidney, and spleen prevented secondary hemorrhage and did not interfere with the natural healing process.[298] Ethyl 2-cyanoacrylate caused leukocytic inflammation and loss of the kidney parenchyma in dogs through histotoxicity.[299] Cyanoacrylate adhesive delays wound healing by preventing the proliferation of fibroblasts and microcirculatory vessels bridging the wounded surfaces.[300] Methyl 2-cyanoacrylate used to close a pancreatic resection in dogs caused inflammation and abscess formation and eventually fibrous tissue formation.[301] This adhesive caused necrosis, abscess formation, and finally dense fibrous tissue when applied to the arterial wall in dogs.[302] Butyl 2-cyanoacrylate caused a high incidence of thrombosis and contraction of the vessel lumen when applied to the femoral artery and vein in the dog.[303] Application of methyl 2-cyanoacrylate to the adventitia caused necrosis of the tunica media and finally thrombotic obliteration of the vessel.[304] Similar results were obtained with butyl and isopropyl cyanoacrylates.[305] Marked inflammation and thrombosis of vessels after application of methyl 2-cyanoacrylate did not occur with n-pentyl 2-cyanoacrylate.[306] The gluing of cartoid and femoral arteries of dogs with methyl 2-cyanoacrylate resulted in stenosis of the arterial lumen, complete obturation of the lumen by thrombi, and aneurysmatic bulging of the arterial wall.[307] Methyl and butyl cyanoacrylate adhesives should not be applied to the myocardium because they are histotoxic and thrombogenic, according to Aaby et al.[308]

Application of butyl 2-cyanoacrylate to the dura, the intact brain surface, and a lesion of the cerebral cortex did not produce inflammation or adhesions to the archnoid or cerebral surface and no damage to the brain parenchyma were observed.[309] Comparison of methyl, N-propyl, N-butyl, N-heptyl, N-octyl, and isobutyl cyanoacrylate adhesives for suturing the radial and peroneal nerves in chimpanzees showed the methyl compound to be the most toxic, producing considerable necrosis in the surrounding tissues. The butyl compound was the most neurotoxic.[310] Methyl 2-cyanoacrylate has been used successfully in skin-grafting in pigs.[311] Butyl 2-cyanoacrylate produces no necrosis or hemorrhage when applied to the pancreas in pigs.[312]

2. Human

Butyl and isobutyl cyanoacrylates can be used as a surface dressing in oral wounds because they are bacteriostatic, reduce pain and infection, and promote healing.[313] Isobutyl cyanoacrylic adhesive has been used for the scaling of emphysematous lung surfaces and sealing of lung surfaces, but it is not effective in closing bronchio pleurocutaneous fistulae.[314] Butyl cyanoacrylic adhesive is useful in trauma patients because it promotes hemostasis, allows immediate postoperative use of anticoagulants, and decreases operating time.[315]

Butyl cyanoacrylate has been used in arterial anastomoses and for reinforcement of bleeding suture lines, but there was less patency in the vessels.[316] This adhesive has been successfully used for closure of arteriotomy wound after embolectomy, thrombendarterectomy, and bypass plastic operations in 127 patients.[317] Ethyl cyanoacrylate is better than methyl cyanoacrylate for blood vessel suturing because it elicits less tissue reaction on adventitia, less thrombus formation on intima, and maintains its tensile strength longer.[318] Methyl 2-cyanoacrylate adhesive has been used as a hemostatic agent in anticoagulated patients undergoing open-heart surgery.[319] In 78 patients, ethyl 2-cyanoacrylate has been used to unite minor arteries and veins. It has neither local nor general toxicity, tissue reaction is moderate, and vascular regeneration progresses normally.[320]

Isobutyl 2-cyanoacrylate adhesive used to close a perforation in an extensively necrotic cornea produced a marked granulomatous keratitis.[321] After removal of the corneal epithelium, a contact lens was attached to Bowman's membrane with cyanoacrylic adhesive in 80 patients. The procedure was well tolerated.[323] In this operation, methyl 2-cyanoacrylate was much more toxic than heptyl 2-cyanoacrylate.[324] Methyl 2-cyanoacrylate adhesive has been used in cataract operations and caused moderate irritation in the eye, with penetration of the material into the eye cavity, partial ungluing of the scleral visor, and bits of the glue cutting through the conjunctiva.[325] Acetone has been used to remove polymerized cyanoacrylate from the eye but it is deleterious to the corneal epithelium. However, permanent damage does not occur.[326] Butyl 2-cyanoacrylate adhesive produced a protracted mild foreign body reaction in the sclera and suprachoroid and a general loosening and atrophy in the choroid and retina, with circumscribed loss of the typical nuclear zones of the sensory retina.[327] Regardless of these adverse reactions, cyanoacrylate adhesive is recommended for retinal surgery.[328] Methyl 2-cyanoacrylate adhesive has been used to glue the eyelashes of the upper lid to the skin of the lower lid to form a temporary tarsorrhaphy for protection of the cornea in cases of facial nerve paralysis.[329] Butyl cyanoacrylate adhesive was used in three cases of perforating injury of the cornea. Disorders of wound healing resulted and toxic keratoiritis with vascular infiltration of the cornea occurred.[330] When cyanoacrylate was used in surgery for retinal detachment, histological examination revealed considerable foreign body reactions due to the adhesive.[331]

Cyanoacrylate adhesives have been used to attach Teflon® sheet to cover skin wounds and promote healing.[332] Ethyl cyanoacrylate adhesive infused into viscerocutaneous fistulas is curative.[333] Methyl 2-cyanoacrylate adhesive is recommended in tracheal anastomosis following resection to remove malignant growths.[334]

Methyl 2-cyanoacrylate adhesive used to attach electrodes or transducers to the skin, caused itching and erythema which cleared rapidly with removal of the test object.[335] Butyl 2-cyanoacrylate adhesive is superior to methyl 2-cyanoacrylate for attaching pacemaker electrodes to the myocardium.[336] Moreover, it caused less damage to the myocardial myofilaments, and less necrosis and fatty degeneration of muscular tissue.[337]

Methyl 2-cyanoacrylate adhesive has been used for sterilization of women because it completely obliterated the lumen of the Fallopian tube.[338] Cyanoacrylate adhesive has been used in circumcisions instead of sutures, but the wound must be dry with good control of bleeding to insure good results.[339] Methyl 2-cyanoacrylate has been used in combination with sutures in urinary bladder surgery, but it caused bladder calculi and prevented muscle regeneration across the site of surgery. Residual adhesive has been found in the tissue for up to 21 months.[340]

D. Teratogenesis
1. Animal

No teratogenic studies have been undertaken in any species with the cyanoacrylic adhesives.

2. Human

No reports are available concerning possible teratogenic properties of cyanoacrylic adhesives in humans.

E. Mutagenesis
1. Animal

No reports of mutagenesis in bacterial or animal species from use of cyanoacrylic are available.

2. Human

No information on possible mutagenesis in humans from use of cyanoacrylate adhesives is available.

F. Pharmacokinetics and Biotransformation
1. Animal

Nothing has been reported on the cyanoacrylate adhesives, but the material remains at the site of application for 21 to 180 days.[341]

2. Human

No data is available.

G. Carcinogenesis
1. Animal

No tumors were found in 13 chimpanzees 11 months after implantation of cyanoacrylate adhesives in soft tissues.[310] Subcutaneous injection of 0.1 or 0.4 mℓ of methyl 2-cyanoacrylate adhesive in dogs did produce tumor in the 2-year observation period. However, fibrosarcomas were produced in rats in 19 months after administration of the larger dose.[342]

2. Human

No reports are available on the carcinogenic potential of cyanoacrylate adhesives in humans.

H. Comment

The cyanoacrylate adhesives may be useful in many types of surgery if their inflammatory and necrotic properties could be eliminated or greatly reduced. The rate and degree of absorption, biotransformation, and elimination should be investigated. Methyl 2-cyanoacrylate adhesive has been showed to be carcinogenic in rats, but no information is available on its higher homologues. Mutagenic potential of these compounds should be investigated.

REFERENCES

Vinyl Chloride

1. **Autian, J.**, Plastics in pharmaceutical practice and related fields. Part I., *J. Pharm. Sci.*, 52, 1, 1963.
2. **Fairhall, L. T.**, *Industrial Toxicology*, 2nd ed., Hafner, New York, 1969, 356.
3. **Albright, L. F.**, Manufacture of vinyl chloride, *Chem. Eng. N.Y.*, 74, 219, 1967.
4. **Albright, L. F.**, Polymerization of vinyl chloride, *Chem. Eng. N.Y.*, 74, 145, 1967.
5. **Albright, L. F.**, Vinyl chloride polymerization by emulsion, bulk and solution processes, *Chem. Eng. N.Y.*, 74, 85, 1967.
6. **Albright, L. F.**, Vinyl chloride polymerization by suspension processes yields polyvinyl chloride resins, *Chem. Eng. N.Y.*, 74, 145, 1967.
7. **Shelly, P. G. and Sills, E. J.**, Monomer storage and protection, *Eng. Chem. Processing*, 65, 29, 1969.
8. **NFPA Committee on Chemicals and Explosives**, *Manual of Hazardous Chemical Reactions*, 4th ed., National Fire Protection Association, Boston, 1971.
9. **Haley, T. J.**, Vinyl chloride: how many unknown problems?, *J. Toxicol. Environ. Health*, 1, 47, 1975.
10. **Baretta, E. D., Stewart, R. D., and Mutchler, J. E.**, Monitoring exposures to vinyl chloride vapor: breath analysis and continuous air sampling, *Am. Ind. Hyg. Assoc. J.*, 30, 537, 1969.
11. OSHA to issue permanent and temporary vinyl chloride standards, *Pesticide Chem. News*, 2, 10, 1974.
12. **Jensen, S., Lange, R., Jernelov, A., and Palmork, K. H.**, Chlorinated byproducts from vinyl chloride production: a new source of marine pollution, *FAO Fish. Rep.*, 99, 1970, 1971.
13. **Boettner, E. A. and Weiss, B.**, An analytical system for identifying the volatile products of plastics, *Am. Ind. Hyg. Assoc. J.*, 28, 535, 1967.
14. **O'Mara, M. M., Crider, L. B., and Daniel, R. L.**, Combustion products from vinyl chloride monomer, *Am. Ind. Hyg. Assoc. J.*, 32, 153, 1971.
15. **Tsuchiya, Y. and Sumi, K.**, Thermal decomposition products of polyvinyl chloride, *J. Appl. Chem.*, 17, 364, 1967.
16. **Popov, L. A. and Yablochkin, V. D.**, Characteristics of gases released by polyvinyl chloride film, *Gig. Sanit.*, 32, 114, 1967.
17. **Foris, A. and Lehman, J. G.**, Gas chromatographic separation of halocarbons on Porapak Q porous polymer beads, *Sep. Sci.*, 4, 225, 1967.
18. **Williams, F. W. and Umstead, M. E.**, Determination of trace contaminants in air by concentration on porous polymer beads, *Anal. Chem.*, 40, 2232, 1968.
19. **Arena, J. A.**, *Poisoning*, 2nd ed., Charles C Thomas, Springfield, 1970.
20. **Meshkova, O. V., Dmitrieva, V. N., and Bezuglgy, V. D.**, Polarographic analysis of waste waters of polyvinyl chloride production, *Sov. Chem. Ind.*, 4, 252, 1971.
21. **Anon.**, Vinyl chloride monitors move on the market, *Chem. Eng. News*, May 27, 1974.
22. **Patty, F. A., Yant, W. P., and Waite, C. P.**, Acute response of guinea pigs to vapors of some new commercial organic compounds. V. Vinyl chloride, *Publ. Health Rep.*, 45, 1963, 1930.
23. **Lehmann, K. B. and Flury, F.**, *Toxicology and Hygiene of Industrial Solvents, Part 1*, Williams & Wilkins, Baltimore, 1938, 1.
24. **Carr, C. J., Krantz, J. C., Jr., and Sauerwald, M. D.**, Anesthesia 27. Narcosis with vinyl chloride, *Anesthesiology*, 8, 359, 1947.
25. **Carr, C. J., Burgison, R. M., Vitch, J. F., and Krantz, J. C., Jr.**, Anesthesia 34. Chemical constitution of hydro-carbons and cardiac automaticity, *J. Pharmacol. Exp. Ther.*, 97, 1, 1949.
26. **Peoples, A. S. and Leake, C. D.**, The anesthetic action of vinyl chloride, *J. Pharmacol. Exp. Ther.*, 48, 284, 1933.
27. **Lester, D., Greenberg, L. A., and Adams, W. R.**, Effects of single and repeated exposure of humans and rats to vinyl chloride, *Am. Ind. Hyg. Assoc. J.*, 24, 265, 1963.
28. **Mastromatteo, E., Fisher, A. M., Christie, H., and Danziger, H.**, Acute inhalation toxicity of vinyl chloride to laboratory animals, *Am. Ind. Hyg. Assoc. J.*, 21, 394, 1960.
29. **Prodan, L., Sucju, I., Pislaru, V., Ilea, E., and Pascu, L.**, Experimental acute toxicity of vinyl chloride (monochloroethene), *Ann. N.Y. Acad. Sci.*, 246, 154, 1975.
30. **Krantz, J. C., Jr., Carr, C. J., Musser, R., and Harne, W. G.**, The effect of chlorinated ethylenes on the perfused leg vessels of the frog, *Arch. Intern. Pharmacodynamie*, 52, 369, 1936.
31. **Smyth, H. F., Jr. and Weil, C. S.**, Chronic oral toxicity to rats of a vinyl chloride-vinyl acetate copolymer, *Toxicol. Appl. Pharmacol.*, 9, 501, 1966.

32. Little, K. and Parkhouse, J., Tissue reactions to polymers, *Lancet*, 1, 857, 1962.
33. Guess, W. L. and Haberman, S., Toxicity profiles of vinyl and polyolefinic plastics and their additives, *J. Biomed. Mater. Res.*, 2, 313, 1968.
34. Autian, J., Rosenbluth, S. A., and Guess, W. L., An evaluation of two urinary bags as to biological activity in animals, *Acta Pharm. Suec.*, 2, 279, 1965.
35. Guess, W. L., Haberman, S., Rowan, D. F., Bower, R. K., and Autian, J., Characterization of subtle toxicity of certain plastic components used in manufacture of polyvinyls, *Am. J. Hosp. Pharm.*, 24, 494, 1967.
36. Haberman, S., Guess, W. L., Rowan, D. F., Bowman, R. O., and Bower, R. K., Effects of plastics and their additives on human serum proteins, antibodies and developing chick embryos, *SPE J.*, 24, 62, 1968.
37. DeHaan, R. L., Toxicity of tissue culture media exposed to polyvinyl chloride tubing, *Nature (London) New Biol.*, 231, 85, 1971.
38. Von Oettingen, W. F., The halogenated aliphatic, olefinic, cyclic, aromatic, and aliphatic-aromatic hydrocarbons including the halogenated insecticides, their toxicity and potential dangers, *U.S. Public Health Serv. Publ.*, 144, 195, 1955.
39. Torkelson, T. R., Oyen, F., and Rowe, V. K., The toxicity of vinyl chloride as determined by repeated exposure of laboratory animals, *Am. Ind. Hyg. Assoc. J.*, 22, 354, 1961.
40. Schotter, W., Toxicology of vinyl chloride, *Chem. Tech.*, 21, 708, 1969.
41. Vazin, A. N. and Plokhova, E. I., Changes of cardiac activity in rats chronically exposed to the effect of vinyl chloride vapours, *Farmakol. Toksikol. Kiev.*, 32, 220, 1969.
42. Vazin, A. N. and Plokhova, E. I., Changes in adrenaline-like substances in rabbit blood following chronic exposure to vinyl chloride fumes, *Gig. Tr. Prof. Zabol.*, 13, 46, 1969.
43. Vazin, A. N. and Plokhova, E. I., Creation of an experimental model of "toxic angioneurosis" developing from the chronic action of vinyl chloride vapors on an organism, *Gig. Tr. Prof. Zabol.*, 12, 47, 1968.
44. Leong, B. K. J. and Torkelson, T. R., Effects of repeated inhalation of vinyl bromide in laboratory animals with recommendations for industrial handling, *Am. Ind. Hyg. Assoc. J.*, 31, 1, 1970.
45. Basalaev, A. V., Vazin, A. N., and Kochetkov, A. G., On the pathogenesis of changes developing due to long-term exposure to the effect of vinyl chloride, *Gig. Truda*, 16, 24, 1972.
46. Vazin, A. N. and Plokhova, E. I., Changes in the rate of inculation of conditional reflexes in rats on prolonged exposure to vinyl chloride vapor in concentrations approaching the maximum permissible concentration, *Gig. Sanit.*, 35, 434, 1970.
47. Vazin, A. N. and Plokhova, E. I., On the pathogenesis of the affection developing secondarily to a chronic exposure of the organism to the action of vinyl chloride, *Farmakol. Toksikol. Kiev.*, 31, 369, 1968.
48. Viola, P. L., Pathology of vinyl chloride, *Med. Lav.*, 61, 174, 1970.
49. Danziger, H., Accidental poisoning by vinyl chloride, *Canad. Med. Assoc. J.*, 82, 828, 1960.
50. Tribukh, S. L., Tikhomirova, N. P., Levin, S. V., and Kozlov, L. A., Working conditions and measures for their sanitation in the production and utilization of vinyl chloride plastics, *Gig. Sanit.*, 10, 38, 1949.
51. Suciu, I., Drejman, I., and Valaskai, M., Investigation of the diseases produced by vinyl chloride, *Med. Interna*, 15, 967, 1963.
52. Antonyuzhenko, V. A., Occupational poisoning by vinyl chloride, *Gig. Tr. Prof. Zabol.*, 12, 50, 1968.
53. Smirnova, N. A. and Granik, N. P., Long-term side effects of acute occupational poisoning by certain hydrocarbons and their derivatives, *Gig. Tr. Prof. Zabol.*, 14, 50, 1970.
54. Kudryavtseva, O. F., Characteristics of electrocardiographic changes in patients with vinyl chloride poisoning, *Gig. Tr. Prof. Zabol.*, 14, 54, 1970.
55. Juhe, S., Lange, C. E., Stein, G., and Veltman, G., Uber die sogenannte, Vinylchlorid-Krankheit (The so-called vinyl chloride illness), *Dtsch. Med. Wochenschr.*, 98, 2034, 1973.
56. Stalova, E. A., Characteristics of the state of thermoregulation in chronic vinyl chloride poisoning, *Gig. Tr. Prof. Zabol.*, 17, 53, 1973.
57. Maricq, H. R., Johnson, M. N., Whetstone, C. L., and LeRoy, E. C., Capillary abnormalities in polyvinyl chloride production workers, *JAMA*, 236, 1368, 1976.
58. Antonyuzhenko, V. A., Golova, I. A., and Aliyeva, N. K., The state of analyzer functions during chronic occupational intoxication with certain substances having a narcotic effect, *Gig. Truda*, 9, 19, 1972.
59. Lange, C.E., Juhe, S., Stein, G., and Veltman, G., Die sogenannte, Vinylchlorid-Krankheit-eine berufsbedingte Systemsklerose? (The so-called vinyl chloride sickness — an occupation-related systemic sclerosis?), *Int. Arch. Arbeitsmed.*, 32, 1, 1974.
60. Kramer, C. G. and Mutchler, J. E., The correlation of clinical and environmental measurements for workers exposed to vinyl chloride, *Am. Ind. Hyg. Assoc. J.*, 33, 19, 1972.

61. **Rumyantseva, Y. P. and Goryacheva, L. A.**, Glucocorticoid function of the adrenal cortex in patients with chronic intoxication with certain unsaturated chlorinated hydrocarbons, *Gig. Truda,* 12, 16, 1968.
62. **Grigorescu, I. and Tiba, G.**, Vinyl chloride: industrial toxicological aspects, *Rev. Chim.,* 17, 499, 1966.
63. **Gabor, S., Lecca-Radu, M., and Manta, I.**, Certain biochemical indexes of the blood in workers exposed to toxic substances (benzene, chlorobenzene and vinylchloride), *Prom. Tokisikol. I. Kliniska Prof. Zabolevanii Khim. Etiol. Sb.,* 221, 1962.
64. **Suciu, I., Drejman, I., and Valaskai, M.**, Investigation of the diseases caused by vinyl chloride, *Med. Lav.,* 58, 261, 1967.
65. **Gabor, S., Lecca-Radu, M., Perda, N., Abrudean, S., Ivanof, L., Anea, Z., and Valaczkay, C.**, Biochemical changes in workers occupied in vinyl chloride synthesis and polymerization, *Igiena,* 13, 409, 1964.
66. **Makk, L., Delmore, F., Creeck, J. L., Jr., Odgen, L. L., II., Fadell, E. H., Songster, C. L., Clanton, J., Johnson, M. N., and Christopherson, W. M.**, Clinical and morphologic features of hepatic angiosarcoma in vinyl chloride workers, *Cancer,* 37, 149, 1976.
67. **Morris, G. E.**, Vinyl plastics. Their dermatological and chemical aspects, *Arch. Ind. Occup. Med.,* 3, 535, 1953.
68. **Key, M. M.**, Occupational dermatitis from plastics, *J. Med. Assoc. Ga.,* 57, 521, 1968.
69. **Fregert, S. and Rorsman, H.**, Hypersensitivity to epoxy resins used as plasticizers and stabilizers in polyvinyl chloride (PVC) resins, *Acta Derm. Venereol.,* 43, 10, 1963.
70. **Pegum, J. S.**, Contact dermatitis from plastics containing triaryl phosphates, *Br. J. Dermatol.,* 78, 626, 1966.
71. **Kalmanovich, F. L.**, Sanitary chemical features of polyvinyl chloride coatings for floors, *Gig. Sanit.,* 33, 107, 1968.
72. **Dyachuk, I. A.**, A contribution to the hygienic evaluation of PVC floor tiles for apartments, *Gig. Sanit.,* 35, 424, 1970.
73. **Deese, D. E. and Joyner, R. E.**, Vinyl acetate: a study of chronic human exposure, *Am. Ind. Hyg. Assoc. J.,* 30, 449, 1969.
74. **Ward, A. M., Udnoon, S., Watkins, J., Walker, A. E., and Darke, C. S.**, Immunological mechanisms in the pathogenesis of vinyl chloride disease, *Br. Med. J.,* 1, 936, 1976.
75. **Cordier, J. M., Fievez, C., Lefevre, M. J., and Sevrin, A.**, Acroosteolyse et lesions cutanees associees chez deuz ouvriers affectes au nettoyage d'autoclaves, *Cahiers Med. Trav.,* 4, 14, 1966.
76. **Wilson, R. H., McCormick, W. E., Tatum, C. F., and Creech, J. F.**, Occupational acroosteolysis. Report of 31 cases, *JAMA,* 201, 577, 1967.
77. **Harris, D. K. and Adams, W. G. F.**, Acroosteolysis occurring in men engaged in the polymerization of vinyl chloride, *Br. Med. J.,* 2, 712, 1967.
78. **Chatelain, A. and Montillon, P.**, An acroosteolysis syndrome of occupational origin and or recent verification in France, *J. Radiol. Electrol.,* 48, 277, 1967.
79. **Anghelescu, M. O., Dobrinescu, E., Hagi-Paraschiv-Dossios, L., Dobrinescu, G., and Ganea, V.**, Clinio-pathogenic considerations of Raynaud's phenomenon in the employees of the vinyl polychloride industry, *Med. Interna,* 21, 473, 1969.
80. **Nitti, G., Petruzzelis, V., and Fasano, V.**, Reographic observations on workers belonging to the plastic materials industry, *Securitas,* 55, 683, 1970.
81. **Basalaev, A. V.**, Experience with the use of large-frame photo-fluorography in examining skeletal bone of persons occupationally dealing with unsaturated hydrocarbons of the ethylene series (olefins) and their chlorine derivatives (vinyl chloride, trichlorethylene), *Gig. Tr. Prof. Zabel.,* 14, 34, 1970.
82. **McCord, C. P.**, A new occupational disease is born, *J. Occup. Med.,* 12, 234, 1970.
83. **Dinman, B. D., Cook, W. A., Whitehouse, W. M., Magnuson, H. J., and Ditcheck, T.**, Occupational acroosteolysis. I. An epidemiological study, *Arch. Environ. Health,* 22, 61, 1971.
84. **Cook, W. A., Giever, P. M., Dinman, B. D., and Magnuson, H. J.**, Occupational acroosteolysis. II. An industrial hygiene study, *Arch. Environ. Health,* 22, 74, 1971.
85. **Dodson, V. N., Dinman, B. D., Whitehouse, W. M., Nasr, A. N. M., and Magnuson, H. G.**, Occupational acroosteolysis. III. A clinical study, *Arch. Environ. Health,* 22, 83, 1971.
86. **Gitsios, C. T.**, Acroosteolysis in PVC workers, *The Medical Bull. (Esso),* 31, 49, 1971.
87. **Juhe, S. and Lange, C. E.**, Sklerodermieartige Hautveranderungen Raynaud-Syndrom und Akoosteolysen bei Arbeitern der PVC-herstellenden Industrie (Scleroderma-like skin changes, Raynaud's syndrome and acroosteolysis in men engaged in the industrial polymerization of vinyl chloride (PVC), *Dtsch. Med. Wochenschr.,* 97, 1922, 1972.
88. **Markowitz, S. S., McDonald, C. J., Fethiere, W., and Kerzner, M. S.**, Occupational acroosteolysis, *Arch. Dermatol.,* 106, 224, 1972.
89. **Meyerson, L. B. and Meier, G. C.**, Cutaneous lesions of acroosteolysis, *Arch. Dermatol.,* 106, 224, 1972.

90. Misgeld, V., Stolpmann, H. J., and Schulte, S., Intoxication by vinyl chloride polymers and/or their additives, *Z. Haut Geschlechtskr*, 48, 425, 1973.
91. John, J. A., Smith, F. A., Leong, B. K. J., and Schwetz, B. A., The effects of maternally inhaled vinyl chloride on embryonal and fetal development in mice, rats, and rabbits, *Toxicol. Appl. Pharmacol.*, 39, 497, 1977.
92. Radike, M., Warkany, J., Bingham, E., O'Toole, B., Larson, E., and Grande, F., Transplacental effects of vinyl chloride in rats, Report USPHS-ES-00159, Center for Study of the Human Environment, Department of Environmental Health, University of Cincinnati.
93. Infante, P. F., Oncogenic and mutagenic risks in communities with polyvinyl chloride production facilities, *Ann. N.Y. Acad. Sci.*, 271, 49, 1976.
94. Infante, P. F., McMichael, A. J., Wagoner, J. K., Waxweiler, R. J., and Falk, H., Genetic risks of vinyl chloride, *Lancet*, 1, 734, 1976.
95. Edmonds, L. D., Falk, H., and Nissim, J. E., Congenital malformations and vinyl chloride, *Lancet*, 2, 1098, 1975.
96. Rannung, U., Johansson, A., Ramel, C., and Wachtmeister, C. A., The mutagenicity of vinyl chloride after metabolic activation, *Ambio*, 3, 194, 1974.
97. Bartsch, H., Malaville, C., and Montesano, R., Human rat and mouse liver-mediated mutagenicity of vinyl chloride in S. typhimurium strains, *Int. J. Cancer*, 15, 429, 1975.
98. McCann, J., Simmon, V., Streitwieser, D., and Ames, B. N., Mutagenicity of chloroacetaldehyde, a possible metabolic product of 1,2-dichloroethane (ethylene dichloride), chloroethanol, and cyclophosphamide (ethylene chlorohydrin) vinyl chloride, *Proc. Natl. Acad. Sci. (U.S.)*, 72, 3190, 1975.
99. Elmore, J. D., Wong, J. L., Laumbach, A. D., and Streips, U. N., Vinyl chloride mutagenicity via the metabolites chlorooxirane and chloroacetaldehyde monomerhydrate, *Biochem. Biophys. Acta*, 442, 405, 1976.
100. Garro, A. J., Guttenplan, J. B., and Milvy, P., Vinyl chloride dependent mutagenesis: effects of liver extracts and free radicals, *Mutat. Res.*, 38, 81, 1976.
101. Greim, H., Bonse, G., Radman, Z., Reichert, D., and Henschler, D., Mutagenicity in vitro and potential carcinogenicity of chlorinated ethylenes as a function of metabolic oxirane formation, *Biochem. Pharmacol.*, 24, 2013, 1975.
102. Laprieno, N., Barale, R., Baroncelli, S., Bauer, C., Bronzetti, G., Cammellini, A., Cercignani, G., Corsi, C., Gervasi, G., Leporini, C., Nieri, R., Rossi, A. M., Stretti, G., and Turchi, G., Evaluation of the genetic effects induced by vinyl chloride monomer (vcm) under mammalian activation in vitro and in vivo studies, *Mutat. Res.*, 40, 85, 1976.
103. Loprieno, N., Barale, R., Baroncelli, S., Bartsch, H., Bronzetti, G., Cammellini, A., Corsi, C., Frezza, D., Nieri, R., Leporini, C., Rosellini, D., and Rossi, A. M., Induction of gene mutations and gene conversions by vinyl chlorine metabolites in yeast, *Cancer Res.*, 36, 253, 1977.
104. Verbugt, F. G., Vinyl chloride mutagenesis in Drosophila melanogaster, *Mutat. Res.*, 48, 327, 1977.
105. Huberman, E., Bartsch, H., and Sach, L., Mutation induction in Chinese hamster V79 cells by two vinyl chloride metabolites, chloroethylene oxide and chloroacetaldehyde, *Int. J. Cancer*, 16, 639, 1975.
106. Anderson, D., Hodge, M. E., and Purchase, I. F. H., Vinyl chloride: dominant lethal studies in male CD-1 mice, *Mutat. Res.*, 40, 359, 1976.
107. Ducatman, A., Hirschhorn, K., and Selikoff, I. J., Vinyl chloride exposure and human chromosome aberrations, *Mutat. Res.*, 31, 163, 1975.
108. Purchase, I. F. H., Richardson, C. R., and Anderson, D., Chromosomal and dominant lethal effects of vinyl chloride, *Lancet*, 2, 410, 1975.
109. Heath, C. W., Dumont, C. R., Gamble, J., and Waxweiler, R. J., Chromosomal damage in men occupationally exposed to vinyl chloride monomer and other chemicals, *Environ. Res.*, 14, 68, 1977.
110. Waxweiler, R. J., Falk, H., McMichael, A., Mallow, J. S., and Grivas, A. S., A cross-sectional epidemiologic survey of vinyl chloride workers, Natl. Inst. Occup. Safety Tech. Inf. Publ. No. 77-177, U.S. Department of Health, Education, and Welfare.
111. Picciano, D. J., Flake, R. E., Gay, P. C., and Kilian, D. J., Vinyl chloride cytogenetics, *J. Occup. Med.*, 19, 527, 1977.
112. Killian, D. J., Picciano, D. J., and Jacobson, C. B., Industrial monitoring: a cytogenetic approach, *Am. N.Y. Acad. Sci.*, 269, 4, 1975.
113. Duprat, P., Metabolic approach to industrial poisoning: blood kinetics and distribution of ^{14}C-vinyl chloride monomer, *Toxicol. Pharmacol. Suppl.*, 122, 1977.
114. Watanabe, P. G., McGowan, G. P., Madrid, E. D., and Gehring, P. J., Fate of (^{14}C) vinyl chloride following inhalation exposure in rats, *Toxicol. Appl. Pharmacol.*, 37, 49, 1976.
115. Bolt, H. M., Laib, R. J., Kappus, H., and Buchter, A., Pharmacokinetics of vinyl chloride in the rat, *Toxicology*, 7, 179, 1977.
116. Watanabe, P. G., McGown, G. R., and Gehring, P. J., Fate of (^{14}C) vinyl chloride after single oral administration in rats, *Toxicol. Appl. Pharmacol.*, 36, 339, 1976.

117. **Watanabe, P. G., Zempel, J. A., and Gehring, P. J.**, Comparison of the fate of vinyl chloride following single and repeated exposure, *Toxicol. Appl. Pharmacol.*, 41, 174, 1977.
118. **Withey, J. R.**, Pharmacodynamics and uptake of vinyl chloride monomer administered by various routes to rats, *J. Toxicol. Environ. Health*, 1, 381, 1976.
119. **Gothe, R., Calleman, C. J., Ehrenberg, L., and Wachmeister, C. A.**, Trapping with 3,4-dichlorobenzenethiol of reactive metabolites formed in vitro from the carcinogen vinyl chloride, *Ambio*, 3, 234, 1974.
120. **Clapp, J. J., Kay, C. M., and Young, L.**, Observations on the metabolism of allyl compounds in the rat, *Biochem. J.*, 114, 6P, 1969.
121. **Hefner, R. E., Jr., Watanabe, P. G., and Gehring, P. J.**, Preliminary studies of the fate of inhaled vinyl chloride monomer in rats, *Ann. N.Y. Acad. Sci.*, 246, 135, 1975.
122. **Van Duuren, B. L.**, On the possible mechanism of carcinogenic action of vinyl chloride, *Ann. N.Y. Acad. Sci.*, 246, 258, 1975.
123. **Montesano, R. and Bartsch, H.**, Mutagenicity and metabolism of vinyl chloride, in *Biological Characterization of Human Tumors*, Proc. 6th Int. Symp. Biol. Characterization of Human Tumours, 3, 242, 1976.
124. **Kappus, H., Bolt, H. M., Buchter, A., and Bolt, W.**, Liver microsomal uptake of (^{14}C) vinyl chloride and transformation to protein alkylating metabolites in vitro, *Toxicol. Appl. Pharmacol.*, 37, 461, 1976.
125. **Jaeger, R. J., Reynolds, E. S., Conolly, R. B., Moslen, M. T., Szabo, S., and Murphy, S. D.**, Acute hepatic injury by vinyl chloride in rats pretreated with phenobarbital, *Nature (London)*, 252, 724, 1974.
126. **Kappus, H., Bolt, H. M., Buchter, A., and Bolt, W.**, Rat liver microsomes catalyze covalent binding of ^{14}C-vinyl chloride to macromolecules, *Nature (London)*, 257, 134, 1975.
127. **Laib, R. J. and Bolt, H. M.**, Alkylation of RNA by vinyl chloride metabolites in vitro and in vivo: formation of 1-N-^{6}ethenoadenosine, *Toxicology*, 8, 185, 1977.
128. **Hathway, D. E.**, Comparative mammalian metabolism of vinyl chloride and vinylidene chloride in relation to oncogenic potential, *Environ. Health Perspect.*, 21, 55, 1977.
129. **Lawrence, W. H., Dillingham, E. O., Turner, J. E., and Autian, J.**, Toxicity profile of chloroacetaldehyde, *J. Pharm. Sci.*, 61, 19, 1972.
130. **Jaeger, R. J.**, Vinyl chloride monomer: comments on its hepatotoxicity and interaction with 1,1-dichloroethylene, *Ann. N.Y. Acad. Sci.*, 246, 150, 1975.
131. **Viola, P. L., Bigotti, A., and Caputo, A.**, Oncogenic response of rat skin, lungs, and bones to vinyl chloride, *Cancer Res.*, 31, 516, 1971.
132. **Caputo, A., Viola, P. L., and Bigotti, A.**, Oncogenicity of vinyl chloride at low concentrations in rats and rabbits, *IRCS*, 2, 1582, 1974.
133. **Maltoni, C. and Lefemine, G.**, Le Potenzialita dei saggi sperimentali nella predizione dei rischi oncongeni ambientali. Un esempio: il chloruro di vinile. (The potential of experimental stages of the prediction of the ambient oncogenic risks. An example: vinyl chloride), *Lincei-Rend. Sc. Fs. Mat. Enat.*, 56, 1, 1974.
134. **Maltoni, C. and Lefemine, G.**, Carcinogenicity bioassays of vinyl chloride. I. Research plan and early results, *Environ. Res.*, 7, 387, 1974.
135. **Maltoni, C. and Lefemine, G.**, Carcinogenicity assays of vinyl chloride, *Ann. N.Y. Acad. Sci.*, 246, 195, 1975.
136. **Maltoni, C.**, Predictive value of carcinogenesis bioassays, *Ann. N.Y. Acad. Sci.*, 271, 431, 1976.
137. **Lee, C. C., Bhandri, J. C., Winston, J. M., House, W. B., Peters, P. J., Dixon, R. L., and Woods, J. S.**, Inhalation toxicity of vinyl chloride and vinylidene chloride, *Environ. Health Perspect.*, 21, 25, 1977.
138. **Lee, C. C., Bhandri, J. C., Winston, J. M., House, W. B., Dixon, R. L., and Woods, J. S.**, Carcinogenicity of vinyl chloride and vinylidene chloride, *J. Toxicol. Environ. Health*, 4, 15, 1978.
139. **Radike, M. J., Stemmer, K. L., Brown, P. G., Larson, E., and Bingham, E.**, Effect of ethanol and vinyl chloride on the induction of liver tumors: preliminary report, *Environ. Health Perspect.*, 21, 153, 1977.
140. **Oppenheimer, B. S., Oppenheimer, E. T., and Stout, A. P.**, Sarcomas induced in rodents by imbedding various plastic films, *Proc. Soc. Exp. Biol. Med.*, 79, 366, 1952.
141. **Oppenheimer, B. S., Oppenheimer, E. T., Stout, A. P., and Danishefsky, I.**, Malignant tumors resulting from embedding plastics in rodents, *Science*, 118, 305, 1953.
142. **Oppenheimer, B. S., Oppenheimer, E. T., Danishefsky, I., Stout, A. P., and Eirich, F. R.**, Further studies of polymers as carcinogenic agents in animals, *Cancer Res.*, 15, 333, 1955.
143. **Russell, F. E., Simmers, M. H., Hirst, A. E., and Pudenz, R. H.**, Tumors associated with embedded polymers, *J. Natl. Cancer Inst.*, 23, 305, 1959.
144. **Kogan, A. K. and Tugarinova, V. N.**, Blastomogenic action of poly (vinyl chloride), *Vopr. Onkol.*, 5, 540, 1959.

145. Kogan, A. K. and Raikhlin, N. T., Dyanmics of morphological changes in the connective tissue capsules developing around the plastic implants during the malignant process, *Arch. Pathol.*, 23, 62, 1961.
146. Stankevich, K. I., Experimental findings on the blastomogenic effect of coumarone and polychloravinyl plates, *Vrache. Delo*, 11, 108, 1962.
147. Shabad, L. M., Plastic carcinogenesis, some experimental data and its possible importance for clinic and prophylaxis of cancer, *Unio Int. Contre Cancer*, 7, 188, 1967.
148. Raikhlin, N. T. and Kogan, A. K., The development and malignant transformation of connective tissue capsules around implants of plastic material, *Vopr. Onkol.*, 7, 13, 1961.
149. Brand, K. G., Buoen, L. C., and Brand, I., Carcinogenesis from polymer implants: new aspects from chromosomal and transplantation studies during premalignancy, *J. Nat. Cancer Inst.*, 39, 663, 1967.
150. Pushkin, G. A., Lesions in the liver and bile ducts in workers producing some kinds of plastics, *Sov. Med.*, 28, 132, 1965.
151. Masteller, H. J., Lelbach, W. K., Mueller, R., Juhe, S., Rohner, H. G., and Veltman, G., Chronic-toxic damage in PVC production workers, *Dtsch. Med. Wochenschr.*, 98, 2311, 1973.
152. Creech, J. L., Jr. and Johnson, M. N., Angiosarcoma of the liver in the manufacture of polyvinyl chloride, *J. Occup. Med.*, 16, 150, 1974.
153. Manucuso, T. F., Cancer and vinyl chloride-polymerization implications, problems and uses, paper presented to the Industrial Union Department AFL-CIO, Feb. 1974.
154. Popper, H. and Thomas, L., Alterations of liver and spleen among workers exposed to vinyl chloride, *Ann. N.Y. Acad. Sci.*, 246, 172, 1975.
155. Lilis, R., Anderson, H., Nicholson, W. J., Damm, S., Fischbein, A. S., and Selikoff, I. J., Prevalence of disease among vinyl chloride and polyvinyl chloride workers, *Ann. N.Y. Acad. Sci.*, 246, 22, 1975.
156. Berk, P. D., Vinyl chloride-associated liver disease, *Ann. Int. Med.*, 84, 717, 1976.
157. Block, J. B., Angiosarcoma of the liver following vinyl chloride exposure, *JAMA*, 229, 53, 1974.
158. Makk, L., Creech, J. L., Jr., Whelan, J. G., and Johnson, M. N., Liver damage and angiosarcoma in vinyl chloride workers, *JAMA*, 230, 64, 1974.
159. Tabershaw, I. R. and Gaffey, W. R., Mortality study of workers in the manufacture of vinyl chloride and its polymers, *J. Occup. Med.*, 16, 509, 1974.
160. Holder, B., Dow Chemical Co. testimony presented at public hearings on proposed standard for occupational exposure to vinyl chloride, U.S. Department of Labor, Washington, D.C., 1974.
161. Monson, R. R., Peters, J. M., and Johnson, M. N., Mortality among vinyl chloride workers, *Environ. Health Perspect.*, 11, 75, 1975.
162. Nicholson, W. J., Hammond, E. C., Seidman, H., and Selikoff, I. J., Mortality experience of a cohort of vinyl chloride-polyvinyl chloride workers, *Ann. N.Y. Acad. Sci.*, 246, 225, 1975.
163. Ott, M. G., Langner, R. R., and Holder, B. B., Vinyl chloride exposure in a controlled industrial environment: a long-term mortality experience in 595 employees, *Arch. Environ. Health*, 30, 333, 1975.
164. Wagoner, J. E., NIOSH presentation before the Environmental Commerce Commission, U.S. Senate, Washington, D.C., 1974.
165. Chiazze, L., Nichols, W. E., and Wong, O., Mortality among employees of PVC fabricators, *J. Occup. Med.*, 19, 623, 1977.
166. Fox, A. J. and Collier, P. F., Mortality experience of workers exposed to vinyl chloride monomer in the manufacture of polyvinyl chloride in Great Britain, *Br. J. Ind. Med.*, 34, 1, 1977.
167. Byren, D., Engholm, G., Englund, A., and Westerholm, P., Mortality and cancer in a group of Swedish VCM and PVC production workers, *Environ. Health Perspect.*, 17, 167, 1976.
168. Delorme, F. and Theriault, G., Ten cases of angiosarcoma of the liver in Shawinigan, Quebec, *J. Occup. Med.*, 29, 338, 1978.
169. Spritas, R. and Kaminski, R., Angiosarcoma of the liver in vinyl chloride/polyvinylchloride workers. Update of the NIOSH register, *J. Occup. Med.*, 20, 427, 1978.
170. Baxter, P. J., Anthony, P. P., MacSween, R. N. M., and Scheuer, P. J., Angiosarcoma of the liver in Great Britain 1963-73, *Br. Med. J.*, 2, 919, 1977.
171. Kuzmak, A. M. and McGaughy, R. E., Quantitative risk assessment for community exposure to vinyl chloride, U.S. Environmental Protection Agency Report, Dec. 5, 1975.
172. Brady, J., Liberatore, F., Harper, P., Greenwald, P., Burnett, W., Davies, J. N. P., Bishop, M., Polan, A., and Vianna, N., Angiosarcoma of the liver: an epidemiologic survey, *J. Natl. Cancer Inst.*, 59, 1383, 1977.

Silicone

173. Gair, T. J. and Thimineur, R. J., Silicone, Modern Plastic Encyclopedia, 1978-1979, New York, 102.
174. Braley, S. A., The chemistry and physical properties of the medical grade silicones, *J. Macromol. Sci. Chem.*, A4(3), 529, 1970.
175. Dow Corning Clean-Grade Elastomers, Bulletin No. 51-209, Dow Corning Corporation, Midland, Mich., August 1973.
176. Dow Corning Silastic 382 Medical-Grade Elastomer, Bulletin, Dow Corning Corporation, Midland, Mich., February 1975.
177. Dow Corning Silastic Medical-Grade Tubing, Bulletin No. 51-247, Dow Corning Corporation, Midland, Mich., September 1975.
178. Dow Corning 360 Medical Fluid, Bulletin, Dow Corning Corporation, Midland, Mich., December 1975.
179. Dow Corning Medical Antifoams, Bulletin No. 51-201A, Dow Corning Corporation, Midland, Mich., February 1977.
180. Hobbs, E. J., Fancher, O. E., and Calandra, J. C., Effect of selected organopoly siloxanes on male rat and rabbit reproductive organs, *Toxicol. Appl. Pharmacol.*, 21, 45, 1972.
181. Smith, A. L., *Analysis of silicones*, John Wiley & Sons, New York, 1974.
182. Horner, H. J., Weiler, J. E., and Angelotti, N. C., Visible and infrared spectroscopic determination of trace amounts of silicones in foods and biological materials, *Anal. Chem.*, 32, 858, 1960.
183. Jankowiak, M. E. and LeVier, R. R., Elimination of phosphorus interference in the colorimetric determination of silicon in biological material, *Anal. Biochem.*, 44, 462, 1971.
184. Neal, P., Campbell, A. D., Firestone, D., and Aldridge, M. H., Low temperature separation of trace amounts of dimethylpolysiloxane from foods, *J. Am. Oil Chem. Soc.*, 46, 561, 1969.
185. Sinclair, A. and Hallam, T. R., Determination of dimethylpolysiloxane in beer and yeast, *Analyst*, 96, 149, 1971.
186. Rowe, V. K., Spencer, H. C., and Bass, S. L., Toxicological studies on certain commercial silicones, *J. Ind. Hyg. Toxicol.*, 30, 332, 1948.
187. Hobbs, E. J., Keplinger, M. L., and Calandra, J. C., Toxicity of polydimethylsiloxanes in certain environmental systems, *Environ. Res.*, 10, 397, 1975.
188. Calandra, J. C., Keplinger, M. L., Hobbs, E. J., and Tyler, L. J., Health and environmental aspects of polydimethylsiloxane fluids. Division of Polymer Chemistry, Inc., *Am. Chem. Soc. Polymer Prepr.*, 17, 1, 1976.
189. Frey, C. F., Thorpe, C., and Brody, G., Silicone fluid in the prevention of intestinal adhesions, *Arch. Surg.*, 95, 253, 1967.
190. Brody, G. L. and Frey, C. F., Peritoneal response to silicone fluid. A histologic study, *Arch. Surg.*, 96, 237, 1968.
191. Ballantyne, D. L., Jr., Howthorne, G., Ben-Hur, N., Seidman, I., and Rees, T. D., Effect of silicone fluid on experimentally induced peritoneal adhesions, *Isr. J. Med. Sci.*, 7, 1046, 1971.
192. Aboulafia, Y. and Polishuk, W. Z., Prevention of peritoneal adhesions by silicone solution, *Arch. Surg.*, 94, 384, 1967.
193. Mellish, R. W. P., Ty, T. C., and Keller, D. J., A study of intestinal healing, *J. Pediatr. Surg.*, 3, 286, 1968.
194. Furman, S. and Denize, A., Serous membrane regeneration: use of intrapleural liquid silicone, *Surgery*, 60, 733, 1966.
195. Nedelman, C. I., Oral and cutaneous tissue reactions to injected fluid silicones, *J. Biomed. Mater. Res.*, 2, 131, 1968.
196. Rees, T. D., Ballantyne, D. L., Jr., Seidman, I., and Hawthorne, G. A., Visceral response to subcutaneous and intraperitoneal injections of silicone in mice, *Plast. Reconstr. Surg.*, 39, 462, 1967.
197. Grasso, P., Fairweather, F. A., and Golberg, L., A short-term study of epithelial and connective tissue reactions to subcutaneous injection of silicone fluid, *Food Cosmet. Toxicol.*, 3, 263, 1965.
198. Ben-Hur, N., Ballantyne, D. L., Jr., Rees, D. T., and Seidman, I., Local and systemic effects of dimethylpolysiloxane fluid in mice, *Plast. Reconstr. Surg.*, 39, 423, 1967.
199. Ballantyne, D. L., Jr., Rees, D. T., and Seidman, I., Silicone fluid: response to massive injections of dimethylpolysiloxane fluid in animals, *Plast. Reconstr. Surg.*, 36, 330, 1964.
200. Bregeon, C., Simard, C., Boasson, M., and Renier, J. C., Utilization d'un silicone, le methylpolysiloxane, comme lubrifiant articulaire, *Ref. Rhum. Mal. Osteo-Articulaires*, 37, 735, 1970.
201. Murray, D. G., The fate of liquid silicone in the rabbit knee joint, *Clin. Orthop.*, 87, 275, 1972.
202. Seidel, W. and Leitz, K., Die Thrombenbildung an Silikonkautschukkugeln von Herzklappenprosthesen, *Langenbecks Arch. Chir.*, 314, 25, 1966.
203. Smith, M. P., Microsurgery in Silastic® vascular grafts and arteriovenous cannulas. Testing with Teflon® tips, Dacron skirt and stainless steel strands, *Am. Surg.*, 32, 531, 1966.
204. Uy, S. and Kammermeyer, K., Nonthrombogenic surface preparation for silicone rubber, *J. Biomed. Mater. Res.*, 3, 587, 1969.

205. Nyilas, E. and Kupski, E. L., Surface microstructural factors and the blood compatibility of a silicone rubber, *J. Biomed. Mater. Res.*, 4, 369, 1970.
206. Streeten, B. W. and Belkowitz, M., Experimental hypotony with Silastic®, *Arch. Ophthamol.*, 79, 503, 1967.
207. Lee, P. F., Donovan, R. H., Mukai, N., Schepens, C. L., and Freeman, H. M., Intravitreous injection of silicone: an experimental study. I. Clinical picture and histology of the eye, *Ann. Ophthalmol.*, 1, 15, 1969.
208. Mukai, N., Lee, P. F., and Schepens, C. L., Intravitreous injection of silicone: an experimental study. II. Histochemistry and electron microscopy, *Ann. Ophthal.*, 4, 273, 1972.
209. Labelle, P. and Okun, E., Ocular tolerance to liquid silicone. An experimental study, *Canad. J. Ophthmal.*, 7, 199, 1972.
210. Hill, R. M., Effects of hydrophilic plastic lenses on corneal respiration, *J. Am. Optom. Assoc.*, 38, 181, 1967.
211. Burns, R. P., Roberts, H., and Rich, L. F., Effect of silicone contact lenses on corneal epithelial metabolism, *Am. J. Ophthalmol.*, 71, 486, 1971.
212. Klopper, P. J., Experimental reconstruction of the trachea with silicone rubber, *Arch. Chir. Neerl.*, 21, 293, 1969.
213. Klopper, P. J. and Haspels, A. A., Experimental replacement of the uteter by silicone rubber, *Arch. Chir. Neerl.*, 21, 243, 1969.
214. Ikeda, M., An experimental study on reconstruction of the esophagus with artificial tubes, *J. Kyoto Prefect. Med.*, 75, 32, 1966.
215. Ashkar, L. and Heller, E., The silastic bladder patch, *J. Urol.*, 98, 679, 1967.
216. Vrind, S. H. M. and Klopper, P. J., Implantation of silicone rubber in the urinary bladder, *Arch. Chir. Neerl.*, 21, 235, 1969.
217. Stanley, T. H. and Lattimer, J. K., Velour lined silicone as an artificial bladder material, *J. Biomed. Mater. Res.*, 6, 533, 1972.
218. Beisner, D. H., Extraocular muscle recessions utilizing silicone tendon protheses, *Arch. Ophthalmol.*, 83, 195, 1970.
219. Bradley, B. E., Vineyard, G. C., Defalco, A., Lawson, D., and Hayes, J. R., An evaluation of silicone rubber tubing as a replacement of the common bile duct, *Am. J. Surg.*, 113, 501, 1967.
220. Schmidt, D., Fichler, H. J., and Schuh, D., Experimentelle Unterschungen zur Vertraglichkeit eines Transplantates aus Silikongumml, *Abt. Exp. Chir.*, 92, 2255, 1967.
221. Van Der Leek, A. J., The use of silastic in vascular surgery, *Arch. Chir. Neerl.*, 22, 127, 1970.
222. Ducker, T. B. and Hayes, G. J., Experimental improvement in the use of Silastic® cuff for peripheral nerve repair, *J. Neurosurg.*, 28, 582, 1968.
223. Roberts, W. C. and Morrow, A. G., Fatal degeneration of the silicone rubber ball of Starr Edwards prosthetic aortic valve, *Am. J. Cardiol.*, 22, 614, 1968.
224. Myhre, E., Dale, J., and Rasmussen, K., Erythrocyte destruction in different types of Starr Edwards aortic ball values, *Circulation*, 42, 515, 1970.
225. Midgley, R. D. and Woolhouse, F. M., Silicone rubber sheathing as an adjunct to neural anastamosis, *Surg. Clin. North Am.*, 48, 1149, 1968.
226. Morin, L. J. and Albert, D. J., Bilateral nephrostomy drainage with silicone rubber catheters, *J. Urol.*, 100, 109, 1968.
227. Kalambaheti, K., Siliconized Foley catheters, *Am. J. Surg.*, 110, 935, 1965.
228. Pearman, R. O., Treatment of organic impotence by implantation of a penile prosthesis, *J. Urol.*, 97, 716, 1967.
229. Reinecke, R. D. and Carroll, J. M., Silicone lacrimal tube implantation, *Trans. Am. Acad. Ophthalmol. Otolaryngol.*, 73, 85, 1969.
230. Rees, T. D., Ballantyne, D. L., and Seidman, I., Eyelid deformaties caused by the injection of silicone fluid, *Br. J. Plast. Surg.*, 24, 125, 1971.
231. Peyman, G. A., Ericson, E. S., and May, D. K., A review of substances and techniques of vitreous replacement, *Surv. Ophthalmol.*, 17, 41, 1972.
232. Okun, E., Intravitreal surgery utilizing liquid silicone. A long term followup, *Trans. Pac. Coast Otoophthalmol. Soc.*, 49, 141, 1968.
233. Linoff, H. A., McLean, J. M., and Nano, H., Scleral Abscess. I. A complication of retinal detachment buckling procedures, *Arch. Ophthalmol.*, 74, 641, 1965.
234. Moreau, P. G., Les silicones intra-oculaires dans les descollements de retine desesperes, *Ann. Ocul.*, 200, 257, 1967.
235. Merz, M. and Czerwinska, W., Silicone rubber in defects of the conjunctiva, *Br. J. Ophthalmol.*, 53, 46, 1969.
236. Lipshutz, H. and Ardizone, R. A., Further observations on the use of silicones in the management of orbital fractures, *J. Trauma*, 5, 617, 1965.

237. Meyer, R. A., Gehrig, J. D., Funk, E. C., and Beder, O. E., Restoring facial contour with implanted silicone rubber. Report of two cases, *Oral Surg.*, 24, 598, 1967.
238. Helal, B. and Karadi, B. S., Artificial lubrication of joints. Use of silicone oil, *Ann. Phys. Med.*, 9, 334, 1968.
239. Meester, W. D. and Swanson, A. B., In vivo testing of silicone rubber joint implants for lipid absorption, *J. Biomed. Mater. Res.*, 6, 193, 1972.
240. O'Leary, J. A., Turner, A. F., and Feldman, M. A. S., Silicone in the prevention of pelvic adhesions, *Am. Surg.*, 35, 622, 1969.
241. Weiner, L. H., Sternberg, T. H., Lehman, R., and Ashey, F. L., Tissue reactions to injected silicone liquids. A report of 3 cases, *Arch. Dermatol.*, 90, 588, 1964.
242. Snyderman, R. K. and Starzynski, T. E., Breast reconstruction, *Surg. Clin. North Am.*, 49, 303, 1969.
243. Amberg, J. R., A hazard of silicone foam diagnostic enema. Report of a case of perforation of the colon, *Am. J. Roentgenol. Radium Ther. Nucl. Med.*, 99, 96, 1967.
244. Ermini, M., Carpino, F., Russo, M., and Benagiano, G., Studies on sustained contraceptive effects with subcutaneous polydimethylsiloxane implants. III. Factors affecting steroid diffusion in vivo and in vitro, *Acta Endocrinol.*, 73, 360, 1973.
245. Weeder, R. S., Brooks, H. W., and Boyer, A. S., Silicone immersion in the care of burns, *Plast. Reconstr. Surg.*, 39, 256, 1967.
246. Batdorf, J. W., Cammack, K. V., and Colquitt, R. D., The silicone dressing management of the burned hand, *Arch. Surg.*, 98, 469, 1969.
247. Cronin, T. D., Use of a Silastic® frame for total and subtotal reconstruction of the external ear: preliminary report, *Plast. Reconstr. Surg.*, 37, 399, 1966.
248. DeBenedictis, N. and Ferro, G. F., Clinical evaluation of a new digestive and antispastic preparation, *Minerva Gastroenterol.*, 18, 35, 1972.
249. Kennedy, G. L., Jr., Keplinger, M. L., Calandra, J. C., and Hobbs, E. J., Reproductive, teratologic and mutagenic studies with some polydimethylsiloxanes, *J. Toxicol. Environ. Health*, 1, 909, 1976.
250. Sakaki, T., Tsurumi, N., Maeda, J., Honda, T., Kawakatsu, K., and Tsutsumino, R., Acid mucopolysacchrides in silicone induced tumors of rats, *Gann*, 63, 167, 1972.
251. Gonzalez Ulloa, M., Stevens, E., Loewe, P., Vargas De La Cruz, J., and Noble, G., Preliminary report on the subcutaneous perfusion of dimethyl polysiloxane to increase volume and alter regional contour, *Br. J. Plast. Surg.*, 20, 424, 1967.
252. Nosanchuk, J. S., Silicone granuloma in breast, *Arch. Surg.*, 97, 583, 1968.
253. Bowers, D. G., Jr. and Radlauer, C. B., Breast cancer after prophylactic subcutaneous mastectomies and reconstruction with silastic prostheses, *Plast. Reconstr. Surg.*, 44, 541, 1969.

Cyanoacrylic Adhesive

254. Baniff, E. H., Cyanoacrylic monomer, U.S. Patent 3,654,340, 1972.
255. Abdurakhimov, N., Alovitdinov, A. B., and Kuchkarov, A. B., Esters of cyanoacetic and cyanoacrylic acids, *Tr. Tashk. Politekh. Inst.*, 107, 22, 1973.
256. Regnault, F. and Neuvellon, M., Alkylα-cyanoacrylate monomers useful as adhesives, especially in surgery, French Patent 2,229,688, 1974.
257. Windholz, M., Ed., *The Merck Index*, 9th ed., Merck, Rahway, N.J., 1976, 2705.
258. Kulkarni, R. K., Johnson, E. C., and Wade, C. W. R., Gas liquid chromatographic analysis of alkyl alcohols, alkyl cyanoacetates and alkyl 2-cyanoacrylates, *Anal. Chem.*, 46, 749, 1974.
259. Lehman, R. A. W., West, R. L., and Leonard, F., Toxicity of alkyl 2-cyanoacrylates. II. Bacterial growth, *Arch. Surg.*, 93, 447, 1966.
260. Noszczyk, W., Szretter Szmid, M., Kulicki, M., and Wichrzycka, E., Experimental closure of wounds with asynthetic glue, *Polprzegl. Chir.*, 41, 882, 1969.
261. Herrmann, J. B. and Woodward, S. C., The effect of cyanoacrylate tissue adhesives upon granulation tissue formation in Ivalon sponge implants in the rat, *Surgery*, 59, 559, 1966.
262. Woodward, S. C., Herrmann, J. B., Cameron, J. L., Brandes, G., Pulaski, E. J., and Leonard, F., Histotoxicity of cyanoacrylate tissue adhesive in the rat, *Ann. Surg.*, 162, 113, 1965.
263. Bhaskar, S. N., Frisch, J., and Margetis, P. M., Tissue response of rat tongue to hexyl, heptyl and octyl cyanoacrylates, *Oral Surg. Oral Med. Oral Pathol.*, 24, 137, 1967.
264. Arthaud, L. E., Lewellen, C. R., and Akers, W. A., The dermal toxicity of isoamyl 2-cyanoacrylate, *J. Biomed. Mater. Res.*, 6, 201, 1972.
265. Bhaskar, S. N., Frish, J., Cutright, D. E., and Margetis, P. M., Effect of butyl cyanoacrylate on the healing of extraction wounds, *Oral Surg. Oral Med. Oral Pathol.*, 24, 606, 1967.
266. Foellmer, W., Rathert, P., and Platzer, T., Comparative histological studies on healing of skin wounds by suture or by adaption with an adhesive (N-butyl-2-cyanoacrylate), *Bruns. Beitr. Klin. Chir.*, 216, 730, 1968.

267. Matsumoto, T., Nemhauser, G. M., Soloway, H. B., Heisterkamp, C., and Aaby, G., Cyanoacrylate tissue adhesives: an experimental and clinical evaluation, *Mil. Med.*, 134, 247, 1969.
268. Pani, K. C., Gladieux, G., Brandes, G., Kulkarni, R. K., and Leonard, F., The degradation of n-butyl alpha cyanoacrylate tissue adhesive, *Surgery*, 63, 481, 1968.
269. Ousterhout, D. K., Larsen, H. W., Margetis, P. M., and Leonard, F., Effects of ingested n butyl alpha cyanoacrylate on the growth of weanling rats, *Oral Surg.*, 27, 275, 1969.
270. Aronson, S. B., McMaster, P. R. B., Moore, T. E., Jr., and Coon, M. A., Toxicity of the cyanoacrylates, *Arch. Ophthalmol.*, 84, 342, 1970.
271. Burger, K., Hubner, R., Wendt, F., and Witter, H., Experimental investigations on the different behavior of several acrylate adhesives applied to the surface of incised wounds of parenchymatous organs, *Z. Exp. Chir.*, 2, 44, 1969.
272. Myers, M. B. and Cherry, G., Effect of methyl cyanoacrylate on the hepatic capsule, *Am. J. Surg.*, 113, 642, 1967.
273. Gottlob, R. and Blumel, G., The toxic action of alkylcyanoacrylate adhesives on vessels. Comparative studies, *J. Surg. Res.*, 7, 362, 1967.
274. Bauchhenss, G. H., Anastomosing of small veins, mechanical and adhesive anastomoses in animal experimentation, *Chir. Praxis*, 12, 43, 1968.
275. Giessmann, H. G., Schlote, H. W., Tuchschere, V., and Zeller, R., The use of tissue adhesive to the eye in the test on rabbits, *Arch. Ophthalmol.*, 183, 294, 1972.
276. Giessmann, H. G., Schlote, H. W., Tuchschere, V., and Zeller, R., The use of tissue adhesive in rabbit eyes. III. Toning conjunctival flaps with adhesives, *Arch. Ophthalmol.*, 184, 309, 1972.
277. Gasset, A. R. and Kaufman, H. E., Epikeratoprosthesis. Replacement of superfacial cornea by methyl methacrylate, *Am. J. Ophthalmol.*, 66, 641, 1968.
278. Freund, J. and Reim, M., Histologic findings in sealed corneal wounds of experimental animals, *Ber. Dtsch. Ophthalmol. Ges.*, 70, 320, 1970.
279. Cardarelli, J. and Basu, P. K., Lamellar corneal transplantation in rabbits using isobutyl cyanoacrylate, *Canad. J. Ophthalmol.*, 4, 179, 1969.
280. Vygantas, C. M. and Kanter, P. J., Experimental buckling with homologous sclera and cyanoacrylate, *Arch. Ophthalmol.*, 91, 126, 1974.
281. Sani, B. P. and Refojo, M. F., $\beta^{14}C$ isobutyl 2-cyanoacrylate adhesive. Determination of absorption in the cornea, *Arch. Ophthalmol.*, 87, 216, 1972.
282. Reim, M. and Vogt, M., Use of adhesive to close corneal wounds in rabbits, *Arch. Klin. Exp. Ophthalmol.*, 179, 53, 1969.
283. Refojo, M. F., Surgical adhesives in ophthalmology, *J. Macromol. Sci. Chem.*, 4, 667, 1970.
284. Lepri, G. and Tota, G., L'Eastman 910 quale sostanza adesiva nelle ferite corneali sperimentali, *Boll. Ocul.*, 44, 343, 1965.
285. Kurokawa, K., Experimental studies on retinopexy using cyanoacrylate in rabbit's eye. II. Clinical observation, *Acta Soc. Ophthalmol., Japan*, 76, 831, 1972.
286. Kublin, K. S. and Refojo, M. F., Closure of ocular lacerations with an adhesive, *J. Am. Vet. Med. Assoc.*, 156, 313, 1970.
287. Carroll, S. E., Amacher, A. L., and Paterson, J. C., The results of using methyl 2-cyanoacrylate monomer in experimental surgery, *Can. J. Surg.*, 9, 435, 1966.
288. Loeffler, K., Untersuchungen zum Wundverschluss mit Methyl-2-cyanoacrylat, *Kleintier-Praxis*, 11, 142, 1966.
289. Huebsch, R. F., Implanting teeth with methyl-2-cyanoacrylate adhesive (dogs), *J. Dent. Res.*, 46, 337, 1967.
290. King, D. R., Reynolds, D. C., and Kruger, G. O., A plastic adhesive for nonsuture sealing of extraction wounds in heparinized dogs, *Oral Surg. Oral Med. Oral Pathol.*, 24, 307, 1967.
291. Olsson, S. E. and Rietz, K. A., Polymer osteosynthesis. II. An experimental study with methyl-2-cyanoacrylate (Eastman 910 adhesive) in bone grafting (dog), *Acta Chir. Scand. Suppl.*, 387, 5, 1966.
292. Quijano, L. M. and Alvarex-Cordero, R., Application of plastic adhesives in acute peritonitis, *J. Chir.*, 94, 521, 1968.
293. Stirling, C. T. and Cohn, I., Jr., Nonsuture small bowel anastomosis. Experimental study of cyanoacrylate monomer, *Am. J. Surg.*, 31, 587, 1965.
294. Sheena, K. S., Healey, J. E., Jr., Gallager, H. S., and McBride, C. M., Nonsuture repair of pelvic fistulae, *Tex. Med.*, 66, 62, 1970.
295. Freese, P., Heinrich, P., and Hinze, M., The treatment of wounds of the liver by means of an acrylic adhesive, *Chirung.*, 30, 483, 1965.
296. Collins, J. A., Pani, K. C., Seidenstein, M. M., Brandes, G., and Leonard, F., Cyanoacrylate adhesives as topical hemostatic aids. I. Experimental evaluation on liver wounds in dogs, *Surgery*, 65, 256, 1969.
297. Ota, K., The use of a binding agent for repair of parenchymatous organs, *Surg. Diagn. Treatm. (Tokyo)*, 6, 883, 1964.

298. Ota, K., Mori, S., Mizuno, K., and Inou, T., Experimental and clinical use of adhesive on parenchymous organs, *Arch. Surg.*, 96, 231, 1968.
299. Neef, H., Preusser, K. P., and Pieper, S., Histomorphological findings after renal bonding, *Zbl. Chir.*, 94, 1629, 1969.
300. Matsumoto, T., Soloway, H. B., Cutright, D. E., and Hamit, H. F., Tissue adhesive and wound healing. Observations of wound healing (tissue adhesive vs sutures) by microscopy and microangiography, *Arch. Surg.*, 98, 266, 1969.
301. Neef, H., Preusser, K. P., Wollenweber, H. D., and Bunge, H. J., Closure of parenchymal pancreas wounds without suturing. Experiment in dogs, *Bruns. Beitr. Klin. Chir.*, 216, 651, 1968.
302. Goetz, R. H., Weissberg, D., and Hoppenstein, R., Vascular necrosis caused by application of methyl 2-cyanoacrylate (Eastman 910 monomer): 7-Month follow up in dogs, *Ann. Surg.*, 163, 242, 1966.
303. Scheetz, W. L. and Matsumoto, T., Cyanoacrylate tissue adhesive: thrombogenic effect, *Am. Surg.*, 36, 418, 1970.
304. Neef, H. and Preusser, K. P., Late damages of the vessel wall by acrylate adhesive, *Z. Exp. Chir.*, 1, 231, 1968.
305. Mori, S., Ota, K., Takada, M., and Inou, T., Comparative studies of cyanoacrylate derivatives in vivo, *J. Biomed. Mater. Res.*, 1, 55, 1967.
306. Herrmann, J. B., Katz, A. R., and Woodward, S. C., Experimental anastomoses of small arteries employing alkyl-cyanoacrylates, *J. Biomed. Mater. Res.*, 1, 395, 1967.
307. Noszczyk, W., Kulicki, M., Szretter Szmid, M., and Wichrzycka, E., Experimental studies on the use of tissue adhesive in vascular surgery, *Pol. Przegl. Chir.*, 41, 1393, 1969.
308. Aaby, G. V., West, R. L., and Jahnke, E. J., Myocardial response to the application of tissue adhesives: comparison of methyl-2-cyanoacrylate and butyl-cyanoacrylate (dog), *Ann. Surg.*, 165, 425, 1967.
309. David, E. and Unger, R. R., An experimental study on the application of acrylate adhesive (butyl 2-cyanoacrylate) to dura mater, brain surface and carotid artery in animals, *Z. Exp. Chir.*, 1, 206, 1968.
310. Lehman, R. A. W., Hayes, G. J., and Leonard, F., Toxicity of alkyl 2-cyanoacrylates. I. Peripheral nerve, *Arch. Surg.*, 93, 441, 1966.
311. Midgley, R. D. and Woolhouse, F. M., Technique for skin graft fixation in experimental wounds of mixed depth, *Ann. R. Coll. Phys. Surg. Canada*, 1, 69, 1968.
312. Hoferichter, J. and Hoferichter, S., Der Effekt von Acrylatklebern bei Pankreasverletzungen, *Bruns. Beitr. Klin. Chir.*, 216, 170, 1968.
313. Bhasker, S. N., Butyl cyanoacrylate as a surface adhesive in human oral wounds, in *Adhesives Biological Systems*, Manly, R. S., Ed., Academic Press, New York, 1970, 201.
314. Keszler, P. and Appel, J., Experimental and clinical observations in the use of adhesive in thoracic surgery, *Tuberk. Tudobet.*, 22, 379, 1969.
315. Heisterkamp, C. H., III, Simmons, R. L., Vernick, J., and Matsumoto, T., An aerosol tissue adhesive, *J. Trauma*, 9, 587, 1969.
316. Matsumoto, T., Pani, K. C., and Hamit, H. F., Use of tissue adhesives for arterial anastomoses, *Arch. Surg.*, 96, 405, 1968.
317. Mussgnug, G. and Alemany, J., Utilization of adhesive for vascular suturing, *Actuelle Chir.*, 4, 145, 1969.
318. Ota, K., Mori, S., Koike, T., and Inou, T., Blood vessel repair utilizing a new plastic adhesive. Experimental and clinical studies, *J. Surg. Res.*, 5, 453, 1965.
319. Braunwald, N. S., A clinical evaluation of methyl-2-cyanoacrylate monomer as a hemostatic agent on the aorta, *Ann. Surg.*, 164, 967, 1966.
320. Gyurko, G., Uniting minor blood vessels with Aron Alpha A adhesive, *Acta Chir. Acad. Sci. Hung.*, 9, 109, 1968.
321. Ferry, A. P. and Barnert, A. H., Granulomatous keratitis resulting from use of cyanoacrylate adhesive for closure of perforating corneal ulcer, *Am. J. Ophthalmol.*, 72, 538, 1971.
322. Ginsberg, S. P. and Polack, F. M., Cyanoacrylate tissue adhesive in ocular disease, *Ophthalmol. Surg.*, 3, 126, 1972.
323. Dohlman, C. H., Carroll, J. M., Richards, J., and Refono, M. F., Further experience with glued-on contact lens (artificial epithelium), *Arch. Ophthalmol.*, 83, 10, 1970.
324. Dohlman, C. H., Payrau, P., and Pouliquen, Y., Artificial corneal epithelium and adhesives, *Arch. Ophthalmol.*, 28, 533, 1966.
325. Bakbardin, Y. V. and Filippenko, V. I., Cyanacrylate glue in sealing off the cararact extraction wound, *Vestn. Oftalmol.*, 3, 66, 1972.
326. Turss, R. and Refojo, M. F., Removal of isobutyl cyanoacrylate adhesive from the corneal with acetone, *Am. J. Ophthalmol.*, 70, 725, 1970.

327. **Spitznas, H., Lossagk, H., Vogel, M., and Joussen, F.,** Intraocular histocompatibility and adhesive strength of butyl 2-cyanoacrylate. Prospective value in retinal detachment surgery, *Albrecht von Graefes Arch. Ophthalmol.,* 187, 102, 1973.
328. **Spitznas, H., Lossagk, H., Vogel, M., and Meyer Schwickerath, G.,** Cyanoacrylate in retinal surgery, *Trans. Am. Acad. Ophthalmol. Otolaryngol.,* 77, OP114, 1973.
329. **Schimek, R. A. and Ballou, G. S.,** Eastman 910 monomer for plastic lid procedures, *Am. J. Ophthalmol.,* 62, 953, 1966.
330. **Muller Jensen, K.,** Initial clinical experience of treatment with adhesives in ophthalmic surgery, *Klin. Monatsbl. Augenheilkd.,* 158, 573, 1971.
331. **Muller Jensen, K. and Gabel, V. P.,** Experience with scleral adhesive pasting in the surgery of retinal detachment, *Klin. Monatsbl. Augenheilkd.,* 159, 597, 1971.
332. **Nakayama, K.,** Skin closure with a new surgical adhesive: Aron Alpha, *Surg. Ther. (Tokyo),* 10, 541, 1964.
333. **Ota, K., Mori, S., and Inou, T.,** Closure of viscerocutaneous fistulas with adhesives, *Arch. Surg.,* 101, 468, 1970.
334. **Sailer, R.,** Tracheal tumors and reconstruction of the trachea with the use of an adhesive harmless to the tissues, *Zbl. Chir.,* 95, 1057, 1970.
335. **Rickles, W. H., Jr. and Seal, H. R.,** Evaluation of four long term transducer adhesive techniques, *Psychophysiology,* 4, 354, 1968.
336. **Nemes, A., Somogyi, E., and Sotonyi, P.,** Histological changes in the myocardium produced by the cementing material used to keep myocardial pacemaker electrode in place, *Morphol. Igazsagugyl Orv. Sz.,* 9, 11, 1969.
337. **Nemes, A., Somogyi, E., and Sotonyi, P.,** Histologic changes in myocardial pacemaker electrodes fixed with adhesive, *Bruns. Beitr. Klin. Chir.,* 217, 277, 1969.
338. **Stevenson, T. C. and Taylor, D. S.,** The effect of methyl cyanoacrylate tissue adhesive on the human Fallopian tube and endometrium, *J. Obstet. Gynaecol. Br. Commonw.,* 79, 1028, 1972.
339. **Takayasu, H., Kitagawa, R., and Nito, H.,** Circumcision for phimosis. Using an adhesive agent, *Operation (Tokyo),* 18, 501, 1964.
340. **Malament, M.,** Experimental bladder closure with a tissue adhesive, *Invest. Urol.,* 3, 429, 1966.
341. **Heiss, W. H.,** Use of synthetic polymeric materials as suture substitutes and their place in pediatric surgery, *Progr. Pediatr. Surg.,* 1, 99, 1970.
342. **Page, R. C.,** Tissue adhesive — eliminates sutures and staples in many types of surgery, *Adhes. Age,* 9, 27, 1966.
343. **Black, M. M. and Van Noort, R.,** Silicone rubber for medical applications, in *Biocompatibility of Clinical Implant Materials,* Vol. 2, Williams, D. F., Ed., CRC Press, Boca Raton, Fla., in press.

Chapter 6

PHARMACOLOGY, TOXICOLOGY, AND CLINICAL ACTIONS OF MONOMERIC AND POLYMERIC METHYLMETHACRYLATE*,**

Ulrich Borchard

TABLE OF CONTENTS

I. Introduction 106

II. Physicochemical Properties of MMA and pMMA 106
    A. Physicochemical Parameters 106
    B. Residual MMA and its Liberation from the Polymer 107
    C. Heat Production During Polymerization 109

III. Toxicology of MMA and pMMA 109
    A. Acute Toxicity of MMA 109
    B. Chronic Toxicity of MMA 110
    C. Teratology 111
    D. Carcinogenicity of pMMA Implantation 111
    E. Localized Actions of MMA and pMMA 112
        1. Investigations on Cultured Tissues 112
        2. Effects on Animal Tissues 112
        3. Localized Effects in Patients 113
            a. Observations Upon Reoperation or Autopsy 113
            b. Allergic Reactions 113

IV. Pharmacology of MMA and Homologous Chemicals 113
    A. Pharmacokinetics of MMA 113
    B. Effect on Nervous System 116
        1. Peripheral Nervous System 116
        2. Central Nervous System 119
    C. Effects on Respiration 122
    D. Cardiovascular Actions 128
    E. Clinical Observations 131

V. Conclusions 133

Acknowledgments 134

References 134

---

* Abbreviations: MMA = methylmethacrylate, pMMA = polymeric methylmethacrylate, EMA = ethyl-, ALMA = allyl-, n-BMA = n-butyl, i-BMA = isobutylester of methacrylic acid, MAM = methacrylamide, NaMA = sodium methacrylate.
** Author's investigations included in this article were supported by the Deutsche Forschungsgemeinschaft, grant number Bo380/3.

I. INTRODUCTION

Monomeric methylmethacrylate (MMA) is widely used in medicine for the production of bone cement (Palacos®, Paladur®, Palaferm®, Palavit®, Simplex®, CMW-bone-cement®) in orthopedic surgery,[1-33] neurosurgery,[34-39] and surgery of the jaw and face.[40-47] In dentistry this compound has been used as a plastic for filling cavities.[48-51] For these applications liquid monomeric MMA and prepolymerized methylmethacrylate powder (pMMA) are mixed to give a paste which, under the influence of activators and catalysts, is hardened by polymerization. This process takes place partially within the prepared bone or dental cavity. Furthermore, MMA allows the conservation of bone, cartilage, and blood vessels.[52-55] One of the most important applications of MMA as introduced by Charnley[13] and McKee and Watson-Farrar[22] is the fixation of metal and plastic prosthesis into the bone. As there is a good blood perfusion of the marrow bone,[56] a considerable quantity of MMA may appear in the circulation,[57] leading to marked influences on the cardiovascular system and respiration as well as the peripheral and central nervous system.[58-61] Free MMA, which may be set free from dental fillings even up to 6 days after polymerization,[62] is able to penetrate the dentinal tubules causing pathological changes of the pulp.[51] In recent years, reports have been published on intraoperative complication due to a slight or severe fall in blood pressure, but with occasional fatality.[33,58-61,63-76]

Fortunately, severe accidents such as cardiovascular collapse and circulatory arrest are rare during insertion of hip prostheses. On the other hand, more or less severe cardiorespiratory changes, such as hypoxemia and hypotension, nearly always appear.[61] These are without persisting consequences in subjects under normal physiological conditions, but they can be fatal in patients with cardiovascular disturbance. Therefore, the study of the pharmacological and toxicological action of MMA has become of great importance.

II. PHYSICOCHEMICAL PROPERTIES OF MMA AND pMMA

A. Physicochemical Parameters

MMA is a clear, volatile liquid which is only slightly soluble in water, but dissolves oils and lipids.[77-83] The physicochemical parameters are summarized in Table 1. Hoffmann[107] in contrast to Mohr[79] and Kreudenstein[51] suggests that MMA precipitates proteins. The high surface tension allows MMA to spread very quickly as a monomolecular film. pMMA which is used in medicinal preparations for the production of bone cement consists of beads which are 5 to 40 μm in diameter,[83,84] with a specific gravity of 1.18 and a molecular weight of 200,000 to 300,000.[49] Liquid MMA is stabilized with 100 ppm hydroquinone. For polymerization of beads of pMMA by MMA an accelerator (usually a tertiary amine) is added which in combination with a peroxide as catalyst starts the exothermic polymerization, according to the following equation:

$$n \begin{pmatrix} CH_3 \\ | \\ C = CH_2 \\ | \\ CO\,OCH_3 \end{pmatrix} \longrightarrow \begin{matrix} CH_3 \\ | \\ -C-CH_2- \\ | \\ CO\,OCH_3 \end{matrix} \begin{matrix} CH_3 \\ | \\ C-CH_2- \\ | \\ CO\,OCH_3 \end{matrix}$$

MMA → pMMA

By scanning electron microscopy it is possible to recognize the beads of pMMA within a structural block.[84-87]

Table 1
PHYSICOCHEMICAL PROPERTIES OF MMA

Molecular weight	100.11
Density (25°C)[77]	0.9374 g/ml
Freezing point	−48.2°C
Boiling point (760 mmHg)	100.3°C
Viscosity (25°C)	0.565 cst
Solubility of MMA in H_2O (20°C)	1.59 w/w %
Solubility of H_2O in MMA (20°C)	1.15 w/w %
Solubility in alcohol, ether	Infinite
Dipole moment	1.79 D
Surface tension	28.2 dyn/cm²
Dielectrical constant	2.9
Heat of polymerisation	11.6 kcal/mol
Vapor pressure (20°C)	29 mmHg
Heat of hydration	28.636 kcal/mol

Table 2
LIBERATION OF MMA FROM THE POLYMER

Author	Year	MMA
Smith and Bains[95]	1956	0.2% in the first 6 hr; 1% in the first 20 hr
Henkel[96]	1961	0.16% extraction with dist. H_2O and subsequent titration
Blumler[97]	1965	0.16% extraction with dist. H_2O and subsequent titration
Koning[98]	1966	1.1% plates extracted 14 days in aqueous solutions
Karalnik[99]	1968	2.1% observed over a period of months
Kutzner et al.[88]	1974	1.11% related to the initial MMA

Note: Values for the MMA content of hardened pMMA range from 2.5 to 11%, but most authors report values of 3 to 5%.[31,97,99,101,156-159]

B. Residual MMA and its Liberation from the Polymer

In clinical use, the plasticity of the polymerizing paste is necessary for intraoperative modeling, but the paste contains considerable amounts of MMA.[57,88] Even after the process of polymerization has ceased, MMA is retained in pores between the polymer beads.[79,89] As the side effects and toxic reactions depend on the amount of MMA liberation,[87] many pharmacological investigations have been carried out on this subject. The results are summarized in Table 2. Most investigators used the polymer in its hardened form, but the greatest amount of MMA is set free during polymerization immediately after implantation. Therefore, MMA liberation from bone cement during polymerization in physiological NaCl[88] and citrous blood[90] has been investigated. Kutzner et al.[88] have reported that 1.11% of the applied MMA was set free within the first 20 min.

Evidence for the liberation of free MMA into the blood stream of rabbits and dogs was given by Homsy et al.[91] and Wenzel et al.[92] Clearance of ^{14}C-labelled MMA from

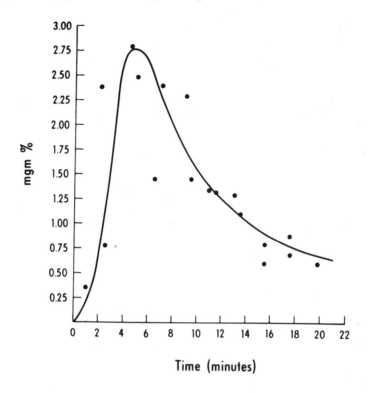

FIGURE 1. ^{14}C-labeled MMA concentration in venous blood. Mean values of four dogs after simulated hip arthroplasty. 1.3 g/kg curing cement were implanted at time zero, 4.5 min after mixing. (From McLaughlin, R. E., et al., *J. Bone Jt. Surg. Am. Vol.*, 55, 1621, 1973. With permission).

the blood was determined in beagles during simulated hip arthroplasty by McLaughlin et al.[182] The blood concentrations of ^{14}C-labeled monomer in the inferior vena cava reached a maximum of 3.5 mg% 3 min after implantation of the cement in the acetabulum and femur (Figure 1). Only 0.5% of the total amount of implanted monomer was detected in the venous circulation. Using ^{14}C-MMA in patients undergoing hip reconstructive surgery, Homsy et al.[93] detected peak concentrations of 1.3 mg MMA/ 100 mℓ blood. MMA is almost completely cleared from the venous circulation in 19 min.[93,182] In 80% of the patients Eggert et al.[57] demonstrated the appearance of free MMA in the vena cava by the use of gas chromatography. The peak value was 16 μg/ mℓ with a mean value of 1 μg/mℓ venous blood. High values of MMA correlate well with a temporary hypotonia. Pahuja et al.[60] measured MMA from mixed venous blood of 34 patients undergoing hip and knee reconstructive surgery. The monomer was detected in nine patients in the range of a trace to 200 mg%. In the patient with 200 mg% morbidity in the early postoperative phase was severe. Pahuja et al.[60] emphasize that the amount of monomer absorbed is much more if the pMMA mix is inserted as a paste as compared to the dough form. Furthermore, the authors demonstrate that the rate of MMA accumulation in the blood depends on the site of insertion. Whereas in total hip series, no MMA was detected following insertion of the cement in the acetabulum, insertion in the femur led to differing MMA blood levels in 5 out of 23 patients. Modig et al.[193] detected MMA concentrations in the pulmonary arterial blood of 13 patients ranging from 0 to 0.25 (mean 0.06) mg/100 mℓ after cementing the acetabulum and from 0 to 0.26 (mean 0.08) mg/100 mℓ after cementing the femur.

Bright et al.[190] observed values of 1 mg/100 mℓ MMA in venous blood of their patients. Bloch et al.[191] found MMA in 10% of the procedures, at levels of 1 to 2 mg% and in one case the level was 10 mg%. Kim and Ritter[192] report a value of 1.1 mg/100 mℓ as the highest level in four out of ten cases. After injection of 0.1 mℓ/kg MMA in dogs blood concentrations of 11.4 mg/100 mℓ, occurring after the 4th min, were necessary to effect significant variations in cardiac output and blood pressure.[94]

This indicates that blood levels generally observed in patients are less than toxic levels reported in animals (Section IV), but until now, exact levels at which pulmonary lesions occur in humans are not known. Local pulmonary hemorrhages were observed in canines at levels as low as 5 mg/100 mℓ, though no long-term morbidity was seen even at levels of 50 mg/100 mℓ.[93]

C. Heat Production During Polymerization

The significance of heat production during polymerization has been discussed for a long time. While Postel et al.,[20] Heidsieck et al.,[55] Roggatz and Ullmann,[100] and Ohnsorge and Goebel[101] report a heat-induced damage to the tissue in contact with bone cement, Münzenberg[102] and Puhl and Schulitz[87] conclude from their experiments that heat production in the vicinity of the vital bone is without damage if the blood perfusion is maintained. The heat given off during 8 to 12 min hardening of a laminated endoprosthesis, made of Dacron soaked in autopolymerizing MMA and used to support unstable spines, can be easily conducted away from the prothesis due to its relative thinness (3 to 5 mm).[33] In an attempt to reduce the exothermic reaction, part of the pMMA needed during implantation was replaced by catgut, assuming that during the polymerization of a smaller amount of MMA less heat would be produced.[103] It was revealed that the peak temperature can be reduced at least 13°C and by a maximum of 23°C using a 15 to 20% filling with catgut fibers. In 350 rats mechanical tests showed only minor reduction of flexion and an increase in the resistance to pressure. The greater elasticity improved the biomechanical properties of bone cement. Under well-defined experimental conditions, Ohnsorge and Kroesen[104] and Kroesen[105] observed that the temperature was 122°C in the center of a polymerizing Palacos sphere and 92°C at the surface. In post-mortem bone, peak temperatures of 58 to 72°C occurred during polymerization.[96] Kuner[106] carried out implantation of Palacos into the marrow cavity of dog's femur and observed peak temperatures in vitro of 87 to 106°C and in vivo of 59.8 to 92.8°C. Further results, which often are reported without acknowledgment of the applied experimental conditions, are summarized in Table 3.

III. TOXICOLOGY OF MMA AND pMMA

A. Acute Toxicity of MMA

Data on acute toxicity of MMA are shown in Table 4. First measurements based on i.v. application in dogs were published in 1969.[91] MMA was injected for 90 s at a dosage which led to blood levels of 5, 10, 15, and 125 mg%. Deichmann[110] injected rabbits with 0.03 to 0.04 mℓ/kg MMA. Respiration was stimulated immediately, indicated by a twofold increase in breathing frequency as compared to control values. But after an additional dose, breathing slowed continuously until it stopped altogether. Spealman et al.[112] also observed that the toxicity of the monomer is mainly due to its action on respiration. The influence on respiration has been confirmed by other authors.[107,117-128] The toxicity of MMA vapor was attributed to its action on the nervous system[125] and the fatigue observed upon inhalation of MMA vapor to its influence on brain function.[120,121]

Lawrence et al.[160] have investigated the acute toxicity (LD_{50}) by i.p. injection of a series of monomeric esters of acrylic and methacrylic acid in mice. Using mathematical

Table 3
PEAK TEMPERATURE DURING POLYMERIZATION OF MMA

Author	Year	Temperature °C
Herrmann[8]	1953	80—90
Hoffmann[107]	1954	60—70
Idelberger[108]	1956	60
Wiltse et al.[12]	1957	79
Unger and Sollmann[38]	1964	70
Otto[58]	1965	73
Scheuba[19]	1965	70—90
Willenberger et al.[21]	1966	>90
Pfeiffer[109]	1967	>100
Homsy et al.[91]	1969	80—100

Table 4
TOXIC DOSAGE OF MMA

Authors	Species	i.v.	Oral	s.c.	i.p.
Deichmann[110]	Rat	—	8.4	13.0[a]	—
Deichmann and Leblanc[111]	Rat	—	10.0[b]	—	—
Spealman et al.[112]	Mouse	—	—	6.3	1.0
	Rat	—	10.0	7.5	1.0
	Guinea pig	—	6.3	6.3	2.0
	Dog	—	5.0	4.5	—
Deichmann and Mergard[113]	Rat	—	8.6—9.1	—	—
Spector[114]	Rat	—	8.4[c]	—	—
	Rabbit	—	6.6—7.5[c]	—	—
Hattemer[115]	Mouse	—	—	12.5[d]	—
Castellino and Colicchio[116]	Mouse	—	—	6.4[c]	0.49
	Rat	—	—	7.7[e]	—
Homsy et al.[91]	Dog	125[f]	—	—	—
Kutzner et al.[117]	Mouse	0.32	—	—	—
Lawrence et al.[160]	Mouse	—	—	—	1.20
Singh et al.[162]	Rat	—	—	—	1.33

Note: Data refer to the LD$_{50}$ (mℓ/kg) except where footnoted.

[a] LD$_{100}$ (mℓ/kg).
[b] Approximate lethal dose (mℓ/kg).
[c] LD$_{50}$ (g/kg).
[d] Tolerated dose (mℓ/kg).
[e] LD$_{100}$ (g/kg).
[f] Lethal dose (mg %).

models, the authors tried to relate structure of the compounds to toxicity. They found that parent acids of the acrylic and methacrylic monomers and of those short chain esters that contain amine or hydroxyl groups were the more toxic compounds of the series; simple alkyls and substituents with additional ester linkages decreased toxicity.

B. Chronic Toxicity of MMA

Studies on chronic oral toxicity of monomeric ethyl and methyl methacrylate have been carried out by Borcelleca et al.[161] The substances were added to the drinking water of rats and beagles in concentrations of 0, 60 to 70, and 2000 ppm for a period of 2 years. Mortality was unaffected. Only the high concentration of the chemicals led to a depression of the body weight. Hematological values and urine concentrations of protein and reducing substances varied within normal limits for all groups of rats. Histopathological findings revealed no lesions attributable to either test material.

C. Teratology

Singh et al.[162] administered the monomers of five methacrylate esters and acrylic acid intraperitoneally to female rats using one tenth, one fifth, and one third of the acute LD_{50} on days 5, 10, and 15 of gestation. MMA produced a number of gross abnormalities (hemangiomas) in the offspring of treated rats, but there were no skeletal malformations observed at any dose in this study. In contrast, n-butyl methacrylate brought about gross and skeletal deformities. The authors conclude from their experiments that methacrylate esters have a deleterious effect on the developing embryo and fetus. The observed incidence of adverse effects as compared to the controls was generally dose-related and dependent on the compounds, the more hydrophilic the compounds the greater tended to be their activity. Each compound at one or more of the doses used showed some or all of the following effects: resorptions, gross and skeletal malformations, fetal death, and decreased fetal size. However, it has to be recognized that the authors applied high i.p. doses of 125, 250, and 417 mg/kg MMA which could lead to high local concentration at the ovaries.

D. Carcinogenicity of pMMA Implantations

In order to determine the relative carcinogenicity of pMMA, Laskin et al.,[183] using the commercial heat-cured polymer, were able to induce 5 tumors in 20 Harlan strain albino Swiss mice after implanting a piece of the polymer, 1 cm² subcutaneously in the lateral abdominal wall for 469 days. Histopathological studies revealed the tumors to be nonmetastatic fibrosarcomas. It is of special interest that the tumors were localized adjacent to, but never adherent to the implant. Oppenheimer et al.[184] carried out investigations with a pure polymer lacking any retarder, catalyst, or coloring materials and with the commercial type polymer. In 4 of the 20 surviving mice, tumors were produced by the commercial polymer after a period from 581 to 658 days. Histopathological observations revealed that, in the absence of tumors, the polymer was encapsulated, whereas, in the presence of tumors, no fibrous capsule was formed. Mitchell et al.[185] implanting pellets 2 by 2 mm in Wistar rats for 27 months and Spealman et al.[112] implanting 0.075 g of the fine granular polymer of two pigmented denture base materials in a gelatine capsule for a period of 1 month, were unable to produce tumorous growth. There is strong evidence given by Stinson,[136] however, that the form and size of the implant is of greater significance in the tumor production than the chemical agents. Large (18 mm) and medium (12 mm) disks implanted into the gluteal muscle of Chester Beatty strain mice induced tumor formation, whereas, small ones (4 mm) did not. Furthermore, the minimum latent period of the large disks (150 days) was much shorter than that of medium disks (278 days). This might be the reason why Spealman et al.[112] did not find tumors in their experiments. While the carcinogenicity of pMMA is well documented in rodents, its significance still remains unknown in man. Pliess and Bornemann[187] have carried out a study of 321 denture patients with sore mouths. Twenty-five of these patients had atypical symptoms such as edema, leukoplasia, hyperkeratosis, parakeratosis, and acanthosis. Four patients showed tumorous growths which reached the submucosa.

By incorporation of ^{14}C into the polymer, Oppenheimer et al.[184] have shown that the polymer is soluble in the body and partly metabolized. The authors conclude from their experiments that either a degradation product of the metabolism of the polymer is the carcinogen, or that the polymer binds to proteins and impairs cellular metabolism, thus producing cancer.

Summarizing the experimental findings, it seems to be well established that prolonged and continuous combined action of the dissolved polymer associated with an ill-fitting denture traumatizing the mucosa may induce tumor growth in man.[155]

E. Localized Actions of MMA and pMMA
1. Investigations on Cultured Tissues

While Strack[129] as well as Virenque and Leroux[130] observed no inhibition of growth and no impairment of cell migration in cultured tissues by pMMA, Debrunner[131] in one series of experiments found an inhibition in growth of fibrocytes by pMMA. The results of Hulliger[132] show a good tolerance of tissue cultures to hardened Palacos, whereas MMA was highly toxic to cells. Schachtschabel and Blencke[133] have applied pulverized plastic (made of MMA) to asynchronous monolayer cultures of fast proliferating Ehrlich ascites tumor cells and slowly proliferating diploid human fibroblasts. They investigated cell growth, rate of DNA synthesis as measured by incorporation of ^{3}H-thymidine into DNA, glucose consumption, and lactate production (rate of glycolysis). Exposure of Ehrlich ascites cells to high concentrations of sonified pMMA (2 mg/mℓ; 1 mℓ medium per culture) resulted in a gradual inhibition of cell growth and rate of DNA synthesis. Protein synthesis, as measured by the incorporation of ^{3}H leucine, was less effected than DNA synthesis. Furthermore, the glycolytic rate of Ehrlich ascites cells was slightly but significantly decreased in the presence of 2 mg/mℓ pMMA. Exposure of the cells to 0.2 or 0.02 mg/mℓ over a period of 46 hr revealed none or only slight inhibitory effects on growth, DNA synthesis, or glycolytic rate. In contrast, 2 mg/mℓ pMMA applied for 92 hr to human fibroblasts did not affect growth, DNA synthesis, or glycolytic rates of human fibroblasts. Therefore, local cellular toxicity seems to be due to MMA but not to pMMA.

2. Effects on Animal Tissues

Numerous experiments have been carried out in order to examine the compatibility of MMA in animal tissues.[6,12,95,101,107,115,134-138] Henrichsen et al.[6] implanted pMMA in the marrow bones of pigs but could not find any change. Dürr[139] applied MMA to the marrow-cavity of dog's femur. The animals were treated with tetracycline to study the structural reorganization by fluorescence microscopy. There was no difference between MMA-treated and control animals. Wiltse et al.[12] after implanting MMA into the marrow cavity of 22 rabbits also found a mild tissue reaction. On the contrary, Yablon[140] was able to show that by application of MMA to endosteal and periosteal regions in the femur of dogs using histologic, fluorescent, and autoradiographic studies, the surface in contact with MMA was necrotic and did not incorporate tetracycline or ^{45}Ca. New bone occurred only on the surface which was not in contact with MMA. Hoppe[134] injected viscous bone cement in rabbits and rats intramuscularly and subcutaneously. Cytotoxic and necrotic effects were attributed to the monomeric MMA. These results were confirmed by Mohr.[79] Intracutaneous application of MMA led to a marked necrosis of the total reaction area followed by peeling and sequestering of the damaged tissue after 10 to 19 days. Necroses were also observed in the human mucosa in contact with MMA by Zschiedrich,[141] Overdiek,[142] and Kreudenstein.[51]

Investigations of MMA toxicity on the nerve were carried out by Mohr.[79,143] After injection of MMA into the sciatic nerve of rabbits, a weak paralysis was induced which was attributed to a toxic neuritis. Decay of the myelin sheath was observed 45 min after injection. Damaged tissue was not repaired but replaced by connective tissue. However, it has to be emphasized that the author applied high concentrations of MMA which do not occur in clinical situations. Kuner[106] reports that toxic damage to tissue is observed only in the first 72 hr of preparation of the paste and disappears after complete hardening. On the other hand, Fischer and Sonnabend[49] found a toxic action of MMA on the pulp of long duration.

3. Localized Effects in Patients
a. Observations upon Reoperation or Autopsy

Since 1950 many reports have been published on the compatibility of pMMA-prostheses in patients.[8,14,17,23,130,131,144-151] Debrunner[131] carried out histological analysis of the tissue in contact with pMMA prostheses in eight reoperated patients. Local reactions were very rare. Charnley[17] confirmed the good local compatibility of MMA and showed that fresh bone fractures of 12 patients were healed reasonably well which was independent of the contact with polymerizing or hardened MMA bone cement.[23] The reaction of the tissue in contact with MMA bone cement seems to depend on the mechanical conditions, the type of tissue at which implantation takes place and the degree of polymerization.[139,148]

From the various data summarized in this chapter, we may conclude that localized toxicity is brought about mainly by the monomeric MMA set free from the pMMA and that pMMA is tolerated well by tissues.

b. Allergic Reactions

Fisher[152] found that MMA was the cause of allergic dermatitis in dentists and dental laboratory technicians who had come into repeated contact with acrylic denture materials. Similar observations in one orthopedic surgeon were reported by Pegum and Medhurst[153] showing that MMA does diffuse through intact surgical rubber gloves. Fries et al.[154] carried out investigations on handlers of bone cement. In 7 out of 13 cases patch test sensitivity to 10% MMA in olive oil was positive, indicating true allergic contact dermatitis characterized by itching, erythema, edema, and vesiculation followed by eczematization. One of the unique and consistent features of the dermatitis from bone cement was parethesia. Also, tenderness was observed outlasting the duration of the eruption. MMA, which is a lipid solvent (cf. Table 1), penetrates s.c. tissue and degreases the skin. As it evaporates very rapidly, it effects allergic reactions only if it is applied under an occlusive or semiocclusive dressing, as under gloves. Contact allergic reactions have also been reported by Strain[155] in patients sensitized to the residual monomer from incompletely polymerized methylmethacrylate dentures. They were classified as (1) allergic eczematous contact type dematitis, as described above or (2) allergic stomatitis, this latter type being less common and showing the following symptoms: soreness, burning sensation, excessive salivation, dryness, and red and edematous mucosa with and without erosion. A typical case of allergic stomatitis has been described by Bradford.[188]

Studies on cutaneous reactions induced by pure MMA have been carried out by Spealman et al.[112] Cotton pellets saturated with MMA and covered with an Elasto patch were applied to the forearm of 50 medical students who had never been treated with MMA. Forty-eight hours later, 25 students showed a mild erythematous reaction. Normal application without cover produced no reaction indicating, as mentioned above, that an occlusive dressing is essential for the effect. Eczematous skin reactions are analogous to those of oral mucosa, which in general reacts and recovers more rapidly.

IV. PHARMACOLOGY OF MMA AND HOMOLOGOUS CHEMICALS

A. Pharmacokinetics of MMA

Distribution of MMA in Wistar rats after i.v. application was investigated by Wenzel

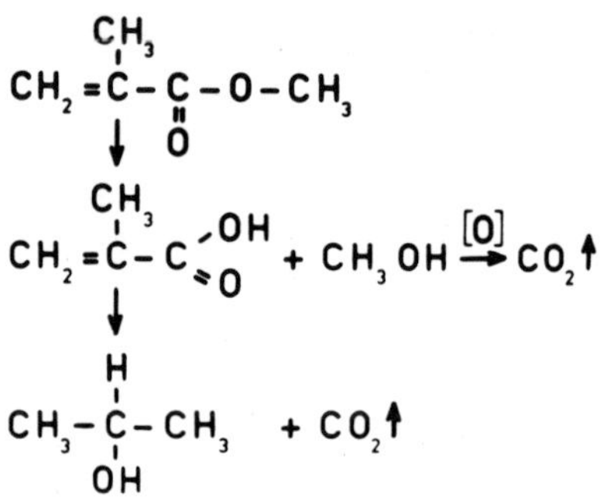

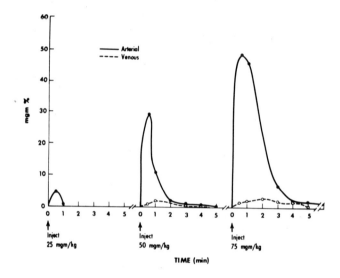

FIGURE 2. Metabolic pathway of MMA in rats.[92]

FIGURE 3. Mean blood levels of ^{14}C labeled MMA in the femoral artery and inferior vena cava of four dogs with i.v. application of three successively increasing doses. (From McLaughlin, R. E., et al., *J. Bone Jt. Surg. Am. Vol.*, 55, 1621, 1973. With permission.)

et al.[92] Using radioactive MMA, the authors carried out autoradiographic studies on whole animals. Five min after the application, the highest activity was found in blood and kidneys. Low concentrations of MMA were in liver and red bone marrow, while in brain and spinal cord no activity could be detected. After 2 hr the total activity had decreased and a shift from the bone marrow to the compact bone could be observed. In the period from 4 to 8 hr, activity was found only in skeletal bones, liver, intestine, and salivary glands. After 24 hr, total injected MMA was eliminated, 5% via urine and feces and more than 90% via respiration. By marking different C atoms, the metabolic pathways were described as illustrated in Figure 2. First, the molecule is split by an unspecific esterase which occurs ubiquitously in the body. Methanol is formed which is oxidized to CO_2. Methacrylic acid is decarboxylated very rapidly yielding CO_2. If the C atom used for the production of CO_2 was marked, the radioactivity was eliminated much more rapidly as compared to marking the methyl group. Furthermore, it

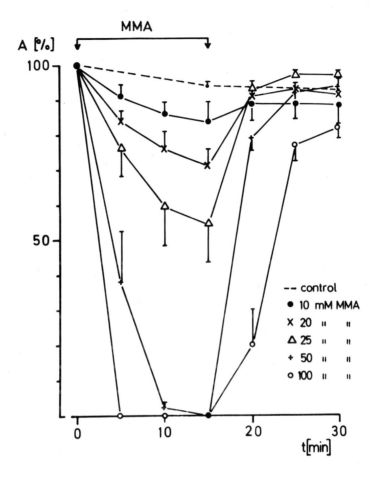

FIGURE 4. Decrease in the amplitude of the compound action potential of the desheathed sciatic nerve at different MMA concentrations, 15 min application. Stimulus duration 0.5 m sec, frequency 50 Hz, and intensity supramaximal. The vertical bars indicate S.E.M. (From Böhling, H. G., Borchard, V., and Drovin, H., *Arch. Toxicol.*, 38, 307, 1977. With permission.)

has been suggested that inhaled MMA is able to alter the pharmacokinetics (biological half-life) of drugs.[164]

Venous blood levels of MMA observed after hip arthroplasty (Figure 1) differ significantly from those after i.v. injection (Figure 3). During arthroplasty, MMA was found only in venous blood with a low clearance, whereas after i.v. injection high arterial levels were measured, occurring and declining rapidly.[182] The following conclusions were drawn: in arthroplasty, MMA directly reaches the inferior vena cava through the veins of the hip and has not yet passed through the major blood filtration systems. Absence of MMA in arterial blood reflects the high clearance in the lungs as well as the small and continuous release of MMA from the curing cement. After injection into the external jugular vein, MMA reaching arterial blood (femoral artery) has already passed through the lungs, while MMA in the venous blood has additionally passed through the peripheral capillary beds.

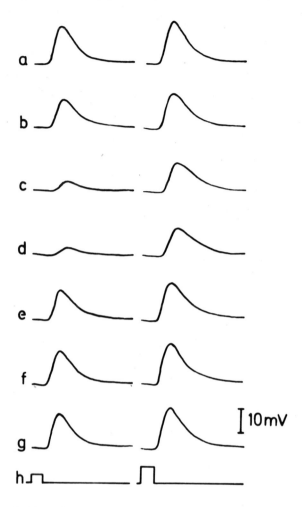

FIGURE 5. Original recordings of the compound action potential of the desheathed sciatic nerve after application of 25 mM MMA. a: control; b, c, d: 5, 10, and 15 min after 25 mM MMA; e, f, g: 5, 10, and 15 min wash; h: stimulus, duration 0.5 m sec, intensity 200 mV = supramaximal (right) and 100 mV (left).

B. Effect on Nervous System
1. Peripheral Nervous System

As already pointed out in Section III, toxic side effects of methylmethacrylate, which may be observed during and after polymerization of MMA, are due to the residual monomeric substance. This is also true for the pharmacological actions of bone cement. Alterations which involve the peripheral nerve have been described by several authors.[51,79,154,165] Mohr[79] summarizes the different types of necrosis which might occur in the pulp, the mucous membrane, and the skin, especially that of the tips of the fingers, which does not cause any pain. This analgesia may be attributed to the influence of MMA on peripheral nerves. A further interesting observation was that the main symptom of handlers of MMA is parathesia.[154] While in experiments of Mohr[79,165] application of pure MMA to the nerve brought about a lysis of the membrane lipids and a destruction of the myelin sheath, our experiments reveal that MMA acts reversibly in a concentration range from 5 to 50 mM.[166] Figure 4 shows that MMA induces a dose-dependent decrease in the amplitude of the compound action potential in the isolated superfused sciatic nerve. At a concentration of 50 mM MMA, the action potential is extinguished after 15 min, but approaches initial values if MMA is removed from the solution. Corresponding compound action potentials are illustrated in Figure 5. Obviously, the effect of MMA is more pronounced in fibers with a low threshold of stimulation. These fibers belong predominantly to the slow conducting type which are activated at low stimulation strength. The increase in effect on compound action potentials at decreased pulse amplitude can also be observed with local anaesthetics or ethylenediamine derivatives.[167] At 25 mM the conduction velocity of the sciatic nerve is decreased by 40% of the control value. At the same time, MMA brings about a dose-related hyperpolarization of the nerve membrane of about 2.5 mM which reaches a maximum after about 4 min, as opposed to the decrease of the compound action potential which continues throughout the application (cf. Figure 4). At 100 mM the hyperpolarization reaches maximal values at about 5 min and then decreases rapidly; and as a sign of the toxic disturbances the nerve membrane is even slightly depolarized after 15 min. Figure 6 illustrates that hyperpolarization by MMA is not influenced by increase in $[K^+]_o$, $[Ca^{2+}]_o$ or decrease in $[Ca^{2+}]_o$, but is reduced by decrease in $[NaCl]_o$. Furthermore, depolarization of the nerve membrane by veratrine or 0 mM $[Ca^{2+}]_o$ is completely reversed by MMA. The depolarization of the nerve membrane in low $[Ca^{2+}]_o$ solutions or with veratrine was attributed to an increase of the sodium permeability.[168] Recently, the action of changes in $[Ca^{2+}]_o$ and other divalent and monovalent ions on the myelinated nerve membrane could be interpreted by the hypothesis of interaction with negative surface charges near the sodium channels.[169] Therefore, it may be concluded from the results of Figure 6 that MMA reduces resting Na^+-permeability of the nerve membrane. As shown in Figure 7 the effect of MMA already occurs at 5 mM and proceeds very quickly. In order to analyze the influence of MMA on the ionic channels of the nerve membrane, we have carried out voltage clamps measurements at single nodes of Ranvier dissected from the sciatic nerve of the frog.[166] Figure 8 shows the influence of MMA on the time course of the sodium and potassium currents (I_{Na} and I_K) through the membrane of the node of Ranvier. Shortly after the beginning of the superfusion there is a fast decrease in I_{Na} and I_K followed by a slow decline of the currents. I_{Na} is decreased to a greater extent than I_K. Removal of MMA from the solution leads to a fast increase in I_{Na} and I_K. But I_{Na} regains a level different from the control by a value which exactly corresponds to the amplitude of the slow decline during MMA application, indicating irreversible damage to the membrane.

As an uncharged molecule, MMA shows the general properties of lipid soluble drugs:[170] at low concentrations these compounds act as membrane stabilizers, and at high and lytic concentrations irreversibly damage both excitable and inexcitable membranes. In the last years it has become of actual interest to study hydrophobic interactions of charged and uncharged molecules with cell membranes and model membranes of lipid bilayers.[171,172] In order to investigate the hydrophobic character of the MMA action on the nerve membrane, we have applied equimolar concentrations of homologous chemicals as well as related substances to the desheathed sciatic nerve.[166] The chemical structures and the abbreviations used are illustrated in Figure 9. Figures 10 and 11 show their influence on the resting potential, compound action potential, and the conduction velocity. Whereas the homologous esters which all have an infinite lipid solubility and are insoluble in water, cause hyperpolarization and a decrease of the amplitude of the compound action potential and of the conduction velocity, the hydrophilic substances MAM and NaMA have no significant influence on the compound action potential and conduction velocity. The depolarization by 25 mM NaMA is due

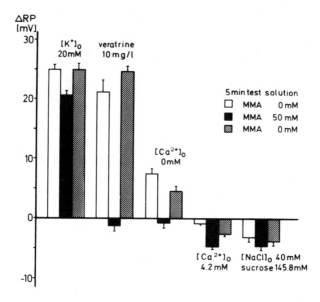

FIGURE 6. Changes in resting potential (ΔRP) of the desheathed sciatic nerve, measured by the single sucrose gap technique. At intervals of 5 min, the nerve was superfused with test solution (white colums), test solution and 50 mM MMA (black colums), and MMA-free test solution (striped colums). Vertical bars indicate S.E.M. $[K^+]_o$, $[Ca^{2+}]_o$, and $[NaCl]_o$: external concentrations of potassium, calcium, and sodium chloride.

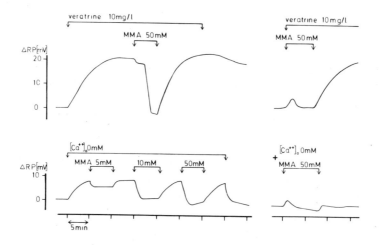

FIGURE 7. Hyperpolarization of the nerve membrane by 50 mM MMA after depolarization with veratrine or with a solution in which Ca^{2+} has been omitted. The test solutions were applied successively (left) or simultaneously (right). (From Böhling, H. G., Borchard, V., and Drovin, H., Arch. Toxicol., 38, 307, 1977. With permission.)

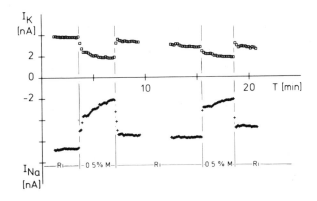

FIGURE 8. Changes in peak sodium and potassium currents of the node of Ranvier after application of 50 mM MMA (0.5% M). (From Böhling, H. G., Borchard, V., and Drovin, H., *Arch. Toxicol.*, 38, 307, 1977. With permission.)

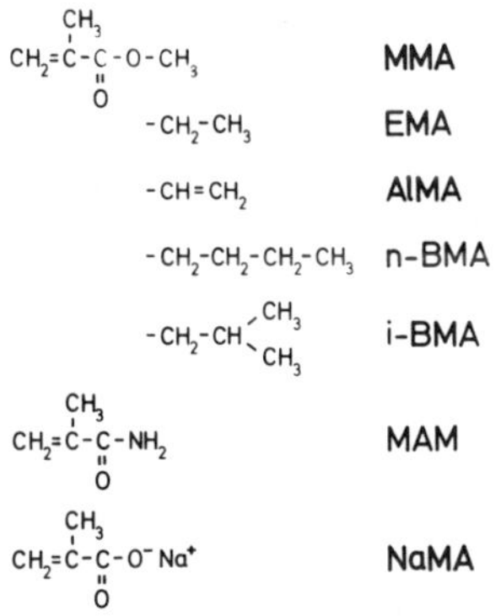

CH$_2$=C(CH$_3$)-C(O)-O-CH$_3$	MMA
-CH$_2$-CH$_3$	EMA
-CH=CH$_2$	AlMA
-CH$_2$-CH$_2$-CH$_2$-CH$_3$	n-BMA
-CH$_2$-CH(CH$_3$)$_2$	i-BMA
CH$_2$=C(CH$_3$)-C(O)-NH$_2$	MAM
CH$_2$=C(CH$_3$)-C(O)-O$^-$Na$^+$	NaMA

FIGURE 9. Chemical structures of homologous methacrylic esters and related substances.

to the increase in [Na$^+$]$_o$. MAM does not alter the resting potential. Analysis of the Na$^+$, K$^+$, and Ca^{2+} content of nerves treated with 50 mM MMA for 60 min showed no significant change in the electrolyte composition of the nerves.[166] This explains why up to this concentration alteration of the function of desheathed nerves is nearly completely reversible.

2. Central Nervous System

It has already been mentioned (Section III) that the toxic actions of MMA on respiration and the fatigue observed upon inhalation of MMA vapor were attributed to its influence on brain function.[117-128] Because of its high lipid solubility, MMA should accumulate in brain tissue. As there are great experimental difficulties in the investigation of drug effects on brain function in vivo, especially in the measurement of concentration response curves, it has become the custom to investigate tissues in vitro. However, studies of isolated brain slices have shown that preparation without damage to cells is very difficult. In order to maintain the function of isolated tissues, rapid

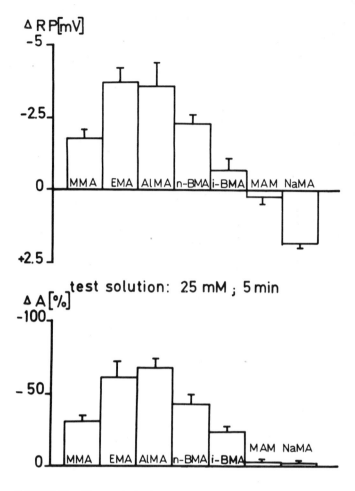

FIGURE 10. Hyperpolarization of the nerve membrane (Δ RP) and decrease of the amplitude of the compound action potential (Δ A) after 5 min application of differing homologous esters of MMA. For abbreviations see Figure 9. The related substance MAM shows no effect, whereas NaMA brings about a small depolarization due to the increase in $[Na^+]_o$. (From Böhling, H. G., Borchard, V., and Drovin, H., *Arch. Toxicol.*, 38, 307, 1977. With permission.)

diffusion of substances from the bathing solution to the intracellular space must be guaranteed. Recently, the author[173,174] has shown that the isolated retina fulfills these conditions and can easily be used as a model for the investigation of the effects of drugs and chemicals on central nervous tissues as, according to the theory of evolution, the retina is derived from the central nervous system. Its preparation is possible without damage to the cellular network and its excitation after light stimulation can be measured very easily by extracellular recording of the exposure potential (ERG). Figure 12 shows original recordings which illustrate the concentration-dependent influence of MMA on the ERG of the isolated frog retina. The ERG consists of the successive waves: (1) fast, downward, (2) fast, upward, and (3) late, slow upward deflection. While the a- and e-waves are probably generated predominantly in the first neuron of the peripheral visual pathway which on excitation is hyperpolarized by an increase in K^+ permeability, the b-wave seems to express the activity of the bipolars, the excitation of which is strongly dependent of $[Na^+]_o$.[175]

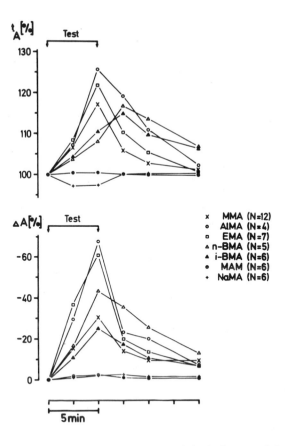

FIGURE 11. Change (% of control) in the latent period t_A, measured from the beginning of the stimulus to peak amplitude of the compound action potential (cf. Figure 4), and change in the amplitude (Δ A) after application of 25 mM test substances for 5 min. t_A is a measure of the mean conduction velocity. Stimulus: duration 0.5 msec, 50 Hz, supramaximal. For abbreviations see Figure 9.

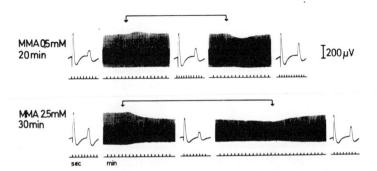

FIGURE 12. Original recordings showing the influence of MMA on the electroretinogram (ERG) of the isolated superfused frog retina, which consists of the consecutive waves: (a) fast downward, (b) fast upward, and (c) late slow upward deflection. Light intensity 20 Lux, stimulus duration 100 msec, frequency 0.1 Hz. Time constant of the amplifier 0.3 sec. Temperature 20°C.

124   Systemic Aspects of Biocompatibility

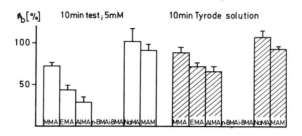

FIGURE 13.   Influence of 5 mM MMA or homologous esters and related substances (cf. Figure 9) on the amplitude of the b-wave ($\varnothing_b$), measured as the difference between the maxima of a- and b-waves. Experimental conditions as in Figure 12. The $\varnothing_b$ values were measured as percent of the control after 10 min test and 10 min wash, respectively.

Low concentrations of 0.5 mM MMA lead to a reversible decrease of the e-wave as illustrated by the dark silhouette curve in Figure 12.[176,177] Above 1 mM there is a decrease in the amplitudes of all waves which, on removal of MMA from the bathing solution, is partly irreversible. The reduced amplitude of the b-wave appears to confirm the general feature of the action of MMA on excitable cells: a decrease in the Na$^+$ conductance of the cell membrane which has already been demonstrated for the nerve membrane. But the striking difference is that decrease in central nervous bioelectrical activity occurs at tenfold lower concentrations as compared to the myelinated nerve. In order to investigate the role which lipophilia plays for the strength of MMA action, homologous and related chemicals were applied to the isolated retina. The result is illustrated in Figure 13. Increase in the length of the alcoholic group of the methacrylic esters corresponds to increased efficiency of action. While the lipophilic butyl derivatives of MMA abolish the b-wave of the ERG irreversibly, the water soluble compounds NaMA and MAM are nearly without effect.

Lawrence and Autian[164] have demonstrated that MMA is able to increase sodium pentobarbital-induced sleeping time in ICR mice considerably. This dose-related increase was attributed to an effect on metabolizing enzymes. But our experimental data presented above, as well as the observation of Karpov[120, 121] that inhalation of MMA vapor brings about fatigue, suggest that there is a direct central nervous action of MMA. The apnoea observed at toxic doses of MMA, as reported by Deichmann[110] and Spealman et al.,[112] and toxic symptoms induced by MMA vapor as described by Filatova[125] emphasize the significance of the central nervous actions of MMA.

Further experimental evidence of the central nervous action of MMA will be presented in the next section, where influence on the breathing center is discussed together with the alteration of respiration.

C. Effects on Respiration

Deichmann,[110] as early as 1941, observed that after injection of 0.03 mℓ/kg MMA in rabbits, respiration was stimulated immediately and stopped after one or two additional doses, upon which the animals died. Further experimental evidence for the influence of MMA on respiration was presented, indicating a peripheral as well as central component of action.[112, 117] We have tried to analyze the mechanism by which MMA affects respiration using an experimental arrangement described in detail by Borchard.[178] MMA or homologous compounds were injected in guinea pigs narcotized

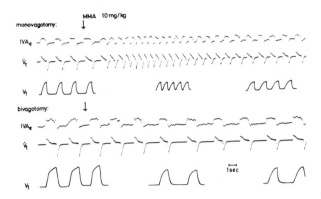

FIGURE 14. Change in respiration of the monovagotomized or bivagotomized guinea pig, induced by 10 mg/kg MMA i.v. IVA_e = integrated efferent vagal activity, $\dot{V}_t$ = velocity of tidal air, V_t = tidal volume.

with 1.25 g/kg i.p. ethyl urethane. The animals were either spontaneously breathing or ventilated by positive pressure respiration. The velocity of tidal air ($\dot{V}_t$) was measured by a Fleisch jet connected to a pneumotachograph[178] and the tidal volume ($\dot{V}_t$) was determined by electronical integration of $\dot{V}_t$. Blood pressure was measured by means of a steel canula in the carotid artery and a pressure transducer. The left vagus nerve was isolated, incubated in paraffin, and cut at its center. Inspiratory afferent vagal activity (IVA_a) and efferent vagal activity (IVA_e) were measured with Ag/AgCl-electrodes from the distal and proximal vagal stump. After rectification of the nerve activity positive spikes were integrated (time constant of the RC circuit 0.5 s). Injections were carried out via a catheter in the jugular vein. Figure 14 illustrates the effect of 10 mg/kg MMA on respiration of a spontaneously breathing guinea pig in mono- and bivagotomy. The tachypnoea and the decrease in ideal volume observed in the monovagotomized animal may be attributed to a bronchospasm because experiments in bivagotomized and ventilated animals (not illustrated here) have shown a reduction of $\dot{V}_t$. The bronchospasm was also described by Kutzner et al.[117] who were able to reduce tachypnoea by the bronchodilator orciprenaline or by lidocaine (Table 5). Figure 14 clearly demonstrates that the tachypnoea is generated by a vagal reflex mechanism, for after bivagotomy an increase in breathing frequency can no longer be observed. On the contrary, bradypnoea is brought about by MMA if the vagal reflex circle is interrupted and activity of the breathing center seems to be diminished as indicated by a decrease in IVA_e and V_t. Respiratory rate is mainly reduced by an increase in the postexspiratory interval. These results confirm central depression of respiration by a direct action on the activity of central inspiratory neurons. pMMA is without action on respiration.

The dose-related tachypnoea induced by MMA in monovagotomized animals commences immediately and is fully reversible up to 5 mg/kg after about 30 sec (Table 6). One minute after injection even a slight bradypnoea is observed which is the predominant effect on bivagotomy as discussed above (Figure 14). This bradypnoea lasts several minutes. High doses of MMA lead to an apnoea. Death from acutely toxic doses by various routes of administration was attributed to respiratory depression.[110,112]

In order to investigate the lipophilic character of the action of MMA on respiration we have applied homologous and related substances in equimolar doses to the sponta-

Table 5
EFFECT OF MMA AND THE COMBINATION OF
MMA AND ORCIPRENALINE ON RESPIRATION
OF THE MONOVAGOTOMIZED GUINEA PIG[179]

Time after injection (sec)	2.4 mg/kg MMA i.v. respiratory rate		2.4 mg/kg MMA + μg/kg orciprenaline i.v. respiratory rate	
	min^{-1}	% of control	min^{-1}	% of control
0	40	100	59	100
5	78	189	71	120
10	50	122	65	110
20	50	122	62	105
30	44	108	61	101

Table 6
DOSE-RELATED CHANGE IN RESPIRATORY
RATE OF THE MONOVAGOTOMIZED
GUINEA PIG BY MMA[179]

MMA (mg/kg i.v.)	N	Respiratory rate % of control (sec after application)			
		5	10	20	30
0.5	15	111.8	107.1	103.2	100.4
0.95	13	123.7	110.3	101.9	92.3
2.4	9	189.1	122.4	122.1	108.2
4.75	9	252.6	174.5	126.6	87.5
9.5	5	446.5	270.1	170.9	136.0

neously breathing guinea pig. As the lipophilic esters show a low water solubility, all substances were dissolved in isotonic NaCl containing 10 w/v % cremophor as a solute which was by itself without influence on respiration. Figures 15 and 16 show characteristic original recordings. The most active compounds were n-BMA and i-BMA, followed by AlMA, MMA, and EMA. After injection there is an elevation of the base line of the curves, illustrating increase in inspiratory activity of the afferent vagus. This increase might be due either to a direct stimulation of lung stretch receptors or by an inflation of the lungs, thus activating the stretch receptors. As orciprenaline reduces the elevation of the base line of the IVA_a curve, bronchospasm is the cause of this effect. Constriction of the respiratory system is demonstrated in Figures 15 and 16 by a decrease in $\dot{V}_t$, despite an increase in respiratory rate. Furthermore, a decrease of the amplitude of the IVA_a curves is observed. This might be due either to decreased degree of inflation or to an endoanaesthetic effect on the excitable membranes of the lung stretch receptors. We were able to demonstrate that on monovagotomized guinea pigs under artificial respiration by application of MMA in comparison to histamine, of veratrine and MMA as well as high doses of MMA, there is a considerable endoanaesthetic effect on lung stretch receptors. This may be expected from the MMA action on the nerve membrane discussed above, but concentrations influencing nerve excitation have to be much higher than those impairing respiration.

Figure 17 illustrates the action of the hydrophilic substances MAM and NaMA on respiration. The increase in respiratory rate does not occur until doses tenfold higher than those of MMA are applied. In bivagotomized animals, a decrease in respiratory

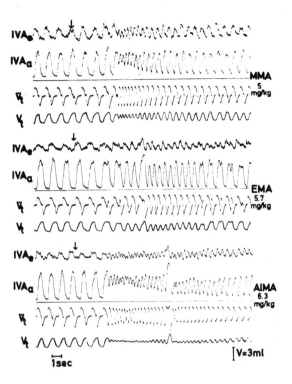

FIGURE 15. Influence of MMA and homologous esters on respiration of the monovagotomized guinea pig. Integrated vagal activity: IVA_e = efferent and IVA_a = afferent. V_t = velocity of tidal air, V_t: tidal volume. For abbreviations see Figure 9.

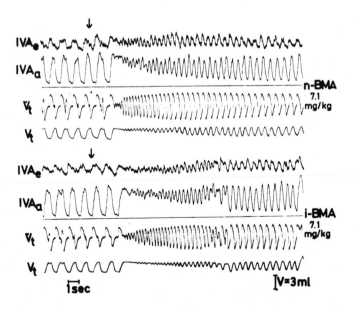

FIGURE 16. For explanation see Figure 15.

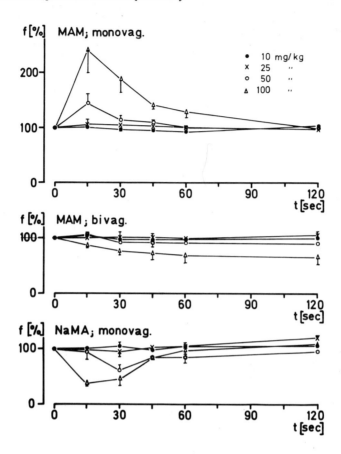

FIGURE 17. Dose-related influence of MAM and NaMA on respiration of the monovagotomized and bivagotomized guinea pig. f = respiratory rate, % change of the control. I.V. application at time 0.

rate is observed at 100 mg/kg, a dose about 20 times higher than a comparably effective dose of MMA. NaMA even at very high doses does not produce tachypnoea but a strong bradypnoea. All effects are reversible within 2 min after injection.

Increase in respiratory rate by MMA has also been reported in mongrel dogs by Mir et al.[180] The authors have investigated MMA and 12 methacrylate esters. The finding that butyl and isobutylmethacrylate showed the strongest effect is consistent with our results. The initial bronchospasm responsible for the tachypnoea induced by a vagal reflex mechanism as well as the late bradypnoea due to a depression of central inspiratory neurons, which we observed under experimental conditions, might be the reason why more or less severe respiratory changes combined with hypoxemia nearly always appear in patients in the early phase of healing after insertion of hip protheses.

The mechanism of bronchospasm has still to be elucidated. MMA leads to a dilatation of the smooth muscle of blood vessels and of the ileum.[181] MMA vapor can produce a direct and reversible inhibition of spontaneous motor activity of rat intestinal smooth muscle strips in vitro and a reduction of intestinal transit after chronic exposure for 6 months.[189] Furthermore, paralytic ileus with pulmonary hypoxia and chest pain was observed in the postoperative period of a patient who showed a venous blood level of 200 mg percent together with a decrease in Pa_{O_2} from 103 to 80 torr.[60] Therefore, a direct myogenic or a cholinergic mechanism of bronchoconstriction seems to

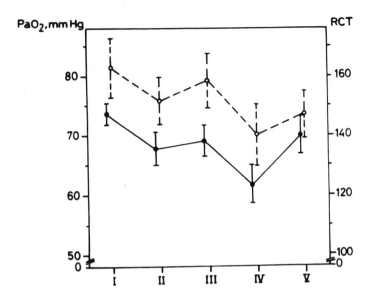

FIGURE 18. Mean recalcification time (RCT, sec) ± SE and mean arterial oxygen tension (Po_2) ± SE during the various periods of operation: before (I) and after (II) insertion of cement and prothesis into acetabulum; before (III) and after (IV) insertion of cement and prosthesis into femur, and 20 to 30 min after impaction of the femoral prosthesis (V). (From Modig, J., et al., *Acta Anaesthesiol. Scand.*, 19, 28, 1975. With permission.)

be unlikely. As MMA induces peripheral vasodilation, and at the same time bronchial constriction, an involvement of receptors for histamine or serotonin has to be discussed.

Modig et al.[197] report that in a 51-year-old female patient the mean pulmonary arterial pressure rose from 25 to 31 mmHg, and the mean arterial blood pressure fell from 100 to 50 mmHg 20 to 25 sec after the insertion of a femoral prothesis into the medullary cavity of the femur by the Charnley technique. Airway resistance increased from 9.0 to 12.3 $dyn.s.cm^{-5}$ at an airflow of 0.5 ℓ/s. Venous admixture during airventilation increased from 17 to 52% leading to an arterial desaturation from 95 to 74% oxygen saturation. These changes indicated peripheral vasodilation, increased pulmonary vascular resistance, and bronchoconstriction, changes which were observed in dogs after experimentally-induced platelet aggregation.[198] Applying ^{51}Cr-labeled platelets and ^{125}I-labeled fibrinogen the day before the operation, Modig et al.[195] could show a transient accumulation of ^{51}Cr over the lung and a strong correlation between deposition of ^{125}I radioactivity in the lung and Pa_{O_2} reduction occurring after the insertion of the femoral prosthesis. In a later study, Modig et al.[193] demonstrated, by measuring the recalcification time, the occurrence of thromboplastic activity in the pulmonary arterial blood during the different phases of insertion of cement and prosthesis into the acetabulum and femur indicating microembolism (Figure 18). The authors conclude from their results that pulmonary dysfunction during certain steps of total hip replacement surgery are associated with the efflux of tissue-thromboplastic products into the pulmonary circulation, thus leading to microembolism. Therefore, it seems to be likely that bronchoconstriction is brought about by microembolism similar to that induced by anaphylatoxin plasma[194] or coli endotoxin which is mediated via the liberation of serotonin, thus effecting changes in respiration and blood pressure comparable to those of MMA.[196] Modig et al.[193] assume that the appearance of bone marrow fat is of minor importance and that the release into the lungs of MMA is probably of no importance. But our results indicate that MMA alone brings about bronchoconstriction and influences lung stretch receptors as well as the center of respiration. Furthermore, it is well documented that MMA blood levels correlate strongly to respiratory or cardiovascular actions.[60,73] As a lipophilic substance (cf. Table 1) MMA might reach much higher concentrations in lipid membranes than in the blood. On the other hand, major cardiovascular alterations are observed during the insertion of cement into the neck of the femur,[199] indicating the influence of pressure developed in the medullary cavity.[200] In addition, cardiovascular changes were also observed using plasticine or paraffin wax.[199]

Many authors have described a decreased pulmonary function induced by MMA injections or occurring during total hip replacement surgery with acrylic bone cement. Animal experiments revealed decreased Pa_{O_2}, elevated Pa_{CO_2}, and a decrease in pH or metabolic acidosis.[94,182] Similar changes in respiratory function were reported in patients undergoing acrylic stabilization of total hip arthroplasty[60,61,124,163,186,193,195,197,200] Decrease in Pa_{O_2} was observed more often upon application of the bone cement in the femoral shaft as compared to the acetabulum. The effect occurred within 30 sec independent of changes in blood pressure and was reversed after about 8 min.[163] Kallos[186] summarizes the possible reasons for decreased Pa_{O_2}: decrease in cardiac output, decreased diffusion capacity, increased ventilation perfusion abnormality, and/or increased shunt. One case of decrease in Pa_{O_2} without change in cardiac output has been reported by Modig et al.[197] Since the decrease in Pa_{O_2} is not totally prevented by breathing high concentrations of oxygen, some of the decrease should be produced by shunting or a decrease in cardiac output.[186] While Park et al.[163] ruled out pulmonary emboli as the cause of decreased oxygen tension after the use of cement in the femur, Kallos[186] suggested that the impairment of oxygenation after the use of bone cement is caused by pulmonary fat and bone marrow emboli, which may bring about the elevation of central venous pressure. The role of fat embolism is also discussed by other authors.[58,65,99,186,201-207] A considerable rise in medullary pressure below the cement during insertion of the prosthesis has been reported.[208] Furthermore, it was demonstrated by Whitenack and Hausberger[209] that in rabbits intramedullary pressures of only 5 to 10 cm of water permitted entry of fat injected into the marrow. Therefore, in the presence of high pressure and MMA as a solute, fat embolism could be facilitated. This is indicated by the results of Sevitt[202] who found fat embolism in 6.8 to 8% of those patients with subcapital fractures treated by primary Thompson's arthroplasty which utilizes acrylic cement, and no evidence of embolism in those given Moore's prosthesis in which cement is not used.

In a number of investigations, morphological alterations such as pulmonary hemorrhage and pulmonary vascular congestion with fluid-filled alveoli combined with increase in hematocrit were observed.[182,190,193] This demonstrates a massive transudation of fluid out of the circulation into the lung tissue and alveoli.

D. Cardiovascular Effects

As several cases of cardiac arrest have been reported during orthopedic reconstructive surgery[32,58,72,78,203-207,212,218-220] in which MMA was suggested as the primary or contributory cause,[65] pharmacological investigations on the direct action of MMA on cardiac tissues have been carried out. The results of our experiments are illustrated in Figure 19. MMA brings about a dose-dependent decrease in force of contraction to a similar extent in left and right atria or papillary muscles of guinea pigs, the ED_{50} ($\mu M/\ell$) being 3.4 and 3.5 or 2.7.

The decrease in contractility reaches steady state values within 10 min. In the atria the reduced force of contraction at 10 mM MMA is quickly reversed after 10 min

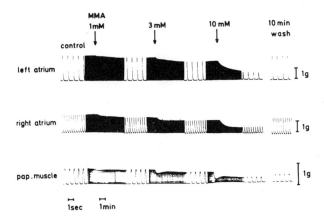

FIGURE 19. Dose-related decrease of the force of contraction of isolated guinea pig heart preparations incubated in Tyrode solution saturated with 5% CO_2 and 95% O_2, pH = 7.4. Original registration. Stimulation at 1 Hz, 3 msec, 20% above threshold at 31°C. MMA was added to the bathing solution cumulatively every 10 min.

wash, whereas in papillary muscles there remains a partly irreversible decrease. These results indicate that mean blood concentrations of 1 mg% observed during hip reconstructive surgery which correspond to 0.1 mM solutions, are without significant effect on cardiac function. Only in those cases reported by Pahuja et al.[60] in which blood levels of 200 mg percent occur, should there be a dramatic drop in contractility. Similar results to ours have been observed for isolated rabbit atria.[213] The experiments of Mir et al.[214] also show a decrease at concentrations between 0.1 and 10 mM, but the authors overestimate the negative inotropic effect because they simultaneously observe a decrease in pulse rate which alone could lead to a diminution of force of contraction (Bowditch stair case phenomenon). This decrease in heart frequency could not be observed in isolated right atria of the guinea pig (cf. Figure 19) or rabbit.[213] Brittain and Ryan[59] (having personally administered or supervized anesthesia in more than 7,000 hip arthroplasties with acrylic bone cement) as well as Park et al.[163] found no changes in the pulse rate of their patients.

Animal experiments have shown a decrease in cardiac output by MMA.[94] This effect was also observed in patients by Convery et al.[215] and Bright,[190] but not by Modig et al.[195] Changes in intraoperative ECG caused by acrylic bone cement are often sinus bradycardia and QRS complexes of diminished amplitude, although these alterations seem to be brought about by embolism and not by MMA.

Many reports have been published about the fall in blood pressure which occurs when acrylic cement is used during prosthetic hip surgery.[32,58,61,67,68,78,186,193,197,200,204,205,210,216-221] Occassional profound hypotension was followed by cardiac arrest.[64,65,72,203,206,207,210,219] It has been suggested that MMA, diffusing from the cement into the blood, is responsible for the decrease in blood pressure.[71,73,216,221] Schuh et al.[221] have determined the frequency and magnitude of circulatory changes following methylmethacrylate implantation in patients (Figure 20). Two of the 52 patients showed no cardiovascular changes. In 29 patients (55%) there was a decrease in blood pressure immediately following cement placement in the acetabular bed, and a second hypotensive episode coincident with placement of methylmethacrylate into the marrow cavity of

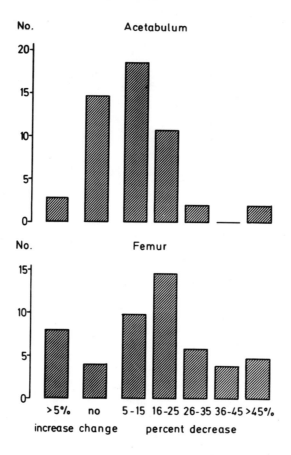

FIGURE 20. Frequency histogram of mean arterial blood pressure following implantation of methylmethacrylate bone cement into the acetabulum (top) and into the femur (bottom) during 52 total-hip-joint replacements in patients. No. = number of observations.[221]

marrow cavity of the femur. One patient showed an increase in blood pressure after each of the two placements, whereas in the remaining 20 patients there were combinations of increase, decrease, or no change of pressure. Hypotension occurred within 1 min after cement insertion, was maximal after 2 min, and regained initial values after 3 min. There was a strong correlation between the age of the patient and the extent of hypotension. Furthermore, there was a greater hypotensive response following femoral as compared to acetabular implantation.

During peripheral vasodilation, Modig et al.[197] observed an increase in pulmonary arterial pressure which paralleled a decrease in $\bar{P}_{AO}$ (Figure 21). There was a simultaneous bronchoconstriction, which was attributed to platelet aggregation in pulmonary circulation (see Section IV C).[193] Kallos[186] has reported a decrease followed by an increase in central venous pressure in his patients, whereas other authors found no change.[68,210] No correlation between changes in blood pressure and changes in arterial oxygen tension was observed.[186]

First reports on the hypotensive action of MMA in animals were published by Deichmann.[110] These results were confirmed by later experimental studies.[73,91,180,181,197,199,222,223] McMaster et al.[223] found an increase in pulmonary vascular bed resistance

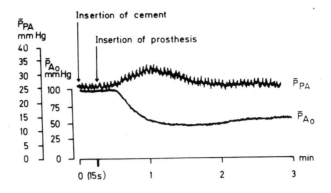

FIGURE 21. Changes in mean pulmonary arterial pressure (P_{PA}) and mean aortic pressure (P_{Ao}) after insertion of the femoral prosthesis. (From Modig, J., et al., *Acta Anaesthesiol. Scand.*, 17, 276, 1973. With permission.)

associated with peripheral vasodilation in dogs. The effects were not influenced by blockade of the beta adrenergic receptors. Decrease in blood pressure in dogs was also observed with homologous esters of MMA.[181] Peebles et al.[73] emphasize that hypertensive, arteriosclerotic, and elderly patients are especially at risk as they compensate poorly for sudden and severe physiological changes. Important experiments on the cardiovascular effects of acrylic cement have been carried out in rabbits and cats by Pelling and Butterworth.[199] The authors could demonstrate that inhibition of respiration, decrease in arterial pressure, and increase in central venous pressure was brought about not only by forcing acrylic bone cement into the medullary cavity of the femur, but also by Plasticine® or soft paraffin wax (Figure 22). Some of the authors' results are summarized in Table 7. The fall in blood pressure was also observed when a femoral venous shunt was used or when femoral venous return was prevented. Therefore, it seems to be unlikely that the acute hypotensive effect is caused by a factor carried in the blood. Minimal cardiovascular values occurred within 12 sec after ramming in the material and were reversed within about 30 sec. In three animals in which blood pressure failed to recover, fat emboli were detected in pulmonary blood vessels. However, microscopical evidence of embolism was not confined to those animals which showed protracted fall in blood pressure. On the other hand, Dustmann et al.[70] demonstrated a connection between the insertion of acrylic bone cement and fat embolism in cats. The authors failed to observe hemodynamic changes after the insertion of Plasticine® as a control substance.

Milne,[78] as well as Phillips et al.,[68] discuss the possibility of a neurovascular reflex mechanism, mediated via receptors similar to the carotid sinus reflex, but an accompanying bradycardia was never observed.[68]

E. Clinical Observations

In the previous sections, it was shown that MMA is able to bring about respiratory and cardiovascular complications, but that it is rather difficult to differentiate them from other complications such as fat or air embolism, platelet aggregation, or neurovascular reflex mechanisms. Charnley[31] states that harm from bone cement is likely to be due to its misuse and that great care should be taken to limit absorption of the monomer. The author reports four cases of cardiac arrest in a total of 3700 hip replacements.

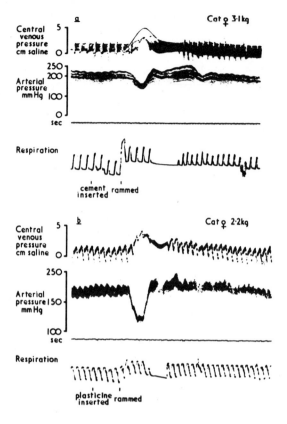

FIGURE 22. Effect of pushing acrylic bone cement (A) or plasticine (B) into femoral medullary cavity of cat. (From Pelling, D., et al., *Br. Med. J.*, 2, 638, 1973. With permission.)

Smith and Turner[32] claim that a hypotensive episode accompanying or immediately following insertion of cement in 2798 patients, occurred with sufficient frequency that it can probably be considered the principle side effect of the cement. In most cases, hypotension lasted only a few minutes. In 83% of the patients a decrease in blood pressure was 9 mmHg or less, in 11% 10 to 19 mmHg, in 5% 20 to 39 mmHg, and in less than 1.7% more than 40 mmHg.

Brittain and Ryan[59] also report that hypotension rarely exceeding 15 mmHg is a common and transient occurrence in about 80% of 7000 patients undergoing hip arthroplasty. The time taken to reach a maximum fall in blood pressure is usually 30 to 60 sec, followed by a rapid return to normal values after about 90 sec. The authors never observed cardiac arrest or ECG changes.

Dandy[204] has analyzed 806 cases of artificial hip joint implantation, taking special interest in the effect of using acrylic cement. Published reports on 4908 total hip replacement operations are summarized excluding mortality attributable to cement, whereas there are 18 reports of cardiac arrest, 11 of them fatal, in patients undergoing Thomson's fracture arthroplasty. A relationship between cardiac arrest and hypotension could not be demonstrated. The author emphasizes the advantageous use of plastic cannula for venting the femoral shaft in order to avoid high pressures, and points out that death and fat embolism are more common when cement is used to secure the prosthesis. Coventry et al.[215] have evaluated 2012 arthroplasties carried out in 1684

Table 7
ARTERIAL BLOOD PRESSURE CHANGES (MEANS ± SE)
DURING THE ACUTE RESPONSE WHICH FOLLOWED
FORCING ACRYLIC CEMENT, PLASTICINE, OR PARAFFIN
WAX INTO THE MEDUALLRY CAVITY OF FEMUR[199]

Species	Material	No. of insertions	Initial b.p. (mm Hg) systolic/diastolic	Fall in b.p. (mm Hg) systolic/diastolic
Rabbit	Cement	19	111 ± 2.9/ 84 ± 2.3	16.3 ± 1.60/17.4 ± 1.50
Rabbit	Paraffin wax	33	120 ± 3.6/ 97 ± 3.7	15.2 ± 1.24/19.5 ± 1.58
Rabbit	Plasticine®	16	122 ± 1.0/ 93 ± 1.6	12.9 ± 2.01/16.0 ± 1.80
Rabbit[a]	Plasticine®	17	120 ± 4.8/ 88 ± 4.5	13.3 ± 1.82/15.0 ± 1.93
Rabbit[b]	Plasticine®	9	116 ± 8.9/ 87 ± 6.9	17.2 ± 4.00/24.2 ± 4.30
Cat	Cement	27	179 ± 5.0/138 ± 4.0	21.3 ± 3.89/16.3 ± 3.19
Cat	Plasticine®	28	166 ± 5.1/131 ± 4.9	22.9 ± 2.76/16.3 ± 2.07

[a] With femoral venous shunt.
[b] With femoral venous return prevented.

patients. The intrahospital mortality was 0.4%. The causes of death were pulmonary embolism in four patients, coronary artery disease in three, cerebrovascular thrombosis in one, and cardiac arrest at time of surgery in another. Based on these findings it was concluded that certain technical considerations are important, including correct use of methylmethacrylate, positioning of prosthestic components, trochanteric wiring, and the configuration of the tissues and cement adjacent to the joint. Cardiac arrest occurred in two patients and there were 44 instances of pulmonary embolism (2.2%), four of them fatal. Desmonts et al.[61] confirm the low rate of cardiovascular collapse or cardiac arrest during arthroplasty. They emphasize the significance of hypoxemia and hypotension during insertion of the cement, particularly in combination with anesthesia or in patients with cardiovascular diseases.

V. CONCLUSIONS

It has been shown in numerous animal experiments that MMA brings about decrease in arterial blood pressure, increase in pulmonary arterial and central venous pressure, decrease in contractility, bronchoconstriction, decrease in intestinal motility, increase in respiratory rate, decrease in activity of the respiration center, and decrease of the excitability and conduction velocity of the peripheral nerve. On a cellular level most of these effects seem to be due to the penetration of MMA as a lipophilic solute into the cell membrane, leading to a decrease in ionic currents. The mechanism by which intravenous MMA brings about bronchoconstriction and increase in pulmonary artery resistance is still unclear, although it resembles the effects brought about by intravasal platelet aggregation. Concentrations needed for the above-mentioned effects are usually much higher than those observed in patients during insertion of acrylic cement. Therefore, direct toxic effects of MMA in patients are rare and may be avoided if the amount of monomer in the cement is kept low. This is made possible if the degree of plasticity of the mixed cement is controlled and if there is an exact choice of time of its introduction. The most common side effects during total hip replacement are hypotension and hypoxemia, although they are also observed if paraffin wax or Plasticine® are used instead of acrylic cement. It has still to be elucidated if fat or air embolism or the diffusion of thromboplastic substances into the circulation combined with platelet aggregation are the main factors for intraoperative accidents. As most operations are carried out on elderly patients, it has to be recognized that preoperative cardiovascular or respiratory disorders may emphasize the toxic side effects brought about by the insertion of acrylic cement.

While pMMA is well tolerated by tissues, local application of MMA may bring about severe cell damage. Furthermore, occlusive contact of MMA with skin and mucosa can result in allergic reactions.

ACKNOWLEDGMENTS

I am greatly indebted to Dr. A. A. L. Fox for critically reading the manuscript and to Mrs. U. Goldstein for technical assistance.

REFERENCES

1. **Burman, M. and Abrahamson, R. H.**, The use of plastics in reconstructive surgery. Lucite in arthroplasty, *Mil. Surg.*, 93, 405, 1943.
2. **Harmon, P. H.**, Arthroplasty of the hip for osteoarthritis utilizing foreign-body cups of plastic, *Surg. Gynecol. Obstet.*, 76, 347, 1943.
3. **Mellen, R. H. and Phalen, G. S.**, Arthroplasty of the elbow by replacement of the distal portion of the humerus with an acrylic prosthesis, *J. Bone Jt. Surg.*, 29, 348, 1947.
4. **Judet, J. and Judet, R.**, Essais de reconstruction prothétique de la hanche après résection de la tête femorale, *J. Chir.*, 65, 17, 1949.
5. **Judet, J. and Judet, R.**, The use of an artificial femoral head for arthroplasty of the hip joint, *J. Bone Jt. Surg. Br. Vol.*, 32, 166, 1950.
6. **Henrichsen, E., Jansen, K., and Krogh-Poulsen, W.**, Experimental investigation of the tissue reaction to acrylic plastics, *Acta Orthop. Scand.*, 22, 141, 1953.
7. **Haboush, E. J.**, A new operation for arthroplasty of the hip based on biomechanics photoelasticity, fast-setting dental acrylic, and other considerations, *Bull. Hosp. Jt. Dis.*, 14, 242, 1953.
8. **Herrmann, K. O.**, Die Verwendung des selbsthärtenden Kunststoffes Palavit zur Substitutionsbehandlung bei Knochenerkrankungen und Knochenbrüchen, *Ärztl. Forschg.*, 7, 543, 1953.
9. **Herrmann, K. O.**, Operative Behandlungsmethoden beim menschlichen Knochenbruch mit Hilfe von Kunststoff, *Erfahrungshkd.*, III, 4, 169, 1954.
10. **Idelberger, M. K.**, Palavit in der operativen Orthopädie, *Verh. Dtsch. Orthop. Ges. Beilageheft Z. Orthop.*, 86, 354, 1955.
11. **Diener, A. and Herrmann, K. O.**, Zur Behandlung von Knochenbrüchen mit Hilfe von Kuststoff, *Erfahrungshkd.*, V, 8, 368, 1956.
12. **Wiltse, L. L., Hall, R. H., and Stenehjem, J. C.**, Experimental studies regarding the possible use of self-curing acrylic in orthopaedic surgery, *J. Bone Jt. Surg. Am. Vol.*, 39, 961, 1957.
13. **Charnley, J.**, Anchorage of the femoral head prothesis to the shaft of the femur, *J. Bone Jt. Surg., Br. Vol.*, 42, 28, 1960.
14. **Muller, M. E.**, Die Verwendung von Kunstharzen in der Knochenchirurgie, *Arch. Orthop. Unfall-Chir.*, 54, 513, 1962.
15. **Herrmann, K. O.**, Die therapeutischen Ergebnisse zweier Fälle mit Kunststoffbehandlung von Frakturen des wachsenden Knochens nach Ablauf von 5 Jahren, *Zentralbl. Chir.*, 88, 1258, 1963.
16. **Huggler, A. H. and Francillon, M. R.**, Zur Hüftarthroplastik nach Charnley, *Schweiz. Med. Wochenschr.*, 93, 297, 1963.
17. **Charnley, J.**, The bonding of prothesis to bone by cement, *J. Bone Jt. Surg. Br. Vol.*, 46, 518, 1964.
18. **Francillon, M. R.**, Osteomien und Gelenkplastiken in der Behandlung der primar chronischen Polyarthritis, *Med. Welt*, 45, 1, 1965.

19. Scheuba, G., Osteosynthese pathologischer Frakturen, *Zentralbl. Chir.*, 90, 1737, 1965.
20. Postel, M., Teinturier, P., and Dubousset, J., Techniques et premiers résultats de l'arthroplastic de la hanche avec la prothèse totale de McKee, *Mém. Acad. Chir.*, 92, 870, 1966.
21. Willenegger, H., Schenk, R., and Bandi, W., Die Anwendung von Leimsubstanzen in der Knochenchirurgie, *Med. Mittlg. (Melsungen)*, 100, 2487, 1966.
22. McKee, G. K. and Watson-Farrar, J., Replacement of arthritic hips by the McKee-Farrar prothesis, *J. Bone Jt. Surg. Br. Vol.*, 48, 245, 1966.
23. Charnley, J., The healing of human fractures in contact with self curing acrylic cement, *Clin. Orthop.*, 47, 157, 1966.
24. Van Dijk, D. J., De behandeling van pathologische femurfracturer met Palacos, *Ned. Tijdschr. Geneeskd.*, 111, 632, 1967.
25. Pfeiffer, R., Ein neues Verfahren der Behandlung frischer Wirbelbrüche unter Anwendung eines Kunstharzes, *Z. Orthop.*, 105, 122, 1968.
26. Huggler, A. H., *Die Alloarthroplastik des Hüftgelenkes mit Femurschaft- und Totalendoprothesen*, Thieme, Stuttgart, 1968.
27. Lange, M., *Orthopädisch-chirurgische Operationslehre. Ergänzungsband: Neueste Operationsverfahren*, Bergmann, München, 1968.
28. Huggler, A. H., On modification of total prothesis, Wiederherstellungschir, *Trauma*, 11, 63, 1969.
29. Witt, A. N., Probleme des Gelenkersatzes, *Munch. Med. Wochenschr.*, 111, 1781, 1969.
30. Buchholz, H. W., Bisherige Erfahrungen mit der totalen Endoprothese am Hüftgelenk, *Orthop. Prax.*, 6, 14, 1970.
31. Charnley, J., *Acrylic Cement in Orthopaedic Surgery*, Livingstone, Edinburgh, 1970.
32. Smith, R. E. and Turner, R. J., Total hip replacement using methylmethacrylate cement. An analysis of data from 3,482 cases, *Clin. Orthop.*, 95, 231, 1973.
33. Hejda, N., Umbach, W., and Janusch, H., Stabilisierung der Wirbelsäule durch laminierte Kunststoffprothesen, *Dtsch. Med. Wochenschr.*, 99(19), 1001, 1974.
34. Woringer, E., Schwieg, B., Brogly, G., and Schneider, J., Nouvelle technique ultra rapide pour la réfection de brèches osseuses craniennes à la résine acrylique. Avantages de la résine acrylique sur la tanfale, *Rev. Neurol.*, 85, 527, 1951.
35. Robinson, R. G. and Mocalister, A. D., Acrylic cranioplasty. A simple one-stage method using a cold curing material, *Br. J. Surg.*, 42, 312, 1954.
36. Rietz, K.-A., The one-stage method of cranioplasty with acrylic plastic, *J. Neurosurg.*, 15, 176, 1958.
37. Ross, P. J. and Jelsma, F., Experiences with acrylic plastic for cranioplasties, *Am. Surg.*, 26, 519, 1960.
38. Unger, R. R. and Sollmann, H., Die Versorgung von Schädelkalottendefekten mit Palacos, *Zentralbl. Chir.*, 89, 849, 1964.
39. Kothe, W. and Lange, C.,G., Die Schädeldachplastik mit schnellhärtenden Acrylaten, *Zentralbl. Chir.*, 92, 497, 1967.
40. Hoppe, W. and Bremer, H., Experimenteller Beitrag zur enossalen Implantation alloplastischen Materials im Kiefernbereich, *Dtsch. Zahnärztl. Z.*, 11, 551, 1956.
41. Immenkamp, A., Beiträge zur maxillo-facialen Chirurgie unter besonderer Berücksichtigung der Korrektur von Fehlbildungen des Mittelgesichts, *Dtsch. Zahnärztl. Z.*, 15, 1073, 1960.
42. Heidsieck, C., Die Therapie veralteter Gesichtsschädelfrakturen, *Dtsch. Zahnärztl. Z.*, 16, 283, 1961.
43. Klewansky, P. and Nicolas, P., Intérêt de l'implantacrylique endosseux dans le traitment des fractures mandibulaires chez l'denté, *Rev. Fr. Odonto-Stomatol.*, 13, 191, 1966.
44. Grasser, H., Die Verwendung von selbsthörtenden Kunststoffen zu Osteosynthesezwecken, *Dtsch. Zahnaerztl. Z.*, 24, 306, 1969.
45. Arnandow, M., Die Implantation künstlicher Zahnwurzeln, *Zahnarztl. Prax.*, 21, 37, 1970.
46. Pape K., Zur plastischen Wiederherstellung des Vorderschadels, *Dtsch. Stomatol.*, 20, 1, 1970.
47. Kreudenstein, Sp. v., T., Alloplastischer Kinnaufbau beim Vogelgesicht mit Verankerung des Kunststoffimplantates im Knochen, *Zahnärztl. Prax.*, 23, 37, 1971.
48. Castagnolq L., Experimentelle Untersuchungen über im Mund selbstpolymerisierende Kunstharzfüllungen, *Schweiz. Monatsschr. Zahnheilkd.*, 60, 880, 1950.
49. Fischer, C. H. and Sonnaband, E., Selbsthörtende Kunststoffe in der konservierenden Zahnheilkunde, *Zahnarztl. Welt*, 6, 7, 1951.
50. Castagnola, L., *Neue Zahnärztliche Kunststoffe*, C. Hanser, München, 1951.
51. Kreudenstein, Sp. v., T., *Kariestherapie mit Schnellhärtendem Kunststoff*, Hanser, München 1952.
52. Hoffmann, K., Konservierung von Knorpel für plastische Operationen im HNO-Bereich, *HNO*, 6, 124, 1957.
53. Carstensen, G., Eine neue Methode der Gefäßkonservierung durch Einbettung in einen schnellhärtenden Kunststoff, *Chirurg.*, 31, 49, 1960.

54. Cain, H. and Carstensen, G., Morphologische Grundlagen für die Verwendung einer neuartigen "Gefäßbank" im Tierexperiment, *Langenbecks Arch. Chir.*, 296, 88, 1961.
55. Heidsieck, C., Bässler, R., and Kohn, J., Experimentelle Untersuchungen zur Frage der Knochenkonservierung in Kunststoff, *Dtsch. Zahnarztl. Z.*, 22, 518, 1967.
56. Peters, G., Biehl, G., and Hanser, U., Experimentelle Untersuchungen über die Wärmeentwicklung im Knochen bei der Polymerisation von Polymethyl-Methacrylat, *Saarl. Ärztebl.*, 23, 637, 1970.
57. Eggert, A., Huland, H., Ruhnke, J., and Seidel, H., Der Übertritt von Methylmethacrylat-Monomer in die Blutbahn des Menschen nach Hüftgelenksersatzoperationen, *Chirurg.*, 45, 236, 1974.
58. Kepes, E. R., Underwood, P. S., and Becsey, L., Intraoperative death associated with acrylic bone cement. Report of 2 cases, *JAMA*, 222, 576, 1972.
59. Brittain, G. J. C. and Ryan, D. J., Hypotension and methylmethacrylate cement, *Br. Med. J.*, 4, 667, 1972.
60. Pahuja, K., Lowe, H., and Chand, K., Blood methylmethacrylate levels in patients having prothetic joint replacement, *Acta Orthop. Scand.*, 45, 737, 1974.
61. Desmonts, J. M., Meilhan, E., and Duvaldestin, P., Modifications hemodynamiques et respiratoires consecutives au scellement des protheses totales de hanche, *Anesth. Analg. Reanim.*, 33, 251, 1976.
62. Sauerwein, E., *Zahnerhaltungskunde*, Thieme, Stuttgart, 1972.
63. Wilson, R. H. and McCormick, W. E., Plastics: the toxicology of synthetic resins, *Arch. Indust. Health*, 21, 536, 1960.
64. Hyland, J. and Robins, R. H., Cardiac arrest and bone cement, *Br. Med. J.*, 4, 176, 1970.
65. Cohen, C. A. and Smith, T. C., The intraoperative hazard of acrylic bone cement, *Anaesthesiology*, 35, 547, 1971.
66. Ling, R. S. M. and James, M. L., Blood pressure and bone cement, *Br. Med. J.*, 2, 404, 1971.
67. Michelinakis, E., Morgan, R. H., and Curtis, P. J., Circulatory arrest and bone cement, *Br. Med. J.*, 3, 639, 1971.
68. Philipps, H., Cole, P. V., and Lettin, A. W. F., Cardiovascular effects of implanted acrylic bone cement, *Br. Med. J.*, 3, 460, 1971.
69. Schulitz, K. P., Koch, H., and Dustmann, H. O., Lebensbedrohliche Sofortkomplikationen durch Fettembolie nach Einsetzen von Totalendoprothesen mit Polymethylmethacrylat, *Arch. Orthop. Unfallchir.*, 71, 307, 1971.
70. Dustmann, H. O., Schulitz, K. P., and Koch, H., Fettembolie nach Anwendung von Knochenzement bei Hüftgelenkersatz, *Arch. Orthop. Unfall-Chir.*, 72, 114, 1971.
71. Ellis, R. H. and Mulvein, J., Total replacement of the hip, *Br. Med. J.*, 2, 528, 1972.
72. Newens, A. F. and Volz, R. G., Severe hypotension during prosthetic hip surgery with acrylic bone cement, *Anaesthesiology*, 36, 298, 1972.
73. Peebles, D. J., Ellis, R. H., Stride, S. D. K., and Simpson, B. R. J., Cardiovascular effects of methylmethacrylate cement, *Br. Med. J.*, 1, 349, 1972.
74. Phillips, H. and Dandy, D., Total replacement of the hip, *Br. Med. J.*, 2, 713, 1972.
75. Gürtner, T., Sommerlad, W., and Vittali, P., Klinische Erfahrungen über intraoperative Herz- und Kreislaufbelastung durch Implantation von Polymethylmethacrylaten. Kongreßbericht I, Internationaler Kongreß für Prothesentechnik und funktionelle Rehabilitation, Wien, 1973.
76. Nicholson, M., Cardiac arrest following acrylic-cement implants, *Anesth. Analg.*, 532, 298, 1973.
77. Lefaux, R., *Chemie und Toxikologie der Kunststoffe*, Krausskopf, Mainz, 1966.
78. Milne, I. S., Hazards of acrylic bone cement. A report of two cases, *Anesthesia*, 28, 538, 1973.
79. Mohr, H. J., Pathologische Anatomie und kausale Genese der durch selbstpolymerisierendes Methacrylat hervorgerufenen Gewebsveränderungen, *Z. Gesamte Exp. Med.*, 130, 41, 1958.
80. Ullmann, *Enzyklopädie der technischen Chemie*, Urban und Schwarzenbeck, Munchen 12, 391, 1960.
81. Beilstein, *Organische Chemie*, Vol. 3, 4th ed., Ergänzungsband, Bd. 2, 2. Teil, Springer, Berlin 1961.
82. Weast, R. C., *Handbook of Chemistry and Physics*, The Chemical Rubber Co., Cleveland, 49, 1d, 1968/69.
83. Römpp, H., *Chemie Lexikon*, Vol. 3, 5th ed., Franckh., Stuttgart, 1962.
84. Ohnsorge, J. and Holm, R., Änderungen der Spongiosafeinstruktur unter dem Einfluß des auspolymerisierenden Knochenzementes, *Z. Orthop.*, 107, 405, 1970.
85. Ohnsorge, J. and Goebel, G., Die Verwendung unterkühlter Metallendoprothesen in der Hüftchirurgie, *Z. Orthop.*, 107, 683, 1970.
86. Hessert, G. R., Bruchfestigkeit und Struktur des Knochenzementes Palacos nach Zusatz von Gentamycin-Sulfat, *Arch. Orthop. Unfall-Chir.*, 69, 289, 1971.
87. Puhl, W. and Schulitz, K. P., Morphologische Untersuchungen über Polymerisation von Knochenzement, *Arch. Orthop. Unfall-Chir.*, 69, 300, 1971.
88. Kutzner, F., Dittmann, E. Ch., and Ohnsorge, J., Restmonomerabgabe von abgehärtetem Knochenzement, *Arch. Orthop. Unfall-Chir.*, 79, 247, 1974.

89. Ketterl, W. and Kierse, H., Mechanische Untersuchungen über Palavit, *Dtsch. Zahnärztl. Z.*, 7, 745, 1952.
90. Schlag, G., Dingeldein, E., Weisse, G., Regele, H., and Sommoggy, St. V., Tierexperimentelle Untersuchungen mit Knochenzement, 1st. Int. Kongr. Prothesentechnik funkt. Rehabil., Wien, 1973.
91. Homsy, C. A., Tullos, H. S., and King, J. W., Evaluation of rapid-cure acrylic for prothesis stabilisation, *Clin. Orthop.*, 67, 196, 1969.
92. Wenzel, H., Garbe, A., and Nowak, H., Experimentelle Untersuchungen zur Pharmakokinetik von Monomethylmethacrylat. 1st Int. Kongr. Prothesentechnik funkt. Rehabil., Wien, 1973.
93. Homsy, C. A., Tullos, H. S., Anderson, M. S., Diferrante, N. M., and King, J. W., Some physiological aspects of prothesis stabilisation with acrylic polymer, *Clin. Orthop.*, 83, 317, 1972.
94. D'Hollander, A. A., Monteny, E., Hooghe, L., Camu, F., Donckerwolcke, M., Wagner, J., and Brauman, H., Cardiovascular effects of methacrylate methyl monomer intravenous injections in dogs, *Acta Anaesthesiol. Belg.*, 27/Suppl., 75, 1976.
95. Smith, D. C. and Bains, M. E. D., The detection and estimation of residual monomer in polymethylmethacrylate, *J. Dent. Res.*, 35, 16, 1956.
96. Henkel, G., Über die Höhe der Restmonomerabgabe bei verschiedenen Kunststoffen, *Dtsch. Zahn-Mund-Kiefer-Heilkd.*, 35, 377, 1961.
97. Blumler, G., Vergleichende Untersuchungen über Art und Menge des Restmonomergehaltes bei dem Prothesenwerkstoff Piacryl/M., *Dtsch. Stomatol.*, 15, 651, 1965.
98. Köning, K., Die Restmonomerabgabe bei kieferorthopädischen Apparaten aus schnellhärtendem Kunststoff, *Dtsch. Stomatol.*, 16, 816, 1966.
99. Karalnik, D. M., The residual monomer content in the acrylic filling material, *Stomatologiia (Moskau)*, 47, 37, 1968.
100. Roggatz, J. and Ullmann, G., Tierexperimentelle Untersuchungen über die Reaktion des Weichteillagers auf flüssiges auspolymerisiertes Palacos, *Arch. Orthop. Unfall-Chir.*, 68, 282, 1970.
101. Ohnsorge, J. and Goebel, G., Oberflächentemperatur des abhärtenden Knochenzements Palacos beim Verankern von Metallendoprothesen im Oberschenkelmarkraum, *Arch. Orthop. Unfall-Chir.*, 67, 89, 1969.
102. Münzenberg, K. J., Submikroskopische Veränderungen des Knochens durch Hitze und Palacos, Dtsch. Orthop. Ges., 57 Congr., Verh. Enke, Stuttgart, 1971.
103. Paulansky, R. and Gabrielova, Z., Snizeni exothermicke reakce u kostniko cementu pomoci catgutu, *Acta Chir. Orthop. Traumatol. Cech.*, 40/5, 388, 1973.
104. Ohnsorge, J. and Kroesen, A., Thermoelektrische Temperaturmessungen des abhärtenden Knochenzementes "Palacos", *Z. Orthop.*, 106, 476, 1969.
105. Kroesen, A., Experimentelle Untersuchungen zur Bestimmung der Oberflächentemperaturen des auspolymerisierenden Knochenzementes Palacos, Inaug. Diss., Köln 1970.
106. Kuner, E. H., Gewebsreaktionen an der Zement-Knochengrenze, Dtsch. Orthop. Ges., 57 Congr., Verh. Enke, Stuttgart, 1971.
107. Hoffmann, G., Feingewebliche Untersuchungen zur Verträglichkeit von Palavit im Tierversuch, *Verh. Dtsch. Orthop. Ges.*, 86, 352, 1954.
108. Idelberger, M. K., Die Kunststoffprothese in der Hüftchirurgie, *Rev. Chir. Orthop.*, 42, 311, 1956.
109. Pfeiffer, R., Fusion der Wirbelsäule mit dem Autopolymerisat Palacos®, *Arch. Orthop. Unfall-Chir.*, 62, 250, 1967.
110. Deichmann, W. B., Toxicity of methyl, ethyl and n-butyl methacrylate, *J. Ind. Hyg.*, 23, 343, 1941.
111. Deichmann, W. B. and Leblanc, T. J., Determination of the approximate lethal dose with about six animals, *J. Ind. Hyg.*, 25, 415, 1943.
112. Spealman, C. R., Main, R. J., Haag, H. B., and Larson, P. S., Monomeric methyl methacrylate. Studies on toxicity, *Ind. Med.*, 14, 292, 1945.
113. Deichmann, W. B. and Mergard, E. G., Comparative evaluation of methods employed to express the degree of toxicity of a compound, *J. Ind. Hyg.*, 30, 373, 1948.
114. Spector, W. S., *Handbook of Toxicology*, Vol. 1, Saunders, Philadelphia, 1956.
115. Hattemer, A. J., Über die Wirkung einiger Kunststoffe im Gewebe, *Dtsch. Zahnärztl. Z.*, 11, 924, 1956.
116. Castellino, N., and Colicchio, G., Ricerche sperimentali sulla tossicita acuta del metacrilato di metile, *Folia Med. (Napoli)*, 52, 337, 1969.
117. Kutzner, F., Dittmann, E. Ch., and Ohnsorge, J., Atemeffekte durch Knochenzement auf Methylmethacrylatbasis, *Z. Orthop.*, 112, 1053, 1974.
118. Treon, J. F., Sigmon, H., Wright, H., and Kitzmiller, K. V., The toxicity of methyl and ethyl acrylate, *J. Ind. Hyg.*, 31, 317, 1949.
119. Harris, D. K., Health problems in the manufacture and use of plastics, *Br. J. Ind. Med.*, 10, 255, 1953.

120. Karpov, B. D., Effect of small concentrations of methyl methacrylate vapors on the inhibition and stimulation process of the cortex and brain, *Tr. Leningr. Sanit. Gig. Med. Inst.*, 14, 43, 1953.
121. Karpov, B. D., Methyl methacrylate from the view-point of labor hygiene, *Gig. Sanit.*, 10, 25, 1954.
122. Oettel, H., Gesundheitsgefährdung durch Kunststoffe? *Naunyn-Schmiedeberg's Arch. Exper. Path.*, 232, 77, 1958.
123. Elkins, H. B., *The Chemistry of Industrial Toxicology*, Chapman and Hall, London, 1959.
124. Turnbull, K. W., Berezowskyj, J. L., Poulsen, J. B., and Root, L. S., General anaesthesia and total hip replacement, *Can. Anaesth. Soc. J.*, 21, 546, 1974.
125. Filatova, V. I., Data for substantiating the maximum permissible concentration of methyl methacrylate in the atmospheric air, *Gig. Sanit.*, 11, 3, 1962.
126. Wirth, W., Hecht, G., and Gloxbuber, Ch., *Toxikologie Fibel,* Thieme, Stuttgart, 1967.
127. Osinteva, V. P., Bepalko, L. E., and Zubets, A. M., Methylacrylate action on the organism of albino rats (morphological investigation), *Farmakol. Toksikol.*, 33, 631, 1970.
128. Daniel, W. W., Coventry, M. B., and Miller, W. E., Pulmonary complications after total hip arthroplasty with Charnley prothesis as revealed by chest roentgenograms, *J. Bone Jt. Surg. Am. Vol.*, 54, 282, 1972.
129. Strack, R., Für und Wider den Kunststoff, *Dtsch. Zahnärztl. Z.*, 2, 370, 1947.
130. Virenque, M. and Leroux, R., Les résines acryliques en prothese et en biologie, *Presse Med.*, 64, 736, 1947.
131. Debrunner, H. U., Die Verträglichkeit von Polymethylmethacrylat (Plexiglas), *Z. Orthop.*, 83, 557, 1953.
132. Hulliger, L., Untersuchungen über die Wirkung von Kunstharzen (Palacos und Ostamer) in Gewebekulturen, *Arch. Orthop. Unfall-Chir.*, 54, 581, 1962.
133. Schachtschabel, D. O. and Blencke, B. A., Effect of pulverized implantation materials (plastic and glass ceramic) on growth and metabolism of mammalian cell cultures, *Eur. Surg. Res. (Basel),* 8/1, 71, 1976.
134. Hoppe, W., Tierexperimentelle Untersuchungen über Gewebsreaktionen auf Injektionen von autopolymerisierendem Kunststoff, *Dtsch. Zahnärztl. Z.*, 11, 837, 1956.
135. Borchard, U. and Böhling, H. G., Investigation of the neuropharmacology of monomeric methylmethacrylate, *Naunyn-Schmiedebergs Arch. Pharmacol.*, Suppl. 293, R 12, 1976.
136. Stinson, N. E., Tissue reaction to polymethylmethacrylate in rats and guinea pigs, *Nature (London),* 188, 678, 1960.
137. Yodh, S. B. and Wright, R. L., Experimental evaluation of synthetic adhesives by intra-arterial injection, *Neurochirurgia (Stuttg.),* 13, 118, 1970.
138. Dürr, W., Experimentelle Untersuchungen zur Beeinflussung des Knochenanbaus durch Methylmethacrylat (Palacos®), *Langenbecks Arch. Klin. Chir.*, 327, 854, 1970.
139. Dürr, W., Knochenanbau und Knochenzement, Med. Habil., F.U. Berlin, 1971.
140. Yablon, I. G., The effect of methylmethacrylate on fracture healing, *Clin. Orthop.*, 114, 358, 1976.
141. Zschiedrich, W., Über Rapid-Paladon und Rapid-Palapont, *Zahnärztl. Rundsch.*, 59, 7, 1950.
142. Overdiek, H. F., Palavit, *Zahnärztl. Welt,* 7, 302, 1952.
143. Mohr, H. J., Gewebsschädigung durch Polymethacrylsäuremethylester, eines für Plastiken, Prothesen und Zahnfullungen gebrauchlichen Kunststoffes, *Verh. Dtsch. Ges. Pathol.*, 39, 212, 1956.
144. Collins, D. H., Tissue changes in human femurs containing plastic appliances, *J. Bone Jt. Surg. Br. Vol.*, 36, 458, 1954.
145. Levy, L. Y., Lipscomband, C. P., and Medonald, H. C., Complications of Judet arthroplasty due to foreign-body-reaction to Nylon-prothesis, *J. Bone Jt. Surg. Am. Vol.*, 36, 1175, 1954.
146. Gierhake, F. W., Histologische Befunde nach Hüftgelenksplastik mittels Judeth-Prothese, *Langenbacks Arch. Klin. Chir.*, 284, 705, 1956.
147. Knight, G., Paraspinal acrylic inlays in the treatment of cervical and lumbar spondylosis and other conditions, *Lancet,* II, 147, 1959.
148. Mittelmeier, H., Gewebereaktionen bei der Allo-Arthroplastik des Huftgelenks, *Langenbecks Arch. Klin. Chir.*, 306, 163, 1964.
149. Willert, H.-G. and Schreiber, A., Unterschiedliche Reaktionen von Knochen- und Weichteillager auf autopolymerisierende Kunststoffimplantate, *Z. Orthop.*, 106, 231, 1969.
150. Cotta, H. and Schulitz, K. P., Komplikationen der Hüftalloarthroplastik durch periartikuläre Gewebereaktionen, *Arch. Orthop. Unfall-Chir.*, 69, 39, 1970.
151. Willert, H.-G. and Puls, P., Die Reaktion des Knochens auf Knochenzement bei der Allo-Arthoplastik der Hüfte, *Arch. Orthop. Unfall-Chir.*, 72, 33, 1972.
152. Fisher, A. A., Allergic sensitization of the skin and oral mucosa to acrylic denture materials, *J. Prosthet. Dent.*, 6, 593, 1956.
153. Pegum, J. S. and Medhurst, E. A., Contact dermatitis from penetration of rubber gloves by acrylic monomer, *Br. Med. J.*, 2, 141, 1971.

154. Fries, I. B., Fisher, A. A., and Salvati, E. A., Contact dermatitis in surgeons from methylmethacrylate bone cement, *J. Bone Jt. Surg. Am. Vol.*, 57, 547, 1975.
155. Strain, J. C., Reactions with acrylic denture base resins, *J. Prosthet. Dent.*, 18, 465, 1967.
156. Axelsson, B., Degree of polymerization of methylmethacrylate in relation to curing time and temperature, *Odontol. Rev.*, 6, 124, 1955.
157. Smith, D. C., The acrylic denture base. Some effects of residual monomer and peroxide, *Br. Dent. J.*, 106, 331, 1959.
158. Koppang, R., Standardized test methods for dental prothetic materials, *Odontol. Tidskr.*, 74, 240, 1966.
159. Koppang, R., Standardization of dental prosthetic materials, *Acta Odontol. Scand.*, 27, 129, 1969.
160. Lawrence, W. H., Bass, G. E., Purcell, W. P., and Autian, J., Use of mathematical models in the study of structure-toxicity relationships of dental compounds. I. Esters of acrylic and methacrylic acids, *J. Dent. Res.*, 51, 526, 1972.
161. Borzelleca, J. F., Larson, P. S., Henningar, C. R., Jr., Huf, E. C., Crawford, E. M., and Smith, R. B., Jr., Studies on the chronic oral toxicity of monomeric ethyl acrylate and methyl methacrylate, *Toxicol. Appl. Pharmacol.*, 6, 29, 1964.
162. Singh, A. R., Lawrence, W. H., and Autian, J., Embryonic-fetal toxicity and teratogenic effects of a group of methacrylate esters in rats, *J. Dent. Res.*, 51, 1632, 1972.
163. Park, W. Y., Balingit, P., Kenmore, P. I., and Macnamara, T. E., Changes in arterial oxygen tension during total hip replacement, *Anaesthesiology*, 39, 642, 1973.
164. Lawrence, W. H. and Autian, J., Possible toxic effects from inhalation of dental ingredients by alteration of drug biological half-life, *J. Dent. Res.*, 51, 57, 1972.
165. Mohr, H. J., Das Verhalten der Methacrylate zum Gewebe, *J. Med. Kosmet.*, 56, 235, 1956.
166. Böhling, H. G., Borchard, U., and Drouin, H., Monomeric methylmethacrylate (MMA) acts on the desheathed myelinated nerve and on the node of Ranvier, *Arch. Toxicol.*, 38, 307, 1977.
167. Borchard, U., Untersuchungen zur Neurotoxizität des Tuberkulostatikums Ethambutol, Diss. Med., Köln, 1976.
168. Ulbricht, W., The effect of veratridine on excitable membranes of nerve and muscle, *Ergeb. Physiol.*, 61, 18, 1969.
169. Hille, B., Woodhull, A., and Shapiro, B. I., Negative surface charge near sodium channels of nerve: divalent ions, monovalent ions, and pH., *Philos. Trans. R. Soc. London*, 270, 301, 1975.
170. Seeman, P., The membrane action of anesthetics and tranquilizers, *Pharmacol. Rev.*, 24, 583, 1972.
171. White, S. H., The lipid bilayer as a solvent for small hydrophobic molecules, *Nature (London)*, 262, 421, 1976.
172. McLaughlin, S., Two different mechanisms by which the hydrophobic adsorption of ions to membranes produces an electrostatic potential, in *Electrical Phenomena at the Biological Membrane Level*, Roux, E., Ed., Elsevier, Amsterdam, 1977.
173. Borchard, U., The isolated retina as a model for the investigation of the effects of drugs on central nervous tissues, in Proc. 33rd Int. Congr. Pharm. Sci., 1973, 24.
174. Borchard, U. and Schneider, K. U., Intoxication, detoxication and copper storage of central nervous tissue at different external Cu(II)-concentrations, *Arch. Toxicol.*, 33, 17, 1974.
175. Borchard, U. and Erasmi, W., Na^+-diffusion in the retinal tissue of the frog, *Vision Res.*, 14, 17, 1974.
176. Borchard, U., and Böhling, H. G., Investigation of the neuropharmacology of monomeric methylmethacrylate, *Naunyn-Schmiedeberg's Arch. Pharmacol. Suppl.*, 293, R 12, 1976.
177. Borchard, U. and Böhling, H. G., Investigation of the neurotoxicity of monomeric methylmethacrylate and homologous chemicals, *Acta Pharmacol. Toxicol.*, 41, (Suppl. 2), 421, 1977.
178. Borchard, U., Untersuchungen zum Wirkungsmechanismus und zur pharmakologischen Charakterisierung von Lokalanästhetika, Habil. Med., Düsseldorf, 1978.
179. Kutzner, F., Tierexperimentelle Untersuchungen zur akuten Toxizität, der Wirkung auf die Atmung und der Restmonomerabgabe des Knochenzementes Palacos®, Diss. Med., Köln, 1974.
180. Mir, G. N., Lawrence, W. H., and Autian, J., Toxicological and pharmacological actions of methacrylate monomers. III. Effects on respiratory and cardiocascular function of anesthetized dogs, *J. Pharm. Sci.*, 63, 376, 1974.
181. Mir, G. N., Lawrence, W. H., and Autian, J., Toxicological and pharmacological actions of methacrylate monomers. II. Effects on isolated guinea pig ileum, *J. Pharm. Sci.*, 62, 1258, 1973.
182. McLaughlin, R. E., Difazio, C. A., Hakala, M., Abbot, B., McPhail, J. A., Mack, W. P., and Sweet, D. E., Blood clearance and acute pulmonary toxicity of methylmethacrylate in dogs after simulated arthroplasty and intravenous injection, *J. Bone Jt. Surg. Am. Vol.*, 55, 1621, 1973.
183. Laskin, D. M., Robinson, I. B., and Weimann, J. P., Experimental production of sarcomas by methylmethacrylate implants, *Soc. Exp. Biol. Med.*, 87, 329, 1954.
184. Oppenheimer, B. S., Oppenheimer, E. T., Danishefsky, I., Stout, A. P., and Eirch, F. R., Study of polymer as carcinogenic agents in animals, *Cancer Res.*, 15, 333, 1955.

185. Mitchel, D. F., Shankwalker, G. B., and Shazer, S., Determining the tumorigenicity of dental materials, *J. Dent. Res.*, 39, 1023, 1960.
186. Kallos, T., Impaired arterial oxygenation associated with use of bone cement in the femoral shaft, *Anesthesiology*, 42, 210, 1975.
187. Pliess, G. and Bornemann, G., Oral lesions and atypical growth of epithelial tissue in patients wearing dentures, *Dent. Abstr.*, 3, 215, 1958.
188. Bradford, E. W., Case of allergy to methyl methacrylate, *Br. Dent. J.*, 84, 195, 1948.
189. Tansy, M. F., Landin, W. E., Perrong, H., and Kendall, F. M., Acute and chronic intestinal motor effects of methyl methacrylate vapor, *J. Dent. Res.*, 55(716), 240, 1976.
190. Bright, D. S., Clark, H. G., and McCollum, D. E., Serum analysis and toxic effects of methylmethacrylate, *Surg. Forum*, 23, 455, 1972.
191. Bloch, B., Hakan, J. K., and Hastings, G. W., Evaluation of cold curing acrylic cement for prothesis stabilization, *Clin. Orthop.*, 72, 239, 1970.
192. Kim, K. C. and Ritter, M. A., Hypotension associated with methyl methacrylate in total hip arthroplasties, *Clin. Orthop.*, 88, 154, 1972.
193. Modig, J., Busch, C., Olernd, S., Saldeen, T., and Waernbaum, G., Arterial hypotension and hypoxaemia during hip replacement: importance of thormboplastic products, fat embolism and acrylic monomers, *Acta Anaesthesiol. Scand.*, 19, 28, 1975.
194. Schumacher, K. A., Classen, H. G., Hagedorn, M., Benner, K. U., Späth, M., and Mittermayer, Ch., Effects of anaphylatoxin plasma in cats: hemodynamic changes induced by platelet aggregation, *Drug Res.*, 24, 122, 1974.
195. Modig, J., Busch, C., Olernd, S., and Saldeen, T., Pulmonary microembolism during intramedullary orthopaedic trauma, *Acta Anaesthesiol. Scand.*, 18, 133, 1974.
196. Borchard, U. and Benner, K. U., unpublished data, 1976.
197. Modig, J., Olernd, S., and Malmberg, P., Sudden pulmonary dysfunction in prothetic hip replacement surgery, *Acta Anaesthesiol. Scand.*, 17, 276, 1973.
198. Rådegran, K. and McAslan, C., Circulatory and ventilatory effects of induced platelet aggregation and their inhibition by acetylsalicylic acid, *Acta Anaesthesiol. Scand.*, 16, 76, 1972.
199. Pelling, D. and Butterworth, K. R., Cardiovascular effects of acrylic bone cement in rabbits and cats, *Br. Med. J.*, 2, 638, 1973.
200. Choffat, P., Delagoutte, J. P., Koff, G., and Leveaux, G., Perturbations clinique et biologiques per opératoires induites par les ciments acryliques, *Rev. Chir. Orthop.*, 61/2 Suppl., 199, 1975.
201. Gresham, G. A. and Kuczinski, A., Cardiac arrest and bone cement, *Br. Med. J.*, 3, 465, 1970.
202. Sevitt, S., Fat embolism in patients with fractured hips, *Br. Med. J.*, 2, 257, 1972.
203. Gresham, G. A., Kuczynski, A., and Rosborough, D., Fatal fat embolism following replacement arthroplasty for transcervical fractures of femur, *Br. Med. J.*, 2, 617, 1971.
204. Dandy, D. J., Fat embolism following prothetic replacement of the femoral head, *Injury*, 3, 85, 1971.
205. Ratliff, A. H. C. and Clement, J. A., Pulmonary embolism and bone cement, *Br. Med. J.*, 2, 532, 1971.
206. Burgess, D. M., Cardiac arrest and bone cement, *Br. Med. J.*, 3, 588, 1970.
207. Harris, N. H., Cardiac arrest and bone cement, *Br. Med. J.*, 3, 523, 1970.
208. Tronzo, R. G., Kallos, T., and Wyche, Q., Elevation of intramedullary pressure when methylmethacrylate is inserted in total hip arthroplasty, *J. Bone Jt. Surg. Am. Vol.*, 56, 714, 1974.
209. Whitenack, S. H. and Hausberger, F. X., Intravasation of fat from the bone marrow cavity, *Am. J. Pathol.*, 65, 335, 1971.
210. Thomas, T. A., Sutherland, I. C., and Waterhouse, T. D., Cold curing acrylic bone cement. A clinical study of the cardiovascular side effects during hip joint replacement, *Anaesthesia*, 26, 298, 1971.
211. Durbin, F. C., Jeffery, C. C., Blundell Jones, G., Ling, R. S. M., Scott, P. J., Woodyard, J. E., and Wrighton, J. D., Cardiac arrest and bone cement, *Br. Med. J.*, 4, 176, 1970.
212. Coventry, M. B., Beckenbangh, R. D., Nolan, D. R., and Alstrup, D. M., 2,012 total hip arthroplasties: a study of postoperative course and early complications, *J. Bone Jt. Surg. Am. Vol.*, 56, 273, 1974.
213. Karcently, L., Dossa, J., Sassine, A., Masse, C., and Baillat, X., Etude expérimentelle des effets cardiaques du métacrylate de méthyle, *Lyon Chir.*, 72/5, 355, 1976.
214. Mir, G. N., Lawrence, W. H., and Autian, J., Toxicological and pharmacological actions of methacrylate monomers. I. Effects on isolated, perfused rabbit heart, *J. Pharm. Sci.*, 62, 778, 1973.
215. Coventry, F. R., Gunn, D. R., Hughes, J. D., and Martin, W. E., The systemic manifestations of polymethylmethacrylate, *J. Bone Jt. Surg. Am. Vol.*, 55(2), 419, 1973.
216. Phillips, H., Lettin, A. W. F., and Cole, P. V., Cardiovascular effects of implanted acrylic cement, *J. Bone Jt. Surg. Br. Vol.*, 55, 210, 1973.

217. **Cadle, D., James, M. L., Ling, R. S. M., Piper, R. F., Pryer, D. L., and Wilmshurst, C. C.**, Cardiovascular responses after methylmethacrylic cement, *Br. Med. J.*, 4, 107, 1972.
218. **Frost, P. M.**, Cardiac arrest and bone cement, *Br. Med. J.*, 3, 524, 1970.
219. **Powell, J. N., McGrath, P. J., Lahiri, S. K., and Hill, P.**, Cardiac arrest associated with bone cement, *Br. Med. J.*, 3, 326, 1970.
220. **Parsons, D. W.**, Cardiac arrest and bone cement, *Br. Med. J.*, 3, 710, 1970.
221. **Schuh, F. T., Schuh, S. M., Viguera, M. G., and Terry, R. N.**, Circulatory changes following implantation of methylmethacrylate bone cement, *Anesthesiology*, 39, 455, 1973.
222. **Ellis, R. H., and Mulvein, J.**, The cardiovascular effects of methylmethacrylate, *J. Bone Jt. Surg. Br. Vol.*, 56, 59, 1974.
223. **McMaster, W. C., Bradley, G., and Wangh, T. R.**, Blood pressure lowering effect of methylmethacrylate monomer, *J. Bone Jt. Surg. Am. Vol.*, 55(2), 419, 1973.

Chapter 7

THE TOXICOLOGY OF ADDITIVES IN MEDICAL PLASTICS

David F. Williams

TABLE OF CONTENTS

I.	Introduction	144
II.	Commercially Used Additives and Their Toxicology	145
	A. Plasticizers	146
	B. Stabilizers	146
	C. Antioxidants	148
	D. Fillers	149
	E. Catalysts	151
III.	Toxicity and Clinical Effects of Phthalic Acid Ester Plasticizers	151
	A. General	151
	B. Di-2-ethylhexylphthalate (DEHP)	152
	1. Extraction of DEHP from PVC Medical Devices	153
	2. Toxicology of DEHP	153
IV.	Conclusions	154
References		154

I. INTRODUCTION

As noted in Chapter 4, modern plastics are complex materials, containing not only the organic macromolecules that impart the basic characteristics, but also a variety of substances, most of which are present by intention but certainly not all, and which modify either the processing or end-use performance. Nonintentionally introduced substances include, primarily, the residues of raw materials used in the preparation of the polymer. The monomers are clearly the most significant of these and many polymer-based structures contain measurable quantities of unreacted monomers. It is also possible for there to be traces of catalysts remaining. The intentionally introduced substances are collectively termed additives. The more important classes of these are listed in Table 1, along with the amounts of each used in the U.S. in 1974.[1] Some of these additives are used to facilitate processing, as with lubricants and blowing agents. The majority are used to confer specific properties to the resulting material, the terms used in Table 1 adequately describing the functions of these additives.

The very nature of the majority of these additives suggests that they may pose toxicological hazards. The principle characteristics of a polymer are derived from its macromolecular structure, but this same structure can also provide difficulties such as an excessive rigidity, a susceptibility to oxidation, and poor rheological characteristics on processing. The only way in which these undesirable features can be obviated, or at least minimized, is to add a lower molecular size substance which can act as a plasticizer to make a rigid material more flexible, as an antioxidant to preferentially participate in oxidation processes by soaking up free radicals, or as a lubricant to facilitate thermoplastic flow. Since it is well known that, in general, toxicity of organic substances increases with decreasing molecular size (or weight), the conversion of a pure high molecular weight polymer to a compounded plastic with appreciable quantities of low molecular weight additives can only increase the toxicological hazard.

The toxicity of the additives has widespread implications in the plastics industry at large, and industrial and occupational toxicologists are concerned about the levels of such additives in the environment at processing plants. It also has significant implications in relation to the use of plastics in medical and paramedical applications when the finished plastic product has contact with tissues of the body.

With this problem in mind, and considering the large variety of polymers now readily available, it would appear an easy solution to simply use pure, additive-free, polymers in all of these clinical applications. That is rarely a feasible proposition, however, on either economic or technical grounds. The total volume of plastics used for medical devices, although increasing, is insignificant in terms of the total volume of plastics used industrially. In only a few situations is it possible to formulate a plastic for a special application where all the additives are specified. It is more usual to select a plastic from a commercially available range on the basis of broad mechanical, physical, and chemical specifications. Only rarely would the economic situation allow the production of a small volume of a special plastic. As implied earlier, additives perform vital functions and in most cases, additive-free material would possess very inferior properties and may be impossible to process.

It is an accepted fact, therefore, that the majority of plastics used in medicine must contain some additives. Even the so-called medical-grade polymeric materials may contain essential additives, the term medical-grade implying only that the additives meet certain requirements and specifications and that usually the material is processed under clean conditions. Silicone rubber, well known for many years as an acceptable implant material, has very poor mechanical properties and has to contain fillers if it is to possess adequate strength and tear resistance. The real significance of the presence of the finely divided silica, used as a filler in a toxicological sense, has only been appreciated

Table 1
MAJOR ADDITIVES IN PLASTICS[1]

	Amount used in 10⁶ kg, in U.S.A., 1974
Plasticizers	750
Flame retardants	170
Colorants	140
Heat stabilizers	42
Lubricants	30
Antioxidants	14
Organic peroxides	12
Blowing agents	6
Antistatic agents	2
Ultraviolet stabilizers	2

in recent years with the possible adverse response to this silica in the tissue surrounding breast prostheses.

If it is necessary to prepare medical plastics that do contain additives, considerable care must be taken over their choice. From the biocompatibility point of view, three factors are important:

1. The rate at which an additive leaches out of the plastic
2. The effect that the loss of additives has on the plastic
3. The toxicity of the additive

The first two of these are, of course, very much dependent on the nature of the polymer and are difficult to discuss in general. The case of polyvinyl chloride (PVC), to be described in detail later, illustrates this point. PVC is a rigid polymer with few uses by itself. However, when compounded with a significant amount of a plasticizer, it becomes flexible and may be used in catheters and similar devices. There has been much discussion about the toxicity of the plasticizer used and the length of time which such catheters can be left in the body. Since a plasticizer is not chemically bonded to the molecular chains of the PVC it can migrate, and certainly fatty substances in the tissues are capable of extracting many of the plasticizers used. As described by Biggs and Baldwin,[2] one of the ways out of the additive-toxicity dilemma with flexible PVC is to use compounds that exhibit a very low degree of plasticizer migration. Since up to 50% of the total volume of flexible PVC may consist of plasticizers, its continued loss will seriously affect the flexibility of a catheter, and it is a well-known fact that such catheters left in tissues for a substantial period of time will become rigid, therefore affecting the performance and biocompatibility. This again clearly indicates the desirability of low rates of additive migration.

The third of the factors mentioned above, the toxicity of additives, is clearly of greatest significance in relation to the biocompatibility, and the remainder of this chapter will be devoted to a discussion of the available information.

II. COMMERCIALLY USED ADDITIVES AND THEIR TOXICOLOGY

In this section different classes of additives are discussed in relation to their known toxicology. The additives discussed are not confined to those known to be used in medical plastics at the moment, but include the general range of commercially used additives. This is done since formulations are continually changing, and the general pattern of toxicology is best seen by a wider review.

It should be borne in mind that the data given in this section is largely derived from experience with industrial toxicology or from experiments with no direct relevance to biomaterials usage. It will become quite apparent from the discussion of phthalate esters in Section III that this type of data only provides the broadest of assessments and cannot indicate whether subtle changes in tissues will arise from the presence of minute quantities of an additive leached from an implanted or extracorporeal plastic device. If a substance is designated even mildly toxic in this section it is highly probable that it would significantly influence the tissue response to, and overall biocompatibility of the plastic material in which it is present. If there is no data available on its toxicity, or if animal experiments involving ingestion of a compound prove negative, there is no guarantee that it will be harmless under the conditions under discussion.

A. Plasticizers

A plasticizer is a substance that is added to a polymer to increase its flexibility. Most thermoplastics are relatively flexible because of the presence of only weak secondary valence bonds between the molecular chains which allow these chains to slide past each other readily. However, some thermoplastics and especially those with highly polar side groups, display stronger intermolecular attractive forces and reduced intermolecular sliding. A plasticizer acts like a solvent, penetrating the polymer and lowering the effective intermolecular cohesion. Cellulose nitrate was the first material to require a plasticizer and camphor was used for this purpose over 100 years ago. Today vinyl resins, and especially PVC, are very widely used materials which depend heavily on plasticizers for their success. About 70% of the total production of plasticizer is used in PVC.[3]

The requirements of a plasticizer are that it should be compatible with the polymer and impart reasonable flexibility. It is especially important in medical devices that the plasticizer should not migrate from the polymer or be extracted by liquids with which it is to come into contact, particularly the fats in tissues which tend to dissolve many organic substances. Good plasticizers generally have a high boiling point, are thermally stable, and resist hydrolysis.

There are many different commercially available plasticizers but the majority are either phosphoric esters, phthalic esters, esters of aliphatic acids such as adipates, sebacates, and ricinoleates, chlorinated aromatic compounds, or derivatives of glycols, all of these being monomeric. Di-2-ethylhexyl phthalate is the most widely used, accounting for 25% of plasticizers production. Tricresyl phosphate is a further widely used monomeric plasticizers. It is also possible to have polymeric plasticizers and some polyesters and butadiene-acrylonitrile copolymers are used for this purpose. There is some interest in these compounds for medical devices since they should be more resistant to extraction than the monomeric type.

Since most plasticizers are organic liquids and since they may comprise 35 or more per cent of a medical plastic, there is naturally some concern over their toxicity. A summary of the more important currently used plasticizers and their known toxicity is given in Table 2.

B. Stabilizers

As described by Gilding in *Fundamental Aspects of Biocompatibility*,[31] polymers are inherently susceptible to environmental degradation. Of particular importance are the effects of heat, oxygen, and light (especially ultraviolet light). While such environmental agencies are not particularly hostile under the conditions of implantation, most medical plastics will encounter these conditions at some stage. Processing of thermoplastics involves heat and will usually involve heat in the presence of oxygen. Moreover, all medical devices have to be sterilized and heat sterilization is very convenient. It is also difficult to avoid exposure to light during storage.

Table 2
TOXICOLOGY OF SOME COMMON PLASTICIZERS

Plasticizer	Toxicology
Ethylene glycol CH_2OHCH_2OH	Oral LD_{50} (rat), 5840 mg/kg; i.v. LD_{50} (mouse) 3000 mg/kg; i.p. LD_{50} (rabbit) 1000 mg/kg. Causes reversible and irreversible changes on exposed tissue, especially skin and mucous membranes. At low levels stimulates CNS, but depresses CNS function at high levels. Produces kidney failure by deposition of calcium oxalate in tubules, but only at very high doses.
Glycerine $CH_2OHCHOHCH_2OH$	Oral LD_{50} (mouse), 470 mg/kg; s.c. LD_{50} (mouse), 10,000 mg/kg; I.U. LD_{50} (mouse), 4250 mg/kg. Can cause tissue damage following oral, s.c., i.v., and i.p. routes but along with esters such as glycerol triacetate, are normally considered nontoxic and used as food additive.
Camphor $C_{10}H_{16}O$	i.p. LD_{50} (rat), 900 mg/kg; s.c. LD_{50} (mouse) 2200 mg/kg. Local irritant, causes tissue changes following i.p. and s.c. routes. Causes nausea, vomiting, and convulsions when ingested.
Castor oil	Generally nontoxic, a food additive permitted for human consumption.
Butyl acetyl ricinoleate $C_{17}H_{32}(O\text{-}COCH_3)COO\text{-}C_4H_9$	Toxicity unknown but generally thought to be very low. LD_{50} of methyl acetylricinoleate is over 50 g/kg.
Dibutyl phthalate $C_6H_4(CO_2C_4H_9)_2$	Oral LD_{50} (rat), 31,000 mg/kg; dermal LD_{50} (guinea pig), 10,000 mg/kg. Only very slightly toxic, having slight irritant effect on skin. Experimental teratogen. Produces GI symtoms at high oral doses. See Section III. Used as packaging material for nonfatty foods.
Ditridecyl phthalate $C_6H_4(COOC_{13}H_{27})_2$	Little data. Appears to have very low toxicity.
Dioctyl adipate (di-2-ethyl-hexyladipate) $(C_8H_{17}CO_2)_2C_4H_8$	Low acute toxicity for animals in LD_{50} (rat) 9.11 ml/kg. No chronic effects observed on oral administration of diet with 100 mg/kg dioctyl phthalate to rats. Used in food packaging materials in some countries.
Dioctyl sebacate (di-2-ethylhexyl sebacate) $(CH_2)_8(CO_2C_8H_{17})_2$	Oral LD_{50} (rat), 1280 mg/kg. Some irritation and reversible or irreversible changes following oral ingestion.
Triphenyl phosphate $PO(OC_6H_5)_3$	Oral LD_{50} (rat) 3000 mg/kg; s.c., LD_{50} (Cat) 100 mg/hg. Highly irritant subcutaneous. Some toxic effects following ingestion. No effect on skin, poorly absorbed. Slight cholinesterase inhibition.
Tricresyl phosphate $(CH_3C_6H_4)_3PO_4$	Oral LD_{50} (human) 1000 mg/kg. Lowest published toxic concentration in humans 6 mg/kg. Ingestion leads to polyneuritis and demyelination which may be fatal. Most toxic of plasticizers.

Table 2 (continued)
TOXICOLOGY OF SOME COMMON PLASTICIZERS

Plasticizer	Toxicology
Chlorinated diphenyls, e.g., monochlorodiphenyl $C_6H_5C_6H_4Cl$	Variable toxicity but all have some effect, usually of two types (1) effect on skin — chloacle; (2) effect on liver — acute yellow atrophy, which may be fatal. The higher the chlorine level in the aromatic derivative molecule, the greater the toxicity. Oral LD_{50} (rat) varies from 1000—12000 mg/kg. Some are experimental carcinogens.
Epoxidized soya bean oil	Little information. Appears nontoxic in low doses but high dietary concentrations in animals show increase in lipid content of blood and modified fat metabolism.

Note: Data obtained from Simonds and Church,[3] Roff and Scott,[4] Lefaux,[5] and Sax.[6]

It is normal for additives, described as stabilizers, to be used to reduce the risk of degradation in susceptible polymers. Heat stabilizers are particularly important in PVC since this polymer readily depolymerizes at elevated temperatures and could not be processed without some protection. The degradation of PVC yields HCl and is an autocatalytic unzipping process. The stabilizers used in this system accept or scavenge the HCl and thus prevent the propagation of the reaction. Alkaline earth stearates were among the first to be used for this purpose and barium-cadmium salts are now among the most important. Examples of the toxicology of these heat stabilizers are given in Table 3.

Light stabilizers protect polymers from photochemical degradation. Since this degradation is largely brought about by ultraviolet light, many stabilizers function by preferentially absorbing light of this wavelength and dissipating the energy through phosphorescence, fluorescence, heat, or simply transferring it to its surroundings. Some examples here include certain salicylates, substituted acrylonitriles, and the o-hydroxy-benzophenones. Alternatively, there are nonabsorbing light stabilizers which photochemically rearrange to become UV-absorbing, which include hexamethylphosphoric triamide and a variety of metal chelates. Of all the common light stabilizers, only a few are listed in either of the works of Sax[6] or Lefaux[5] so that their toxicology is unknown. Phenyl salicylate is used in some plastics. According to Lefaux this is not a very harmful substance and it does not impart toxicity to films in which it is used in concentrations of 1 to 2%. However, Sax describes it as a highly toxic substance with an LD_{Lo} (human) of only 50 mg/kg. Clearly the information available on the toxicity of light stabilizers is very limited.

C. Antioxidants

Many polymers are susceptible to attack by oxygen, the rate of reaction being increased as the temperature is raised. As far as medical devices are concerned, oxidation is most likely during processing, sterilization, and to a lesser extent, storage. Oxidation within the body at physiological temperatures is unlikely to be a problem, although it is known that pure polypropylene will oxidize slowly after implantation by a mechanism that is very similar to auto-oxidation in air at elevated temperatures.[7] The adverse effects of oxidation can be significantly reduced by the presence of antioxidants, which inhibit or retard oxidation processes. Oxidation normally proceeds by a free radical

Table 3
TOXICOLOGY OF SOME HEAT STABILIZERS

Stabilizer	Toxicology
Calcium derivatives of fatty acids e.g., stearate and ricinoleate	No toxic effects. Widely used in food packaging. Not very efficient stabilizers.
Barium derivatives of fatty acids, e.g., stearate, laurate, ricinoleate, 2-ethylhexoate	Not used for food packaging. Not very efficient stabilizers but used synergistically with cadmium derivatives. Toxicology consistent with soluble barium compounds, yielding soluble and toxic barium chloride on accepting Cl⁻ in PVC.
Cadmium derivatives of fatty acids, e.g., stearate, laurate	Oral toxicity of cadmium compounds is high. Severely irritant to tissues, experimental carcinogens.
Lead compounds including normal salts (orthosilicate, stearate) and basic salts which contain reactive uncoordinated PbO (basic lead carbonate, tribasic lead sulphate, dibasic lead stearate)	Lead and its compounds are highly toxic and all lead stabilizers must be considered as potentially harmful if they gain access to the tissues. Actual effects will vary, depending on solubility. Lead stearates, both normal and dibasic are insoluble in water and are recommended for use in drinking water supply pipes. See Sax for details of lead toxicology.
Organotin compounds, e.g., dibutyltin dilaurate, dibutyltin maleate	Many organotin compounds are toxic and irritant especially trialkyl derivatives. Dialkyl derivatives are far less toxic. Dioctytin compounds are virtually insoluble and unabsorbed, hence relatively nontoxic. Dibutyltin derivatives are more irritant: oral LD_{50} (rat) dibutyltin dilaurate, 243 mg/kg.
Organic phosphites, e.g., triphenyl phosphite	Generally have high toxicity: Triphenyl phosphite oral LD_{50} (rat), 1600 mg/kg; i.p. LD_{50} (rat), 250 mg/kg s.c. LD_{LO} (rat), 2000 mg/kg. Higher derivatives, especially with long alkyl chains (e.g., nonyldiphenyl phosphite) considerably less toxic.

Note: Data obtained from Simonds and Church,[3] Roff and Scott,[4] Lefaux,[5] and Sax.[6]

chain reaction in which free radicals (R°) present or generated within a polymer react with oxygen to form peroxy radicals (ROO°) which extract hydrogen atoms to form a hydro-peroxide (ROOH) and another polymer-free radical. The cycle then repeats itself. Antioxidants generally either react preferentially with the peroxy radicals or decompose the hydroperoxides and prevents the formation of additional free radicals. The former type are usually phenols or amines, while the peroxide decomposers are either sulfur compounds or metal complexes such as dithio-phosphates. Some common antioxidants are included in Table 4 along with their toxicology. Since, by definition, these compounds are reactive, a certain degree of toxicity should be expected.

D. Fillers

Virtually all commercially available plastics contain fillers to some extent. This may be for economic reasons, fillers usually being far cheaper than the polymer, or to improve mechanical properties. There are very many different fillers which may be used, the choice being dependent upon the particular requirements. Some are very low cost cellulosic materials such as wood chips, cotton, and sisal. Other mineral fillers

Table 4
TOXICOLOGY OF SOME COMMON ANTIOXIDANTS

Antioxidant	Toxicology
Di-tert-butyl-p-cresol	Oral LD_{50} (rat) 3510 mg/kg, (mouse) 1040 mg/kg. Experimental teratogen, carcinogen, has some irreversible effects after ingestion. However, used as a food additive (less than 0.01%) and as antioxidant in food packaging platics.
Butyl hydroxy anisole	Oral LD_{Lo} (rat), 1000 mg/kg. Low toxicity. No effect on rats kept for 2 years on diet with 0.5% BHA. Used as food additive.
4,4'-thiobis (6-tert-butyl-m-cresol)	Used as antioxidant for polyolefins, including high density polyethylene. i.p. LD_{Lo} (mice), 50 mg/kg — high i.p. toxicity. Oral toxicity much lower.
Dilauryl thiodipropionate	Little information but generally considered to be of low toxicity.
Triphenyl phosphite	Oral LD_{50} (rat), 1600 mg/kg i.p. LD_{50} (rat), 250 mg/kg high acute toxicity.

Note: Data obtained from Simonds and Church,[3] Roff and Scott,[4] Lefaux,[5] and Sax.[6]

Table 5
TOXICOLOGY OF SOME COMMON FILLERS

Filler	Toxicology
Cellulosics, ligeneous compounds, proteinaceous compounds, synthetic polymers (Nylon®, Dacron®)	Not expected to be toxic or irritant although some may be allergens.
Carbon (graphite, carbon black or carbon fibers)	Chemically nontoxic but in finely divided form may be irritant. Under some circumstances may be contaminated with carcinogenic hydrocarbons.
Asbestos	Recognized carcinogen. Prolonged inhalation can cause cancer of the lung, pleura and peritoneum. Irritant to s.c. tissues.
Quartz and other siliceous compounds	As with asbestos, causes premonary irritation on inhalation, but generally thought to be nonirritant by other routes. Used as food additive for animal and humans.
Silicates (mica, clay)	Similar to quartz, initiating pulmonary fibrosis. Soluble alkaline silicates act like mild alkalines producing irritation.

Note: Data obtained from Simonds and Church,[3] Roff and Scott,[4] Lefaux,[5] and Sax.[6]

such as quartz, glass, silicates, or even ceramic whiskers are being used in situations where improved mechanical properties are required. Some examples are included in Table 5 along with their toxicology. A very good example of the potentially irritant behavior of what might seem to be an inert filler is seen here with the case of silica in medical grade silicone rubber. A further interesting example is the use of asbestos fibers in periodontal dressings which, although they increase the cohesiveness of the dressing, result in marked epithelial damage. This has been discussed by Heaney in *Biocompatibility of Dental Materials*.[32]

Table 6
TOXICOLOGY OF SOME COMMON CATALYSTS

Catalyst	Toxicology
Benzoyl peroxide	Irritant to skin and mucous membranes. High toxicity via i.p. and inhalation routes.
Azobis isobutyronitrile	Oral LD_{50} (mouse), 700 mg/kg i.p. LD_{50} (mouse) 25 mg/kg. High toxicity via i.p., moderate toxicity via oral routes. Decomposes to give highly toxic nitriles.
Boron trifluoride	Inhalation LC_{50} (mouse) 3460 mg/kg. Highly toxic via inhalation. Strong irritant.
Titanium tetrachloride	Inhalation LC_{Lo} (mice) 10 mg/m. Highly irritant to skin, eyes mucous membranes.
Antimony trioxide	Details unknown but antimony compounds are usually highly toxic via oral, inhalation and i.p. routes.
Triethylamine	Oral LD_{50} (rat), 460 mg/kg; dermal LD_{50} (rabbit) 570 mg/kg. High toxicity via oral, inhalation routes. Dermal irritation.

Note: Data obtained from Simonds and Church,[3] Roff and Scott,[4] Lefaux,[5] and Sax.[6]

E. Catalysts

Polymerization processes are chemical reactions which do not usually occur spontaneously. Instead, these reactions require the presence of an initiator and possibly other agents to control the kinetics of polymerization and the degree, if any, of cross-linking. These substances are generally called catalysts even though they do not follow the laws of conventional catalysts.

Addition polymerization is normally initiated by a free radical mechanism requiring the presence of a substance which can supply free radicals under the appropriate conditions. Peroxides are very widely used for this purpose, especially benzoyl peroxide, but aliphatic azo compounds are also common. Some ionic species can also initiate addition polymerization, an important cationic catalyst being boron trifluoride. Ziegler catalysts, such as titanium tetrachloride are anionic catalysts and are used in the preparation of polyethylene. Examples of these and other types of catalysts are given in Table 6. Although these are very reactive chemicals and, therefore, are likely to be toxic, they are used in small amounts and the probability of toxicity arising from their presence in polymerized material is low.

III. TOXICITY AND CLINICAL EFFECTS OF PHTHALIC ACID ESTER PLASTICIZERS

A. General

As noted in Section II, phthalic acid esters are among the most widely used plasticizers in the plastics industry. They were first used to replace camphor for plasticizing cellulose nitrate some 60 years ago. During the 1930s when PVC was developed and replaced cellulose nitrate in many applications, the phthalates found even greater use, the flexible PVC containing up to 50% of these substances.

There are many different types of phthalate available. By 1958, five phthalates (diethyl phthalyl, butylphthalyl butyl glycolate, ethylphthalyl ethyl glycolate, diisooctyl phthalate, and di-2-ethylhexyl phthalate) were permitted for use in plastic food packaging material while a further 18 have since been allowed.[8]

These phthalates are derived from phthalic anhydride, the structure of which is given

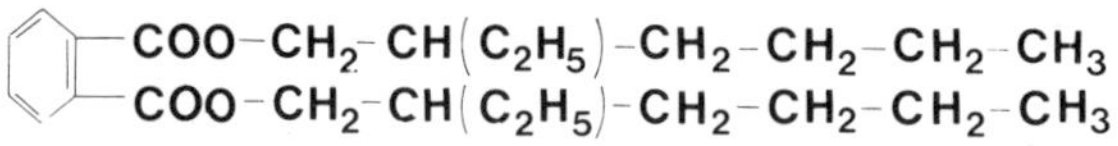

FIGURE 1. (a) Structure of phthalic anhydride, (b) esterification of phthalic anhydride, and (c) structure of di-2-ethylhexylphthalate

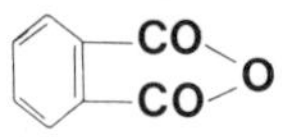

in Figure 1a. Esterification of the phthalic anhydride is then undertaken with the appropriate alcohol, as shown in Figure 1b. The most widely used phthalate is di-2-ethylhexyl phthalate (DEHP), the structure being shown in Figure 1c. It is this plasticizer that is discussed at greatest length in this section since it is most commonly employed, especially in PVC hemodialysis tubing and PVC catheters. Some others are relevant to medical polymers, however, the butylphthalyl butyl glycolate plasticizing acrylic resins in certain dental materials, for example.

It has generally been assumed in the past, on the basis of available data that these phthalate esters have very low toxicity. In a review of the toxicology of plastics, Lefaux[5] concludes that the majority are either nontoxic or only very slightly toxic, the latter description being given to the DEHP that is discussed in detail later. Even in the latest edition of Sax's *Dangerous Properties of Industrial Materials*[6] the oral LD_{50} (rat) of DEHP is given as high as 3100 mg/kg and the TD_{Lo} (lowest published toxic concentration) in man to produce gastrointestinal symptoms as 143 mg/kg. Acute toxicity was described as low to none, the only significant adverse effect to be noted being that it is an experimental teratogen. Thus in terms of industrial toxicity these esters are relatively harmless.

As with a number of other situations, however, the juxtaposition of the material with human tissues, following the use of these phthalate esters as plasticizers in biomedical plastics, can lead to far more significant effects than the industrial toxicology would predict. Much of our knowledge of this subject emanates from the work of Jaeger and his colleagues.[9-12] It has now been shown that phthalate ester plasticizers are leached from flexible PVC biomedical devices[10,11,13,14] and can be readily identified in the tissues of patients with whom they come into contact. Most of the work has involved DEHP, but several reports have dealt with other phthalate esters and find similar results, with the lower chain alkyl esters generally proving most toxic.

B. Di-2-Ethylhexyl Phthalate (DEHP)

Since DEHP is the most widely used phthalate ester plasticizer in biomedical devices, it will be discussed in some detail. Rubin and Jaeger[12] have discussed the toxicity of DEHP and commented on the general data which indicates a low order of toxicity. The majority of available toxicological data deals primarily with gross, overt effects, including morbidity, mortality, changes in body and organ weights, and histological changes. The data is normally obtained following ingestion of the phthalates, but, as reviewed by Krauskopf,[15] such studies do not normally detect subtle toxicological effects and it is these that are potentially important from the biomedical device point of view.

Rubin and Jaeger[12] have described some pharmacological and toxicological effects of DEHP. They found, for example, that DEHP can alter the pharmacological response to a barbiturate and can alter behavior patterns. DEHP was also shown to either depress or stimulate reticuloendothial function, depending on the amount administered, to increase the microaggregation of platelets in stored blood and at a concentration of only 4 μg/ml to be lethal to 97 to 98% of chick heart cells maintained in tissue culture. All these observations imply that there may be many subtle effects arising from the accumulation of DEHP in tissues.

1. Extraction of DEHP from PVC Medical Devices

Numerous studies have demonstrated that DEHP is extracted from PVC medical devices in vitro by various solutions. Again the work of Jaeger and Rubin is important since they have shown the significance of proteinaceous and other physiological fluids in this extraction.[16] Pumping of saline solution through PVC tubing in an isolated perfusion system for 4 hr did not extract any DEHP at all, but 0.7 mg was extracted by 4% bovine serum albumin. This amount was increased threefold to 2 mg, when perfusate was used instead of albumin, indicating the importance of cells, lipids, and other proteins. Other studies have shown migration from hemodialysis tubing, including those of Easterling et al.,[17] Lewis et al.,[18] and Ono et al.[19] Jaeger and Rubin[16] have shown that the rate of DEHP accumulation in blood is 0.25 mg/100ml/day over a period of 21 days storage in PVC bags.

This migration leads to raised levels of DEHP in patients on hemodialysis or who receive blood stored in PVC bags. Lewis et al.[18] report a mean concentration of DEHP in patients after a single hemodialysis session to be 750 μg/l serum with much higher levels after multiple treatment sessions. Rubin and Schiffer[20] found that the serum level of DEPH in patients receiving transfusions of blood that had been stored in PVC bags was 0.02 mg/dl plasma for each mg of DEHP administered per square meter of surface area. Jaeger and Rubin[16] performed autopsies on a number of patients who had received multiple transfusions, especially involving cardiopulmonary bypass and often, although not always, found significant levels in the lung, liver, and spleen. In one patient, for example, after cardiopulmonary bypass the lungs were found to contain 91.5 μg/g (dry weight) of DEHP, the liver 69.5 μg/g, and the spleen 25.3 μg/g. No DEHP was detected in the systemic circulaton in this or in the majority of patients, indicating that some organs efficiently remove DEHP. This aspect has also been discussed by Fayez et al. in relation to the distribution of DEHP in blood and tissues.[21]

2. Toxicology of DEHP

In spite of a large amount of experimental data, the actual toxicological hazards associated with DEPH release from PVC medical devices is not well understood. Certainly toxic effects of DEHP have been noted in animals. Singh et al.[22] studied the effect of several different phthalate ester plasticizers on developing rat fetuses, all having harmful effects, the more toxic at doses as low as 0.1 of the acute LD_{50}. Using techniques of enzyme histochemistry, Salthouse et al.[23] have demonstrated considerably greater tissue responses to PVC plasticized with numerous phthalates than to controls. Dimethoxyethyl phthalate has also been shown to inhibit cell growth in replicating mouse fibroblast L cells in culture.[24] Peters and Cook[25] showed that intraperitoneal injection of DEHP in pregnant rats affected both implantation and parturition.

Schulz and Rubin[26] have demonstrated that after i.v. injections of DEHP in rats, the phthalate disappears rapidly from the bloodstream and accumulates in the liver. The DEHP appears to be extensively metabolized by the rat to water-soluble metabolites which are excreted in the urine and feces. Nazir et al.[27] have further demonstrated the localization of DEHP in bovine heart muscle mitochondria. Stein et al.[28] did not observe any accumulation of DEHP in the liver of rats given a diet supplemented with this phthalate, but did see a significant hypertrophy of the liver. Supplementation of the diets with DEHP did, however, result in deposition in the heart.

It would seem from the available evidence that DEHP, along with several other phthalate ester plasticizers, can have subtle effects on the tissues of patients who come into contact with plasticized PVC. The extent of the problem is far from clear, and it is not suggested that clinical toxicity arising from this source is of serious proportions. There has been some concern expressed about the role of DEHP in necrotizing enterocolitis as discussed by Hillman et al.[29] and also by Hastings in *Biocompatibility in Clinical Practices*.[30] In this case, PVC is used for exchange transfusions in neonates, a proportion of whom develop necrotizing enterocolitis, a syndrome consisting of vomiting, abdominal distension, gastrointestinal bleeding, and apnea. Although the etiology is far from clear, a causative factor may be DEHP, and certainly this substance has been found in tissues of some of these neonates at post-mortem.

IV. CONCLUSIONS

The purpose of this brief chapter has been to highlight the possibility of additives affecting the biocompatibility of certain polymers. Very little hard data is available on this subject, but it is clear that whenever the additive is capable of migrating to the polymer surface it is capable of modifying the host response. The exact effect will depend on the quantity of the addition in the polymer, the rate of migration through the polymer, and the specific toxicity. On this basis, plasticizers, which may be present in large amounts and which by definition are loosely bound to the polymer, offer the greatest risks. Other substances are present in only small amounts and may be bound firmly to the polymer so that the risks are much smaller.

REFERENCES

1. **Liepins, R. and Pearce, E. M.**, Chemistry and toxicity of flame retardants for plastics, *Environ. Health Perspect.*, 17, 55, 1976.
2. **Biggs, M. S. and Baldwin, J.**, A flexible PVC compound for long term contact with human tissue, Proc. Conf. Plastics in Medicine and Surgery, Enschede, Netherlands, Plastics and Rubber Institute, London, 1979.
3. **Simonds, H. R. and Church, J. M.**, *The Encyclopedia of Basic Materials for Plastics*, Reinhold, New York, 1967.
4. **Roff, W. J. and Scott, J. R.**, *Fibres, Films, Pastics and Rubbers*, Butterworths, London, 1971.
5. **Lefaux, R.**, *Practical Toxicology of Plastics*, Iliffe, London, 1968.
6. **Sax, N. I.**, *Dangerous Properties of Industrial Materials*, 5th ed. Van Nostrand, New York, 1979.
7. **Liebert, T. C., Chartoff, R. P., Cosgrove, S. L., and McCuskey, R. S.**, Subcutaneous implantation of polypropylene filaments, *J. Biomed. Mater. Res.*, 10, 939, 1976.
8. **Shibko, S. I. and Blumenthal, H.**, Toxicology of Phthalic acid esters used in food-packaging material, *Environ. Health Perspect.*, 3, 131, 1973.
9. **Jaeger, R. J.**, Studies on the Extraction, Accumulation and Meatbolism of Phthalate Ester Plasticized from Polyvinylchloride Medical Devices, Ph.D. thesis, John Hopkins University, Baltimore, 1971.
10. **Jaeger, R. J. and Rubin, R. J.**, Plasticizers from plastic devices: extraction, metabolism and accumulation by biological systems, *Science*, 170, 460, 1970.
11. **Jaeger, R. J. and Rubin, R. J.**, Migration of a phthalate ester plasticizer from polyvinylchloride blood bags into stored human blood and its localisation in human tissues, *N. Engl. J. Med.*, 287, 1114, 1972.
12. **Rubin, R. J. and Jaeger, R. J.**, Some pharmacologic and toxicologic effects of Di-2-ethylhexylphthalate and other plasticizers, *Environ. Health Perspect.*, 3, 53, 1973.

13. Marcel, Y. L. and Noel, S. P., Contamination of blood stored in plastic packs, *Lancet*, 1, 35, 1970.
14. Marcel, Y. L. and Noel, S. P., A plasticizer in lipid extracts of human blood, *Chem. Phys. Lipids*, 4, 418, 1970.
15. Kraushopf, L. G., Studies on toxicity of phthalates via ingestion, *Environ. Health Perspect.*, 3, 61, 1973.
16. Jaeger, R. J. and Rubin, R. J., Extraction, localisation and metabolism of Di-2-ethylhexylphthalate from PVC plastic medical devices, *Environ. Health Perspect.*, 3, 95, 1973.
17. Easterling, R. E., Johnson, E., and Napier, E. A., Plasma extraction of plasticizers from medical grade PVC tubing, *Proc. Soc. Exp. Biol. Med.*, 147, 572, 1974.
18. Lewis, L. M., Flechter, T. W., Kerkay, J., Pearson, K. H., and Nahamoto, S., Bis (2-ethylhexyl) phthalate concentrations in the serum of haemodialysis patients, *Clin. Chem.*, 24, 741, 1978.
19. Ono, K., Tatskawa, R., and Wakimoto, T., Migration of plasticizer from haemodialysis blood tubing, *JAMA*, 234, 948, 1975.
20. Rubin, R. J. and Schiffer, C. A., Fate in humans of the plasticizer DEHP arising from transfusion of platelets stored in vinyl plastic bags, *Transfusion*, 16, 330, 1976.
21. Fayez, S., Herbert, R., and Marlin, A. M., The release of plasticizer from polyvinyl chloride haemodialysis tubing, *J. Pharm. Pharmacol.*, 29, 407, 1977.
22. Singh, A. R., Lawrence, W. H., and Autian, J., Teratogenicity of phthalate esters in rats, *J. Pharm. Sci.*, 61, 51, 1972.
23. Salthouse, T. N., Matlaga, B. F., and O'Leary, K., Microspectrophotometry of macrophage lysosomal enzyme activity: a measure of polymer implant tissue toxicity, *Toxicol. Appl. Pharmacol.*, 25, 201, 1973.
24. Pillingham, E. O. and Aution, J., Teratogenicity, mutagenicity and cellular toxicity of phthalate esters, *Environ. Health Perspect.*, 3, 81, 1973.
25. Peters, J. W. and Cook, R. M., Effects of phthalate esters on reproduction of rats, *Environ. Health Perspect.*, 3, 91, 1973.
26. Schulz, C. O. and Rubin, R. J., Distribution, metabolism and excretion of DEHP in the rat, *Environ. Health Perspect.*, 3, 123, 1973.
27. Nazir, D. J., Beroza, M., and Nair, P. P., DEHP in bovine heart muscle mitochondria, *Environ. Health Perspect.*, 3, 141, 1973.
28. Stein, M. S., Caasi, P. I., and Nair, P. P., Some aspects of DEHP and its action on lipid metabolism, *Environ. Health Perspect.*, 3, 149, 1973.
29. Hillman, L. S., Goodwin, S. L., and Sherman, W. R., Identification and measurement of plasticizer in neonatal tissue after umbilical catheters and blood products, *N. Engl. J. Med.*, 292, 381, 1975.
30. Hastings, G. W., Catheters in neonatology, in *Biocompatibility in Clinical Practice*, Vol. 1, Williams, D. F., Ed., CRC Press, Boca Raton, Fla., in press.
31. Gilding, D. K., Degradation of polymers: mechanisms and implications for biomedical applications, in *Fundamental Aspects of Biocompatibility*, Williams, D. F., Ed., CRC Press, Boca Raton, Fla., in press.
32. Heaney, T., Periodontal dressing materials, in *Biocompatibility in Dental Materials*, Smith, D. C. and Williams, D. F., Eds., CRC Press, Boca Raton, Fla., in press.

Chapter 8

SOLUBLE POLYMERS IN MEDICINE

Jindrich Kopeček

TABLE OF CONTENTS

I.	Introduction		158
II.	Possibilities of Therapeutical Uses of Soluble Polymers		158
	A.	Blood Plasma Expanders	159
		1. Dextran	159
		2. Hydroxyethyl Starch	160
		3. Gelatin and Its Derivatives	161
		4. Polyvinylpyrrolidone	162
		5. Other Blood Plasma Expanders	164
	B.	Synthetic Polymers as Drugs	165
	C.	Carriers of Biologically Active Compounds	166
III.	Biocompatibility of Soluble Polymers		166
IV.	Some Possibilities for the Preparation of Synthetic Biodegradable Polymers		167
	A.	Polymer Analogous Reactions Leading to Enzymatically Degradable Bonds	168
	B.	Degradability of Bonds in Side Chains of Soluble Synthetic Polymers	170
	C.	Destructibility of Cross-links	171
References			174

I. INTRODUCTION

Polymeric compounds in their insoluble (cross-linked) form are used in medicine for a number of applications, including the substitution of organs, in plastic surgery, or in auxiliary surgical procedures.[1-5] This review deals with polymers which are soluble in water or in saline, and which may be applied as blood plasma expanders, polymeric drugs, or drug carriers (generally, carriers of biologically active compounds). This field, too, has been dealt with in numerous reviews.[6-10] Many polymers, and especially those which can be metabolized in the organism, are used in clinical practice, mainly as blood plasma exapnders[11-14] or as polymeric drugs.[15] Several other proposed applications have only reached the experimental stage and their application in human medicine is hindered by a great number of unsolved problems, such as the effect of polymer structure on the immunological response of the organism, the relationship between the structure of the polymer and its elimination from the organism, the deposition of the polymer in individual organs, and changes in the polymer properties (with respect to living matter) due to the binding of the biologically active compound. Even though these problems cannot be expected to be solved in the immediate future, their solution would lead the way to many new developments which could certainly play an important role in the life and health of mankind.

Bearing in mind the scope of this chapter, all the aspects just outlined cannot, of course, be treated with the same attention. We therefore concentrate on the properties of synthetic soluble polymers and on the possibility of modifying their structures by the introduction of biodegradable bonds. The other problems, have been reported in the literature and are mentioned only briefly.

II. POSSIBILITIES OF THERAPEUTICAL USES OF SOLUBLE POLYMERS

Soluble polymers or copolymers may be applied in medicine either unmodified or as polymeric carriers of biologically active compounds. Unmodified polymers are active as such, without binding a biologically active compound. The group of compounds thus defined has two main applications, namely,[6] as blood plasma expanders and as polymers active as drugs, i.e., polymers, the pharmacological activity of which can be attributed to their molecular weight (e.g., polyvinylpyridine-N-oxide).

Polymers suited for these purposes must meet many requirements. For blood plasma expanders, these conditions have been summarized by Grönvall:[16]

1. The molecular weight must allow an adequate colloid-osmotic effect to be achieved.
2. The colloid-osmotic pressure of the solution should be close to that of plasma.
3. The viscosity of the solution should be close to that of blood.
4. The polymer must not, in the long-term, be deposited in the organism, but must be fully eliminated, or biodegraded. The term biodegradability is used here in a qualified way;[17] it also includes nonbiodegradable polymers small enough to pass the renal threshold.
5. The polymer solution must not possess toxic properties and must not provoke allergic or pyrogenic reactions.
6. The polymer solution must not be antigenic and must not cause sensitization.
7. The polymer solution must be easy to sterilize and to store.

Polymers which meet the above requirements may also be applied as drug carriers. The drug can be attached to the polymer carrier[6] by ionic or covalent bonds. Under physiological conditions, the latter may be nondegradable, hydrolyzable, or undergo enzymatic cleavage.

Results obtained by many authors indicate that the molecular weight is an important parameter of the biological activity of the polymer.[8] It is predominantly reflected in the rate of elimination of the polymer from blood circulation and in the possible deposition of the polymer in some organs. The approach to the synthesis of polymers suitable for use in blood circulation depends on the polymers initially used, that is, natural or synthetic. If the initial polymers are natural, such as starch or gelatin, and consequently contain many biodegradable bonds, the structure of the polymer in many applications must be modified so as to reduce the rate of biodegradation. If, on the other hand, a synthetic polymer is used, i.e., a polymer which does not contain bonds degradable in the organism (e.g., poly(vinyl alcohol), polyvinylpyrrolidone, poly(vinylpyridine-N-oxide), the main condition for their use is an adequate molecular weight distribution. The reported data show[7] that the majority of synthetic polymers with molecular weights over 80,000 to 100,000 cannot be excreted by kidneys although this is not always the case.[44] These polymers can be slowly eliminated via the liver into the intestine. Compounds having a molecular weight over 1000 cannot pass the blood-brain barrier and cannot consequently penetrate into the cerebrospinal fluid. Synthetic polymers are polydisperse. An experimentally proved finding can be derived therefrom,[18-20] namely that after some time not only a change (loss) in the amount of the polymer in the organism takes place, but its molecular weight distribution also changes. The lower fractions are quickly eliminated from the organism while the higher fractions remain for a longer time.

It would be useful for a number of applications if synthetic polymers could be used, the molecular weight of which lies beyond the critical value for passing through the kidneys, so that the polymer would remain in the blood circulation for a longer time and would be excreted in the normal way. This aim may be achieved by binding relatively short synthetic polymer chains to each other using bonds susceptible to enzymatic attack or in other words, by additional cross-linking of synthetic polymers to a level short of the gel point.[21-24] The advantage of a polymer thus prepared is that its molecular structure contains only the smallest number of bonds susceptible to enzymatic attack and that the rate of scission can be controlled by changing the structure, amount, and length of the cross-links. In addition, the investigation of these problems allows us to examine systematically an important and rather undeveloped field, namely, the relationship between structure and susceptibility to enzymatic cleavage for polymeric substrates. These problems are discussed in detail at the end of this review.

A. Blood Plasma Expanders
1. Dextran

Dextrans are polysaccharides consisting of α-D-glucose units joined predominantly by 1 → 6 glucosidic linkages. Many species of Lactobacilla are able to convert sucrose into dextran; the use of a particular bacterial strain, B-512, of *Leuconostoc mesenteroides* produces a dextran with a minimum number of side chains (i.e., 1 → 3 linkages). This dextran of high linearity has proved especially suitable for clinical use,[25] but its molecular weight must be adjusted in advance by partial acid hydrolysis followed by fractionation. The molecular weight distributions of two basic products in clinical use, Rheomacrodex® and Macrodex,® are shown in Figure 1.

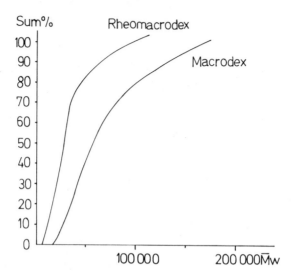

FIGURE 1. Molecular weight distributions for Rheomacrodex® and Macrodex®. (From *Blood Flow Improvement*, Pharmacia AB, Uppsala, 1968. With permission.)

Both types of dextrans can be used[11-14,25,26] for volume substitution and thrombosis prophylaxis; Rheomacrodex® (M_w = 40,000) can also be used for the treatment of microcirculation disturbances, and Macrodex® (M_w = 70,000) can be employed in hemodilution. The limiting molecular weight[14] passing through the kidneys is M_w = 50,000. The greater part of parenterally introduced dextran is excreted by the kidneys. The fate of the residue was explained by experiments with ^{14}C-labelled dextran.[11,14] Ninety percent of the administered amount is excreted within 10 days, of which 64% is excreted as urine, 26% is excreted as $^{14}CO_2$ and about 2% is excreted in feces. Practically all the radioactive material is eliminated within 2 weeks; 30% of it is excreted as $^{14}CO_2$. Dextran is enzymatically degraded in the organism (with dextranase), so that the amount of glucose in blood increases after the application. On the other hand, ^{14}C is also incorporated into amino acids and compounds synthesized from them by the organism. The rates of elimination[27] of the two basic types of dextran are shown in Figure 2. Allergic reactions are very rare in the application of dextran.[26,30,49] An effect on blood coagulation can be observed at doses higher than 1 g/kg of body weight. It was also observed that the results of some laboratory methods were affected by the infusion of dextran.[14] It can be said, however, that dextran is still the colloid substitute most widely used, and that the number of side reactions[99] observed following the application of a vast number of doses was minimal. The properties and applications of dextrans have also been reviewed in Reference 59.

2. Hydroxyethyl Starch

The use of hydroxyethyl starch (HES) as a blood plasma expander was suggested by Wiedersheim.[28] The reason for the choice of HES as an artificial colloidal agent was that the branched component of starch, amylopectin, is very similar in structure to glycogen, the reserve polysaccharide of animals, and therefore was likely to be compatible with the body tissue.[29] The use of native starch is prevented by the presence of amylases in blood, which degrade the starch too quickly. Substitution reduces the rate of degradation.[29] The most commonly accepted model for HES is one in which the

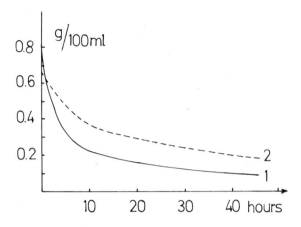

FIGURE 2. Dextran concentration in human serum after infusion of Rheomacrodex® (500 mℓ, 10%, curve 1), or Macrodex® (500 mℓ, 10%, curve 2). (From Artuson, G., Granath, K., Thoren, L., and Wallenius, G., *Acta Chir. Scand.*, 127, 543, 1964. With permission.)

substituent hydroxyethylgroup is formed on carbon atom 6 of the glucose ring.[14] It should be assumed, however, that substitution also takes place on atoms 2 and 3.[29]

Commercially available products are prepared by acid hydrolysis of waxy starches from soya and corn with subsequent substitution of hydroxyethyl groups by reaction with ethylene oxide. To obtain a satisfactory rate of degradation, the commercial solutions have hydroxyethyl units attached to 70% of the glucose units. The average molecular weight is about 450,000, which corresponds to 6 hr of intravascular half-life.[26]

HES is widely used in the U.S.A., having the same properties as dextran,[31] but better storage life and is cheaply prepared.

3. Gelatin and Its Derivatives

Gelatin is prepared by the hydrolysis of animal collagens.[11-14,26] The separation of peptide chains is followed by cross-linking with glyoxal, succinic acid anhydride, or diisocyanates. Three basic types can be distinguished, according to the method of preparation, namely:

1. Oxypolygelatin[32] (OPG)
2. Modified fluid gelatin[33] (MFG)
3. Urea-linked gelatin[34]

Data on molecular weight differ. In all cases the preparations are polydisperse with an average molecular weight of about 30,000. The polymers are biodegradable, having an intravascular half-life of only 2 to 3 hr.[26] Anaphylactic reactions following the use of gelatin solutions are roughly as frequent as with other substitute solutions,[26,30,49] even though the mechanism of these reactions is different and depends on the structure of the sample.[26]

4. Polyvinylpyrrolidone

Polyvinylpyrrolidone (PVP) is the synthetic polymer that has been most completely investigated with respect to its fate in the organism. It has the structure:

$$\left[-CH_2-CH- \right]_x$$
$$\begin{array}{c} | \\ N \\ / \ \backslash \\ CH_2 \quad CO \\ | \quad | \\ CH_2-CH_2 \end{array}$$

A detailed review of results obtained in the study of interactions of this polymer with the living organism has been published recently by Rypáček.[35] Hecht and Weese[36] introduced this polymer into medicine. Its therapeutical applications and clinical experience have been dealt with in detail in some reviews.[11-14] Since the polymer is synthetic and undegradable, the problems connected with its elimination from and deposition in the organism must be considered. PVP is eliminated mainly through the kidneys, the rate of elimination depending on many factors, and especially on its molecular weight distribution. Ammon and Depner[37] investigated the amount of PVP excreted in urine as a function of molecular weight (Figure 3). Two stages are discerned in the elimination: the first steep stage represents the polymer eliminated directly from blood circulation. The second stage is slower and controlled by the rate of transition of the polymer from the lymph and interstitial space into the plasma.[37] The results were confirmed and extended by other authors.[18-20,38-40] An important question that remains open is the route by which polymers with molecular weights above 50 to 60,000 are excreted into the urine; these polymers appear in the urine, although they should not pass through glomerular filtration. Hespe,[39,40] using his own work and that of Ravin et al.,[41] Hecht and Scholtan,[42] Marek et al.,[43] and Gärtner,[44] made the following subdivision with regard to the relation of molecular weight and urinary excretion rate in the rat:

1. PVP molecules with molecular weight up to 25,000 are eliminated by glomerular filtration, which is a rapid process (a matter of days)
2. PVP molecules with molecular weight between 25,000 and 110,000 are eliminated by postglomerular filtration, which is a far less rapid process (a matter of weeks)
3. For PVP molecules with molecular weight above 110,000 uptake in the reticuloendothelial system seems to dominate, which can lead to very prolonged storage, but elimination via urinary excretion still seems to be possible, presumably through postglomerular endothelial gaps.

Gärtner et al.,[44] found, using ^{131}I-PVP, that postglomerular (intertubular) capillaries are permeable up to a molecular weight of about 650,000, which makes possible an exchange of plasma proteins and molecules of synthetic polymers between the plasma of the capillaries and the interstitium. Another route by which PVP may be eliminated from the organism is through the liver and its biliary system into the intestine.[7] Macromolecules get into the intestine either by passing through the intestinal mucosa or with gall.[35,39,45] A similar course of elimination has also been observed with other synthetic polymers.[6,9,10,46]

Polymer and especially fractions having higher molecular weights may be deposited

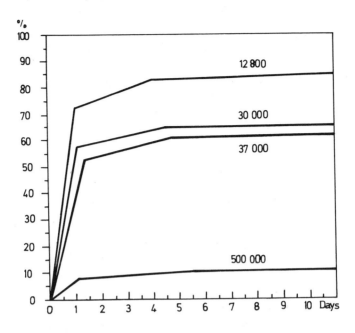

FIGURE 3. Elimination of polyvinylpyrrolidone fractions with different molecular weight from rabbits. (From Ammon, R. and Depner, E., *Z. Ges. Exp. Med.*, 128, 607, 1957. With permission.)

in various organs, predominantly in cells possessing phagocytic activity. These are mainly cells of the reticulo-endothelial system (RES), also called the mononuclear phagocytic system (MPS). Some authors assert[14] that such deposition is only transient and without any functional changes, others assume it to be irreversible, accompanied by a decrease in phagocytic activity. An excessive accumulation of macromolecules at large doses and high molecular weight in the RES organs and in mesenchymal tissues leads to a number of pathological and morphological changes, which are collectively called the macromolecular syndrome[47] or PVP-disease.[48] It follows from papers already published[39] that the deposition of PVP in RES may be prevented by the elimination of a fraction having a molecular weight higher than 25,000. Endocytosis (or pinocytosis, in the case where the polymer enters in solution) is in some cases essential for the medical application of a polymer, in other cases it is not necessary, while on other occasions it might be disadvantageous.

For all these reasons, elucidation of the fate of synthetic polymers in the organism and of their elimination mechanism necessitates an investigation of pinocytosis on isolated cells, monocellular organisms cultured in vitro, or on organ cultures. Using a quantitative study of the pinocytosis of ^{125}I-PVP by rat yolk sac cultured in vitro or in rat peritoneal macrophages, Lloyd et al.,[50-52] have attempted to explain the basic obscure problems of pinocytosis: do pinocytically ingested solutes mainly enter in free solution or are they adsorbed to the plasma membrane? By what means and to what extent can pinocytosis be stimulated and inhibited? Lloyd et al.,[50] assume that PVPs uptake into yolk sac tissue is presumably accomplished by micropinocytosis and that the site of intracellular accumulation may be supposed to be the vacuolar system. The rate of pinocytosis of ^{125}I-PVP was constant over the concentration range 0.15 to 24.0 µg/mℓ medium; it appears, therefore, that within this range ^{125}I-PVP neither stimulates nor inhibits pinocytosis. From the results it seems[51] that pinocytic ingestion rather than reversible surface adsorption was responsible for the accumulation of the polymer.

This conclusion is corroborated by results obtained in an investigation of the pinocytosis of PVP in rat peritoneal macrophages.[52]

PVP as a blood plasma expander was most widely used for therapeutical purposes during World War II involving about 500,000 patients.[21] Nowadays, low-molecular PVP (e.g., Periston® N-M_w = 12,500) is used in some countries for the detoxication of organisms.[53-57]

5. Other Blood Plasma Expanders

Of other polymers suggested as blood plasma expanders, the condensation products of ethylene oxide and propyleneglycol,[60] poly-α,β-[N-(2-hydroxyethyl)-D,L-aspartamide],[35,61-63] and poly [N-(2-hydroxypropyl)methacrylamide][64-70] may be mentioned. The latter polymer has the structure

$$\left[-CH_2 - \underset{\underset{\underset{\underset{OH}{|}}{NH-CH_2-CH-CH_2}}{\underset{|}{CO}}}{\overset{\overset{CH_3}{|}}{C}} - \right]_x$$

and has been investigated within a broad study of the relationship between the structure of synthetic polymers and their biocompatibility.[6]

The properties of this polymer with regard to its application as a blood plasma expander are being investigated and characterized by Štěrba et al. of the Institute of Haematology and Blood Transfusion in Prague. Six standard batches of Duxon (the working name for poly[N-(2-hydroxypropyl)-methacrylamide]) have been investigated.[65] The samples were nontoxic and apyrogenic, and did not affect cell cultures as measured by the growth of LEP and HeLa cells. The colloid-osmotic capacity of the samples was also favorable.[65] Preliminary results obtained by immunological tests carried out by Korčáková, et al.[67] showed that Duxon did not cause formation of antibodies. ^{14}C-labelled Duxon was also used to examine the elimination of the polymer from blood circulation and its deposition in individual organs.[66] At M_w = 28,000 with $M_w/M_n \doteq 2$ a single application causes a decrease in the polymer concentration in blood to half its original value at 8 hr; within 9 days, the polymer disappears completely from the blood system. These values obviously vary depending on molecular weight.[64,66] Deposition of the polymer in individual organs was investigated after repeated application of the polymer to test animals (for 10 days, 10 mℓ of 4% Duxon was administered intravenously to rabbits once a day). Two polymers were compared (M_w = 11,000, M_w = 33,000; $M_w/M_n \doteq 2$). In agreement with studies dealing with other synthetic polymers,[39-45] it was observed that the deposition of the polymer with a higher molecular weight was more pronounced, mainly in organs possessing phagocytic activity. Molecules having a lower molecular weight which exhibit a good therapeutical effect, are not permanently deposited in the organism.

In particular, one should mention attempts to prepare samples which would possess properties of the whole blood, or in other words, which, unlike blood plasma expanders, would additionally possess the function of red blood corpuscles. There exist several trends in this field:[71]

1. Efforts to attach hemoglobin to synthetic polymers[17,73,74]
2. Attempts to prepare whole erythrocytes by the microencapsulation of hemoglobin solutions into gas-permeable polymeric capsules[72,75,76]

3. The use of aqueous emulsions of perfluorinated hydrocarbons[75,77-79] which are able to dissolve substantial volumes of oxygen and CO_2

It seems probable that in the future at least some of these extremely interesting compounds may be applied for practical purposes.

B. Synthetic Polymers as Drugs

Many synthetic polymers possess biological activity which can be attributed only to their structure. It was found that some synthetic polymers could be employed as drugs in the treatment of silicosis; these include PVPs[80] and poly(2-vinyl-pyridine-1-oxide),[15,81,82] the latter being the most thoroughly investigated. It has the structure

$$\left[-CH_2-CH-\right]_x$$
(with pyridine N-O substituent)

It was shown that this polymer prevented cytotoxic influence of quartz dust. Its effect is explained[81] by the adsorption of the weakly basic polymer on the weakly acid quartz surface. The dependence of the activity of the sample on polymeric structure is proved by the finding[9] that the lowest molecular weight which guarantees the activity of the polymer is 30,000.

Polyelectrolytes constitute another group of polymers possessing biological activity.[83-87] Anionic and cationic polyelectrolytes of both natural and synthetic origin have been found to exhibit inhibitory effects on viruses, bacteria, tumors, and enzymes. Polyanions, in particular, have received considerable attention in the areas of oncology and virology. The structures of the polyanions studied[83-86] include, pyran copolymer (divinyl-ether + maleic anhydride), polyacrylic acid,[46,86] polymethacrylic acid,[86] acrylic acid-maleic anhydride copolymer, and maleic anhydride-furan copolymer. Some of the polymers mentioned also act as interferon inducers.[85] The investigation of the antiviral activity of a number of fractions of poly(acrylic acid)s and poly-(methacrylic acid)s with different tacticities, different molecular weights, and molecular weight distributions, showed[46,86] that isotactic polyacrylic acids exhibit a considerably higher antiviral activity than atactic ones. Independent of the tacticity, poly(acrylic acid) samples with molecular weights smaller than 5000 show no significant effects. The optimum efficacy is obtained between 6,000 and 15,000. Isotactic poly(acrylic acid) samples with narrow molecular weight distributions are more active than those with broad distributions. The atactic poly(acrylic acid)s with lower efficacy do not show this relationship. Atactic poly(methacrylic acid)s are not antiviral-active in vivo; isotactic ones have a just detactable efficacy which is appreciably lower than the activity of atactic poly(acrylic acid).

Positively charged polyelectrolytes (cationic polymers) have several biomedical applications which include antiviral, antibacterial, and antineoplastic activities.[83,84] These properties are primarily related to their ability to bind to (cell) surfaces with negative electrolytic charges. An example can be seen in copolymers of N-vinylpyrrolidone with 2-methacryloyloxyethyl-trialkylammonium iodide (or chloride), which possess antimicrobial activity depending both on the alkyl structure and on the concentration of quaternary groups in the polymeric molecules.[87]

Polymers with structures conducive to surface activity also exhibit pharmacological activity.[9] Polymeric detergents of the Triton® type have a strong metastases inhibition activity.[88] These and similar compounds are also active in experimental tuberculosis.[89]

C. Carriers of Biologically Active Compounds

Soluble polymers, both synthetic and natural, can be used as carriers of biologically active compounds (drugs, hormones, enzymes). By binding a biologically active compound (BAC) onto a polymer, the following main effects may be achieved:[7]

1. Depot effect, i.e., retarded absorption and/or retarded excretion
2. Reduction of toxicity of the sample
3. Influence on the solubility of the sample
4. Influence of pharmacokinetics and pharmacodynamics. These may be varied within a broad range by controlling the rate of release of the active component, by specific interaction with cell membranes and by different course of endocytosis
5. Possibility to combine drugs along the polymer chain
6. Possibility to obtain specific effects, and even a change in the drug effects by binding onto the polymer.

If a particular polymeric carrier is considered, the effects of the drug, or generally of BAC, depends on the way in which the latter is bound onto the polymer. The following types of binding of BAC onto the polymeric matrix are possible:

1. Ionic bond (and other types of noncovalent bonds)
2. Covalent bond
   a. Indestructible under physiological conditions
   b. Chemically hydrolyzable under physiological conditions
   c. Enzymatically hydrolyzable under physiological conditions

Concrete cases of bonds between biologically active compounds and soluble polymeric carriers go beyond the scope of this review. The reader can find them in papers already published.[6-10]

The major achievements[90] in the development of new chemotherapeutic agents up to a few years ago were limited to the synthesis of new chemical entities and the evaluation of their pharmacological activity. Today new concepts are presented capable of transporting drug molecules from the site of application directly to the site of action.[91-96] It is desirable that the polymeric carrier should contain, in addition to the active group (drug) yet another group (e.g., antibody), which would guarantee that the desired conjugate of drug and carrier would be capable of locating and binding to the cells against which they were intended to react. Attempts are made to use this concept in the treatment of cancer. Cancerostatic drugs, such as Daunomycin or Adriamycin®, are either attached on antibodies,[94,95,97] or a conjugate antibody-polymer-drug[97] is prepared, and selective endocytosis (the so-called piggyback endocytosis[7,91]) by tumor cells is studied. Antibodies are not the only possible carriers, however,[97] and many other macromolecules and low molecular weight compounds with an affinity for specific cells or organs will eventually be used as carriers for affinity therapy studies,[93,97] including plasma proteins, lectins, nucleic acids, hormones, polysaccharides, cells, and liposomes.

III. BIOCOMPATIBILITY OF SOLUBLE POLYMERS

The problem of biocompatibility of soluble polymers should be regarded similarly to that of the harmlessness of new drugs. Procedures prescribed in particular countries for the testing of polymers or polymeric drugs are based on rules controlling the testing of new drugs, taking into account the specificity ensuing from the high molecular weight of polymeric molecules and from their polydispersity.

Among other things, the following properties are examined in the study of biocompatibility of soluble polymers and polymeric drugs:

1. Acute toxicity — after single administration — at least two types of animals, one of which must be nonrodent, separately for the two sexes
2. Chronic toxicity — using repeated administration

The evaluation of toxic effects is based on observations of changes in behavior and growth, on analyses of blood and urine, on functional tests, autopsy results, histological and biochemical examinations; sometimes these results are supplemented by tests on cell cultures. Additional tests include: pyrogenicity test (a rise in body temperature after parenteral application), tests of damages of reproduction, carcinogenicity, effects on blood clotting, effects on the determination of blood groups, and the influence on microcirculation. With blood plasma expanders, the colloid osmotic capacity must be investigated; with polymeric drugs, pharmacokinetic and pharmacodynamic investigations must be carried out, and determination of the mechanism of action of the drug must be attempted.

For all polymeric samples, an important factor is elimination from the organism or deposition in individual organs. This problem has been already dealt with in preceding chapters. The elimination of a polymer having the given molecular weight is affected in a decisive manner by its biodegradability. The existing findings concerning the relationship between structure and biodegradability can be found in reference 58.

Immunological problems involved in the application of polymers also deserve attention. Polymers may act as antigens, hapten carriers, and haptens.[10] The immunity response of the organism is specific towards the type of macromolecules which cause immunity, and may be realized both in the humoral, via production of specific antibodies, and in the cellular, via production of sensitized phagocytic cells, way.[99] The provocation of immunity response may be reflected in various unfavorable phenomena,[30,35,49,99] such as instantaneous hypersensitivity with the danger of anaphylactic shock or in a gradual sensitization. Synthetic water-soluble polymers are weak immunogens, which lead to the equilibrium between immunological stimulation and immunological paralysis.[98] Small doses of weak immunogens stimulate antibody formation, while larger doses induce immunological paralysis. For this reason, in studying these problems one should discuss not only the effect of polymer structure, but also the influence of the dose and of the method of application. The effects of detailed polymer structure[100] and of molecular weight[101] are also important.

Quite understandably, most of the tests just described have been carried out on all the polymers used in the clinical practice. The results of these tests can be found in the references indicated for the individual polymers.

IV. SOME POSSIBILITIES FOR THE PREPARATION OF SYNTHETIC BIODEGRADABLE POLYMERS

It may be inferred from the discussion of the elimination of polymers from blood circulation in Section II that in human medicine one must use either metabolizable (biodegradable) polymers or synthetic polymers having relatively low molecular weight. In both cases, the duration of intravascular half-life can be counted in hours.[29] For a number of polymer applications it would be suitable if synthetic polymers could be used, the molecular weight of which goes beyond the critical level for passage through the kidneys, so that the polymer would remain in the body for a longer time and would then be excreted by the normal mechanism. In our view, this aim can best be achieved by joining relatively short synthetic chains to each other using enzymatically degradable bonds, or in other words, by additional cross-linking below the gel point with formation of enzymatically degradable bonds.[21-23]

An advantage of polymers thus prepared would consist in the possibility of controlling the number of enzymatically degradable sites. One may also assume the possibility of controlling the rate of splitting through the length and structure of the cross-link.[6] The process of devising a solution to this problem can be divided into several stages:

1. Polymer analogous reactions leading to enzymatically degradable bonds
2. Relationship between the structures of polymers and their susceptibilities to degradation
    a. In the side chain
    b. In the main chain.

Results of these studies may be used as a basis for the synthesis of novel polymers suitable for use as blood plasma expanders, of biodegradable carriers of biologically active compounds, and for binding BAC on to polymers through an enzymatically degradable bond.

A. Polymer Analogous Reactions Leading to Enzymatically Degradable Bonds

In solving this problem, we start with copolymers of N-(2-hydroxypropyl) methacrylamide with p-nitrophenyl esters of N-methacryloylated oligopeptides.[100-103]

$$-CH_2-\underset{\underset{\underset{\underset{CH_3}{|}}{CH-OH}}{\underset{|}{CH_2}}}{\underset{|}{\underset{|}{C}}{\underset{|}{NH}}}-CH_2-\underset{\underset{\underset{\underset{NH}{|}}{CO}}{\underset{|}{CH_2}}}{\underset{|}{\underset{|}{C}}{\underset{|}{NH}}}-CH_2-\underset{\underset{\underset{\underset{CH_3}{|}}{CH-OH}}{\underset{|}{CH_2}}}{\underset{|}{\underset{|}{C}}{\underset{|}{NH}}}-$$

with side chain continuing: CH–CH$_2$–C$_6$H$_5$, CO, O, C$_6$H$_4$–NO$_2$

Abbreviated formula: Gly–Phe–ONp

$$R-\bar{N}H_2 + \underset{R'}{\overset{O}{\underset{|}{C}}}-OR'' \rightleftharpoons \left[R-\underset{H}{\overset{H}{\underset{|}{N}}}\underset{R'}{\overset{\bar{O}}{\underset{|}{C}}}\oplus \blacktriangleleft OR''\right] \rightleftharpoons \left[R-\underset{H}{\overset{H}{\underset{|}{N}}}\oplus \underset{R'}{\overset{\bar{O}}{\underset{|}{C}}}\blacktriangleleft OR''\right]$$

$$\rightleftharpoons \left[R-\underset{H}{\overset{H}{\underset{|}{N}}}\oplus \underset{R'}{\overset{\bar{O}}{\underset{|}{C}}}\oplus + {}^{\ominus}OR''\right] \rightleftharpoons R-NH-CO-R' + HOR''$$

These polymers react readily with compounds containing an aliphatic amino group in the molecule (BAC very frequently possess such a group), with formation of the amide bond. If this bond originates in an amino acid specific for a certain enzyme, an enzymatically cleavable bond is formed. Specific acids for chymotrypsin are L-phenylalanine, L-tyrosine and L-leucine; with trypsin, for example, it is L-lysine.

[Structural diagram of methacrylamide copolymer with pendant peptide-drug model; the dashed bond marked with * indicates:]

* Bond cleavable with chymotrypsin (originating in L-phenylalanine)

← Drug model

If a degradable carrier is to be prepared, the copolymers of N-(2-hydroxypropyl)methacrylamide with p-nitrophenyl esters of N-methacryloylated derivatives of oligopeptides should be reacted with diamines. In such a case, three possibilities should be contemplated: cyclization, formation of free $-NH_2$ groups, and cross-linking (cross-linking below the gel point).

[Reaction scheme showing Gly-Phe-ONp side chains reacting with $NH_2-(CH_2)_6-NH_2$ via routes a (cyclization), b (formation of free NH_2 groups), and c (crosslinking), giving products a, b, c.]

The course of the reaction just mentioned can be controlled by choosing suitable reaction conditions. The participation of the cross-linking reaction is pronounced only[100] if the ratio of p-nitrophenyl (ONp) and NH_2 groups is 1:1. At a higher level virtually all the diamine is consumed by the formation of free $-NH_2$ groups or by cyclization. The polymer containing free NH_2 groups can be reacted with that containing ONp groups with formation of cross-linked polymers not containing small rings.

Hence, it may be stated in summary that additional cross-linking below the gel point of the copolymers under study may be carried out in two ways:

1. In one state — i.e., by reacting the copolymer containing ONp groups with diamine
2. In two stages — by preparing a polymeric amine and reacting the latter with the copolymer containing ONp groups

The limitation of intramolecular cyclization in the two-stage cross-linking leads to a greater increase in molecular weight.

B. Degradability of Bonds in Side Chains of Soluble Synthetic Polymers

An investigation of bond-splitting in side chains of synthetic polymers provides us with the necessary material for designing the structure of biodegradable polymeric carriers containing degradable bonds in the main chain, and at the same time contributes to the study of attachment of BAC to polymeric carriers via enzymatically cleavable bonds. The necessary condition is that the degradable bond should originate in the amino acid specific for the given enzyme. The splitting in the organism may occur both in blood circulation and after penetration of the polymer into the cell by means of endocytosis[7] through lysosomal enzymes. It is obvious that the polymer molecule can hinder the formation of the enzyme-substrate complex; in such case, the simplest approach consists in binding the compound onto the polymer chain by means of a spacer. Jatzkewitz[104,105] was the first to suggest this method of binding in 1954. He bound the psychopharmaceutic drug mescaline on the copolymer of vinylpyrrolidone with acrylic acid using a dipeptidic link, glycylleucine.

$$-CH_2-CH-CH_2-CH-CH_2-CH-CH_2-CH-$$

(structural formula showing vinylpyrrolidone units, COOH unit, and unit bearing CO—Gly—Leu—Mes side chain)

$$Mes = -NH-CH_2-CH_2-\text{C}_6\text{H}_2(OCH_3)_3$$

Mescaline alone and glycyl-L-leucylmescaline are eliminated from the organism of the test animal within about 24 hr, but the polymeric pharmaceutic drug provided gradual release of mescaline over 17 days. Obviously, the pharmaceutical drug is eliminated in this case by the action of proteolytic enzymes.

Model experiments using p-nitroanilides (as drug models) were performed by Fu and Morawetz,[106] and at the Institute of Macromolecular Chemistry of the Czechoslovak Academy of Sciences in Prague.[107] The results of the investigation of the dependence on structure of the rate of cleavage of polymeric substrates with chymotrypsin obtained in these two cases shows that the rate of splitting of L-phenylalanine-p-nitroanilide residues is a function of their spacing from the polymer chain backbone.

The finding that the results of splitting of side chains could not be extrapolated to cross-links[101] led to a detailed study of the relationship between the structure and degradability of polymeric substrates. The possibility of cleavage is strongly dependent not only on steric factors, but also on interactions along the chain which is to be split. Our reasonings regarding the relationship between the structure and properties are based on the papers of Schechter and Berger.[108-111]

The active site of an enzyme performs the twofold function of binding a substrate and catalyzing a reaction. The efficiency of these actions determines the overall activity of the enzyme towards the particular substrate, i.e., the specificity of the enzyme.[101]

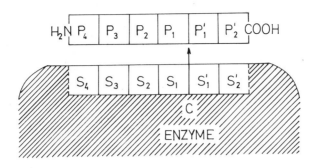

FIGURE 4. Schematic representation of the possible enzyme-substrate complex.[108-111]

Since the active site of proteolytic enzymes is rather large,[111,112] capable of combining with a number of amino acid residues, it is convenient to subdivide the binding site into subsites. A subsite is defined as a region on the enzyme surface which interacts with one amino acid residue of the substrate (Figure 4). The substrates are lined up on the enzyme in such a way that the CO-NH group being hydrolyzed always occupies the same place (the catalytic site); the amino acid residues occupy adjacent subsites, those towards the NH_2 end occupying subsites S_1, S_2, etc., those towards the COOH end occupying subsites S_1', S_2' and so on. The effects of structure on properties may be conveniently described as S_1-P_1; S_2-P_2; S_3-P_3, interactions.

Chymotrypsin was the enzyme chosen for model studies; its active site has at least four subsites towards the NH_2 end of the substrate[113] and at least two subsites towards the COOH end of the substrate.[114] The effect of S_1-P_1 and S_2-P_2 interactions was investigated by using a series of copolymers of N-(2-hydroxypropyl)methacrylamide prepared by us, which contained

```
     Gly
      |
      X           X = Gly, Ala, β-Ala, Val, Leu, Ileu, Phe
      |
      Y           Y = Phe, Tyr
    - -|- -
     NAp
```

where Gly . . . glycine; Ala . . . L-alanine; β-Ala . . . β-alanine; Val . . . L-valine; Leu . . . L-leucine; Ilue . . . L-isoleucine; Phe . . . L-phenylalanine; Tyr . . . L-tyrosine; NAp . . . p-nitroanilide, the latter modeling biologically active compounds.

In position P_3, glycine was chosen for all substrates, because subsite S_3 of chymotrypsin is occupied by Gly-216.[113] Results of kinetic measurements[115] (determination of k_{cat} and K_M) show that in position P_1, Tyr is more advantageous than Phe. In position P_2, hydrophobic amino acids (Val, Ileu, Leu) are more suitable; the splitting proceeds more slowly if X = Gly or β-Ala. This is due to the fact that both Gly and β-Ala interact only with difficulty with Ileu-99.[113] The former is not capable of hydrophobic interactions, while the structure of the latter is not favorable to their formation (absence of α-CH_3 group).

C. Destructibility of Cross-links

When developing structures for copolymers of N-(2-hydroxypropyl)methacrylamide containing cross-links, (remembering that the cross-linking occurs below the gel point, i.e., the resulting polymers are soluble) suitable for degradation with chymotrypsin, results obtained in investigations of the splitting of polymer side-chains were used. In the initial stages of the investigation it was found[101] that the splitting of those structures which were cleaved in side-chains (bonds originating from sequences -Gly-Phe; -Acap-Phe; -Acap-Leu) was very low in the case of cross-links. In all the polymers listed below (a-d), less than 10% of cross-links were split.[101]

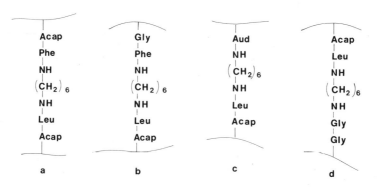

where Acap ... ε-aminocaproic acid; Aud ... ω-aminoundecanoic acid; Phe ... L-phenylalanine; Leu ... L-leucine; Gly ... glycine.

The search for the reasons for the insufficient and nonspecific splitting of cross-links by chymotrypsin in vitro was based on the following assumptions:

1. The effects of steric hindrance of polymer chains are regarded as decisive. With relatively short cross-links, steric hindrance is responsible for the inability of the substrate to assume a suitable conformation at the active site of chymotrypsin.
2. However, for the length of cross-links used in this study, steric hindrances were not regarded as the decisive factor for cross-links. Insufficient splitting is also due to the fact that the structure of cross-links is not optimal with respect to S-P interactions.

The possibility of eliminating steric hindrance was established by splitting copolymers cross-linked with the oxidized B-chain of insulin.[101] In this polymer, complete splitting of cross-links took place (100% splitting). This helped to demonstrate that, with a sufficient length of cross-link, the steric hindrance of the polymer chains can be neglected. On the other hand, owing to the different lengths of the cross-links, we cannot deduce from this whether or not the steric hindrance of polymer chains is the decisive factor affecting the splitting of substrates containing sequences -Gly-Phe-, -Acap-Phe-, or -Acap-Leu-. To answer this question, we tried to optimize the structure of the cross-links under study.[101] The degradability was raised considerably by incorporating the sequence -Gly-Phe-Phe-. This structure takes into account interactions S_1-P_1, S_2-P_2, S_3-P_3 and corresponds to the structure of the tripeptide 23-25 of the oxidized B-chain of insulin (one of the bonds broken by chymotrypsin originates in 25th amino acid). Arguments leading to the choice of this structure are described in detail in the final part of Reference 101. The result clearly indicates that in the cleavage of cross-links, one must very carefully choose a structure which makes possible interactions with the highest possible number of subsites of the active site of the enzyme; this is much more important with cross-links than it is with low molecular weight substrates or side chains. The more important conclusion, however, is to identify and develop the right method and technique for preparing enzymatically degradable polymers.

These studies were extended to include a detailed investigation on the contribution[116] of S_1-P_1, S_2-P_2, S_3-P_3, S_4-P_4, and S_1'-P_1' subsite interactions to the degradability of the polymers studied. We prepared and subjected to cleavage, polymeric substrates in which the cross-links connecting the poly[N-(2-hydroxypropyl)methacrylamide] chains contained an oligopeptidic sequence of 3 to 5 amino acids, cleavable by chymotrypsin: -Gly-X-Y- (Gly . . . glycine; X . . . glycine, L-alanine, β-alanine, L-valine, L-leucine, L-isoleucine, L-phenylalanine; Y . . . L-phenylalanine, L-tyrosine); -Gly-Gly-Phe-Y- (Phe . . . L-phenylalanine); -W-Gly-Val-Y- (W . . . L-alanine, glycine; Val . . . L-valine); and -Gly-Phe-Y-W.

Several examples may be given to illustrate the possible ways of preparing cleavable cross-links taking into account S-P interactions:[116]

```
    Gly            Gly            Gly            Gly
    |              |              |              |
    Ala            Val            Phe            Gly
    |              |              |              |
    Phe            Phe            Phe            Phe
    |              |              |              |
    NH             NH             NH             Phe
    |              |              |              |
    (CH₂)₆         (CH₂)₆         (CH₂)₆         NH
    |              |              |              |
    NH             NH             NH             (CH₂)₆
    |              |              |              |
    Phe            Phe            Phe            NH
    |              |              |              |
    Ala            Val            Phe            Phe
    |              |              |              |
    Gly            Gly            Gly            Phe
                                                 |
                                                 Gly
                                                 |
     e  10%        f  45%         g  70%         Gly

                                                 h  100%
```

where the symbols have the meanings specified above. Figures in brackets give percentage of cross-links which undergo cleavage.

A comparison of structures a to h allows us to estimate qualitatively the effect of the detailed structure of the cross-links on their degradability with chymotrypsin in vitro (conditions of splitting cf.[101,116]). Differences in the degradability of polymers e,f,g may be assigned predominantly to structural effects (influence on S_2-P_2 interactions); the difference between g and h is mainly due to steric effects. Polymer h undergoes 100% splitting, the same as the polymer cross-linked with the oxidized insulin B-chain; the cross-link in polymer h is much shorter, however. This finding is an important guideline for the determination of the influence of the polymer chain on the possibility of formation of the enzyme-substrate complex.

An attempt was made to explain the effect of steric hindrance of polymer chains on enzymatic degradability by studying the hydrolysis of polymeric substrates catalyzed by polymer-bound chymotrypsin.[115] Preparation of trypsin[117] and papain[118] cleavable polymers can be regarded as a proof of the possibility of extending the suggested procedure for the synthesis of enzymatically degradable polymers to other enzymes.[118] Recently it was shown that the oligopeptide sequences in N-(2-hydroxypropyl)methacrylamide copolymers are cleaved as in vitro by intracellular (lysosomal) enzymes[119] as in vivo.[120]

REFERENCES

1. **Lee, H. and Neville, K.**, *Handbook of Biomedical Plastics,* Pasadena Technology Press, Pasadena, 1971.
2. **Williams, D., Ed.**, *Biocompatibility of Implant Materials,* Pitman Medical Publishing Co., London, 1976.
3. **Gregor, H. P., Ed.**, *Biomedical Application of Polymers,* Plenum Press, New York, 1975.
4. **Kronenthal, R. L., Oser, Z., and Martin, E., Eds.**, *Polymers in Medicine and Surgery,* Plenum Press, New York, 1975.
5. **Williams, D. F. and Roaf, R.**, *Implants in Surgery,* W.B. Saunders Co., London, 1973.
6. **Kopeček, J.**, Soluble biomedical polymers, *Polym. Med. (Wroclaw),* 7, 191, 1977.
7. **Ringsdorf, H.**, Structure and properties of pharmacologically active polymers, *J. Polym. Sci. Polym. Symp.,* 51, 135, 1975.
8. **Donaruma, L. G.**, Synthetic biologically active polymers, *Prog. Polym. Sci.,* 4, 1, 1974.
9. **Batz, H. G.**, Polymeric drugs, *Adv. Polym. Sci.,* 23, 25, 1977.
10. **Drobník, J.**, *The Use of Polymers in Pharmacology,* Institute of Macromolecular Chemistry Press, Prague, 1977.
11. **Appel, W. and Bickert, E.**, Substitution of plasma by high-molecular compounds (in German), *Angew. Chem.,* 80, 719, 1968.
12. **Lutz, H.**, *Plasma Substitutes,* 2nd ed., Thieme, Stuttgart, 1975.
13. **Labrude, P. and Vigneron, C.**, Blood substitutes. General review and perspectives, *Lyon Pharm.,* 26, 629, 1975.
14. **Gruber, U. F.**, *Blood Substitution (in German),* Springer-Verlag, Berlin, 1968.
15. **Schlipköter, H. W. and Brockhaus, A.**, Prevention of experimental silicosis by subcutaneous application of poly(vinylpyridine-N-oxide) (in German), *Klin. Wochenschr.,* 39, 1182, 1961.
16. **Grönvall, A.**, *Dextran and its use in Colloidal Infusion Solutions,* Almquist and Wiksell, Uppsala, 1957.
17. **Zaffaroni, A. and Bonsen, P.**, Controlled chemotherapy through macromolecules, in *Polymeric Drugs,* Donaruma, L. G. and Vogl, O., Eds., Academic Press, New York, 1978.
18. **Hardwicke, J., Hulme, B., Jones, J. H., and Rickets, C. R.**, Measurement of glomerular permeability to polydisperse radioactively labelled macromolecules in normal rabbits, *Clin. Sci.,* 34, 505, 1968.
19. **Hulme, B. and Hardwicke, J.**, Human glomerular permeability to macromolecules in health and disease, *Clin. Sci.,* 34, 515, 1968.
20. **Hulme, B., Dykes, P. W., Appleyard, I., and Arkell, R. W.**, Retention and storage sites of radioactive polyvinylpyrrolidone, *J. Nucl. Med.,* 9, 389, 1968.
21. **Kopeček, J.**, Interaction of the solutions of hydrophilic polymers with blood, Lecture D9/2 presented on the 2nd Disc. Conf. Macromolecular Matrices and Carriers Biol. Functions, Prague, 1972.
22. **Kopeček, J.**, Reactive copolymers of N-(2-hydroxypropyl)methacrylamide with derivatives of L-leucine and L-phenylalanine. I. Preparation, characterization and reaction with diamines, *Makromol. Chem.,* 178, 2169, 1977.
23. **Kopeček, J. and Rejmanová, P.**, Reactive copolymers of N-(2-hydroxypropyl)methacrylamide with derivatives of L-leucine and L-phenylalanine. II. Reactions with polymeric amine and stability of crosslinks towards chymotrypsin in vitro, *J. Polym. Sci. Polym. Symp.,* 66, 15, 1979.
24. **Kopeček, J.**, Polymeranalogous reactions leading to enzymatically degradable polymers, presented at Int. Symp. Macromolecular Chemistry, Tashkent, U.S.S.R., October 17 to 21, 1978.
25. *Blood Flow Improvement,* Pharmacia AB, Uppsala, 1968.
26. **Doenicke, A., Grote, B., and Lorenz, W.**, Blood and blood substitutes, *Br. J. Anaesth.,* 49, 681, 1977.
27. **Arturson, G., Granath, K., Thorén, L., and Wallenius, G.**, The renal excretion of low molecular weight dextran, *Acta Chir. Scand.,* 127, 543, 1964.
28. **Wiedersheim, M.**, An investigation of oxyethylstarch as a new plasma volume expander in animals, *Arch. Int. Pharmacodyn. Ther.,* 111, 353, 1957.
29. **Banks, W., Greenwood, C. T., and Muir, D. D.**, Studies on hydroxyethyl starch. I. A review of the chemistry of hydroxyethyl starch, with reference to its use as a blood plasma volume expander, *Stärke,* 24, 181, 1972.
30. **Ring, J. and Messmer, K.**, Infusion therapy with colloidal blood volume expanders, *Anaesthesist,* 26, 279, 1977.
31. **Hartel, W.**, Blood volume expanders. Possibilities and indications, *Fortschr. Med.,* 89, 903, 1971.
32. **Campbell, D. H., Koepfli, J. B., Pauling, L., Abrahamson, N., Dandliker, W., Feigen, G. A., Lanni, F., and Rosen, le A.**, The preparation and properties of a modified gelatin (oxypolygelatin) as an oncotic substitute for serum albumin, *Tex. Rep. Biol. Med.,* 9, 235, 1951.

33. Tourtelotte, D. and Williams, H. E., Acylated gelatins and their preparation, U.S. Patent 2,827,419, 1958.
34. Schmidt-Thome, J., Mager, A., and Schöne, H. H., Chemistry of a new blood plasma expander (in German), *Arzneim. Forsch.*, 12, 378, 1962.
35. Rypáček, F., Fluorescent labelling of polymers for biomedical applications (in Czech), Ph.D. thesis, Institute of Macromolecular Chemistry, Prague, Czechoslovakia, 1978.
36. Hecht, G. and Weese, H., Periston a new blood plasma expander (in German), *Münch. Med. Wochenschr.*, 90, 11, 1943.
37. Ammon, R. and Depner, E., Elimination and deposition of different types of polyvinylpyrrolidone in the organism (in German), *Z. Ges. Exp. Med.*, 128, 607, 1957.
38. Vogel, G., Ströcker, H., and Höller, M., Penetration of polyvinylpyrrolidone through the plasma-lymph barrier in rats, *Pflügers Arch. Ges. Physiol.*, 279, 187, 1964.
39. Hespe, W., Meier, A. M., and Blankwater, Y. J., Excretion and distribution studies in rats with two forms of ^{14}C-labelled polyvinylpyrrolidone with a relatively low mean molecular weight after intravenous administration, *Arzneim. Forsch.*, 27, 1158, 1977.
40. Hespe, W., Blankwater, Y. J., and Wieriks, J., A combined study on the distribution of tetracycline and polyvinylpyrrolidone in rats, *Arzneim. Forsch.*, 25, 1561, 1975.
41. Ravin, H. A., Seligman, A. M., and Fine, J., Polyvinylpyrrolidone as a plasma expander. Studies on its excretion, distribution and metabolism, *N. Engl. J. Med.*, 247, 921, 1952.
42. Hecht, G. and Scholtan, W., About elimination of polyvinylpyrrolidone through normal kidneys (in German), *Z. Ges. Exp. Med.*, 130, 577, 1959.
43. Marek, H., Matzkowski, H., and Koch, H., Autoradiographical investigation of the time dependent deposition of ^{3}H-polyvinylpyrrolidone in organs of the rat (in German), *Zschr. Inn. Med.*, 23, 233, 1968.
44. Gärtner, K., Vogel, G., and Ulbrich, M., Investigation of penetration of macromolecules (polyvinylpyrrolidone) through glomerular and postglomerular capillaries into urine and into lymph in the kidney and investigation of the amount of exchange of ^{131}I-albumin between intra- and extravasal space in the kidney (in German), *Pflügers Arch. Ges. Physiol.*, 298, 305, 1968.
45. Steele, R., Van Slyke, D., and Plazin, J., The fate of intravenously administered polyvinylpyrrolidone, *Ann. N.Y. Acad. Sci.*, 55, 479, 1952.
46. Mück, K. F., Christ, O., and Kellner, H. M., Behaviour of isotactic poly(acrylic acid)s in organism. Distribution and excretion with mice (in German), *Makromol. Chem.*, 178, 2785, 1977.
47. Hueper, W. C., Macromolecular substances as pathogenic agents, *Arch. Pathol.*, 33, 267, 1942.
48. Cabanne, F., Michielis, R., Duserre, P., Bastien, H., and Justrab, E., Polyvinylic disease (in French), *Ann. Anat. Pathol.*, 14, 419, 1969.
49. Ring, J. and Messmer, K., Anaphylactic reactions after infusion of blood volume expanders (in German), *Pädiatr. Prax.*, 17, 283, 1976.
50. Williams, K. E., Kidston, E. M., Beck, F., and Lloyd, J. B., Quantitative studies of pinocytosis. I. Kinetics of uptake of ^{125}I-polyvinylpyrrolidone by rat yolk sac cultured in vitro, *J. Cell Biol.*, 64, 113, 1975.
51. Roberts, A. V. S., Williams, K. E., and Lloyd, J. B., The pinocytosis of ^{125}I-labelled polyvinylpyrrolidone, ^{14}C-sucrose and colloidal ^{198}Au by rat yolk sac cultured in vitro, *Biochem. J.*, 168, 239, 1977.
52. Pratten, M. K., Williams, K. E., and Lloyd, J. B., A quantitative study of pinocytosis and intracellular proteolysis in rat peritoneal macrophages, *Biochem. J.*, 168, 365, 1977.
53. Beuchelt, H., The effect of Periston N (in German), in *Medizin und Chemie Bayer Leverkusen*, Vol. 5, Verlag Chemie, Weinheim, 1956.
54. Schubert, R., About elimination of compounds bound to Periston (bilirubin, indigocarmin) and something about the elimination of Periston itself (in German), *Klin. Wochenschr.*, 26, 143, 1948.
55. Schubert, R. and Seybold, G., The elimination of granular deposited compounds from certain cells by Kollidon of different molecular weight, *Ärztl. Forsch.*, 6, 1, 1952.
56. Ansel, R., Intorp, H., and Weisschedel, E., Investigation of the effect of polyvinylpyrrolidone on tetanus toxin, *Arzneim. Forsch.*, 13, 949, 1963.
57. Vasiliev, P. S., Suzdaleva, V. V., Guldbadamova, N. M., Kiseleva, A. A., and Vidavskaya, G. M., Comparative evaluation of PVP preparations of different molecular weights, *Bibl. Haematol.*, 38(Suppl. 2), 777, 1971.
58. Kronenthal, R. L., Biodegradable polymers in medicine and surgery, in *Polymers in Medicine and Surgery*, Kronenthal, R. L., Oser, Z., and Martin, E., Eds., Plenum Press, New York, 1975, 119.
59. Derrick, J. R. and Guest, M. M., Eds., *Dextrans. Current Concepts of Basic Actions and Clinical Applications*, Charles C. Thomas, Springfield, 1968.
60. Hymes, A. C., Physiological salt solutions of ethylene oxide propylene glycol condensation products as blood plasma substitutes, U.S. Patent 3,590,125, 1969.

61. Neri, P., Antoni, G., and Benvenuti, F., Method of preparation of α,β-poly(hydroxyalkylaspartamides) and their therapeutical application (in German), *Ger. Offenlegungsschrift*, 2, 032, 470, 1971.
62. Neri, P., Antoni, C., Benvenuti, F., Cocola, F., and Gazzei, G., Synthesis of α,β-poly(2-hydroxyethyl)-D,L-aspartamide, a new plasma expander, *J. Med. Chem.*, 16, 893, 1973.
63. Antoni, G., Neri, P., Pedersen, T. G., and Hesen, M. O., Hydrodynamic properties of a new plasma expander: polyhydroxyethylaspartamide, *Biopolymers*, 13, 1721, 1974.
64. Kopeček, J., Šprincl, L., and Lím, D., New types of synthetic infusion solutions. I. Investigation of the effect of solutions of some hydrophilic polymers on blood, *J. Biomed. Mater. Res.*, 7, 197, 1973.
65. Štěrba, O., Paluska, E., Jozova, O., Špunda, J., Nezvalová, J., Šprincl, L., Kopeček, J., and Činátl, J., New types of synthetic infusion solutions. II. Basic biological properties of poly-[N-(2-hydroxypropyl)methacrylamide] (in Czech), *Čas. Lék. Česk.*, 114, 1268, 1975.
66. Šprincl, L., Exner, J., Štěrba, O., and Kopeček, J., New types of synthetic infusion solutions. III. Elimination and retention of poly[N-(2-hydroxypropyl)methacrylamide[in test organism, *J. Biomed. Mater. Res.*, 10, 953, 1976.
67. Korčáková, L., Paluska, E., Hašková, V., and Kopeček, J., A simple test for antigenicity of colloidal infusion solutions — the draining lymph node activation, *Z. Immunitaetsforsch.*, 151, 219, 1976.
68. Kopeček, J. and Bažilová, H., Poly[N-(2-hydroxypropyl)methacrylamide]. I. Radical polymerization and copolymerization, *Eur. Polym. J.*, 9, 7, 1973.
69. Bohdanecký, M., Bažilová, H., and Kopeček, J., Poly[N-(2-hydroxypropyl)methacrylamide]. II. Hydrodynamic properties of diluted polymer solutions, *Eur. Polym. J.*, 10, 405, 1974.
70. Strohalm, J. and Kopeček, J., Poly[N-(2-hydroxypropyl)methacrylamide]. IV. Heterogeneous polymerization, *Angew. Makromol. Chem.*, 70, 109, 1978.
71. Rosenberg, G. J., Perspectives on the problem of blood substitutes, *Bibl. Haematol.*, 38 (Part II), 737, 1971.
72. Chang, T. M. S., *Artificial Cells*, Charles C. Thomas, Springfield, Ill., 1972.
73. Bayer, E. and Holzbach, G., Synthetic hemopolymers for reversible uptake of molecular oxygen (in German), *Angew. Chem.*, 89, 120, 1977.
74. Tsuchida, E., Honda, K., and Sata, H., Hydrophobic environmental effects on oxygenation of polymeric hemochrome in aqueous solutions, *Makromol. Chem.*, 176, 2251, 1975.
75. Sloviter, H. A. and Kamimoto, T., Erythrocyte substitute for perfusion of brain, *Nature (London)*, 216, 458, 1967.
76. Sloviter, H. A., Erythrocyte substitutes, *Med. Clin. N. Am.*, 54, 787, 1970.
77. Geyer, R. P., Whole animal perfusion with fluorocarbon dispersions, *Fed. Proc. Fed. Am. Exp. Biol.*, 29, 1758, 1970.
78. Geyer, R. P., A fluorocarbon-polyol mixture for essentially total replacement of blood in vivo, *Bibl. Haematol.*, 38 (Part II), 802, 1971.
79. Riess, J. C. and LeBlanc, M., Perfluorocompounds as blood substitutes (in German), *Angew. Chem.*, 90, 654, 1978.
80. Schlipköter, H. W. and Brockhaus, A., Effect of polyvinylpyridine on experimental silicose (in German), *Dtsch. Med. Wochenschr.*, 85, 920, 1960.
81. Klosterköter, W. and Gono, F., *Results of the Investigation in the Field of Supression of Dust Effects and Silicose in the Mining Industry*, Vol. 7, Bösman, Detmold, 1969.
82. Holt, P. F., Lindsay, H., and Beck, E. G., Some derivatives of poly(vinylpyridine-1-oxides) and their effect on the cytotoxicity of quartz in macrophage cultures, *Br. J. Pharmacol.*, 38, 192, 1970.
83. Donaruma, L. G., Ottenbrite, R. M., and Regelson, W., Biological activity, *Encyclopedia Polymer Science Technology*, Suppl. Vol. 2, Interscience, New York, 1977, 113.
84. Ottenbrite, R. M., Regelson, W., Kaplan, A., Carchman, R., Morahan, P., and Munson, A., Biological activity of polycarboxylic acid polymers, in *Polymeric Drugs*, Donaruma, L. G. and Vogl, O., Eds., Academic Press, New York, 1978, 263.
85. Levy, H. B., Polymers as interferon inducers, in *Polymeric Drugs*, Donaruma, L. G. and Vogl, O., Eds., Academic Press, New York, 1978, 305.
86. Mück, K. F., Rolly, H., and Burg, K., Preparation and antiviral activity of poly(acrylic acid) and poly(methacrylic acid), *Makromol. Chem.*, 178, 2773, 1977.
87. Solovskij, M. V., Panarin, E. F., Vershinina, T. M., and Kropachev, V. A., Synthesis and antimicrobial properties of monomeric and polymeric quaternary ammonium salts based on aminoalkyl esters of methacrylic acid (in Russian), *Khim. Farm. Zh.*, 8, 20, 1974.
88. Franchi, G., Morsca, L., Reyers-Degli-Innocenti, I., and Garattini, S., Triton WR 1339 (TWR), an inhibitor of cancer dissemination and metastases, *Eur. J. Cancer*, 7, 533, 1971.
89. Cornforth, J. W., Morgan, E. D., Potts, K. T., and Rees, R. J. W., Preparation of antituberculous polyoxyethylene ethers of homogeneous structure, *Tetrahedron*, 29, 1659, 1973.
90. Samour, C. M., Polymeric drugs in the chemotherapy of microbial infections, in *Polymeric Drugs*, Donaruma, L. G. and Vogl, O., Eds., Academic Press, New York, 1978, 161.

91. DeDuve, C., de Barsy, T., Poole, B., Trouet, A., Tulkens, P., and van Hoof, F., Lysosomotropic agents, *Biochem. Pharmacol.*, 23, 2495, 1974.
92. Trouet, A., Deprez-de Campeneere, D., and deDuve, D., Chemotherapy through lysosomes with a DNA-daunorubicin complex, *Nature (London) New Biol.*, 239, 110, 1972.
93. Gregoriadis, G., Targeting of drugs, *Nature (London)*, 265, 407, 1977.
94. Hurwitz, E., Levy, R., Maron, R., Wilchek, M., Arnon, R., and Sela, M., The covalent binding of daunomycin and adriamycin to antibodies with retention of both drug and antibody activities, *Cancer Res.*, 35, 1175, 1975.
95. Levy, R., Hurwitz, E., Maron, R., Arnon, R., and Sela, M., The specific cytotoxic effects of daunomycin conjugated to antitumour antibodies, *Cancer Res.*, 35, 1182, 1975.
96. Goldberg, E. P., Polymeric affinity drugs for cardiovascular cancer and urolithiasis therapy, in *Polymeric Drugs*, Donaruma, L. G. and Vogl, O., Eds., Academic Press, New York, 1978, 239.
97. Wilchek, M., Affinity therapy, presented at the 4th Enzyme Engineering Conf., Bad Neuenahr, Federal Republic of Germany, 1977.
98. Gill, T. J., III, and Kunz, H. W., The immunogenicity of vinyl polymers, *Proc. Natl. Acad. Sci. U.S.A.*, 61, 490, 1968.
99. Richter, A. W., Immunological in vivo and in vitro studies of the dextran-antidextran system, Monograph, Department Clin. Chem., University Hospital, Uppsala, and the Research Division, Pharmacia AB, Uppsala, 1973.
100. Gill, T. J., III., Kunz, H. W., and Papermaster, D. S., Studies on synthetic polypeptide antigens, XVIII. The role of composition, charge, and optical isomerism in the immunogenicity of synthetic polypeptides, *J. Biol. Chem.*, 242, 3308, 1967.
101. Dintzis, H. M., Dintzis, R. Z., and Vogelstein, B., Molecular determinants of immunogenicity: the immunon model of immune response, *Proc. Natl. Acad. Sci. U.S.A.*, 73, 3671, 1976.
102. Rejmanová, P., Labský, J., and Kopeček, J., Aminolyses of monomeric and polymeric p-nitrophenyl esters of methacryloylated amino acids, *Makromol. Chem.*, 178, 2159, 1977.
103. Chytrý, V., Vrána, A., and Kopeček, J., Synthesis and activity of a polymer which contains insulin covalently bound on a copolymer of N-(2-hydroxypropyl)methacrylamide and N-methacryloyl-4-nitrophenyl ester, *Makromol. Chem.*, 179, 329, 1978.
104. Jatzkewitz, H., Binding of physiologically active compounds to blood plasma expander (in German), *Hoppe-Seyler's Z. Physiol. Chem.*, 297, 149, 1954.
105. Jatzkewitz, H., Peptamine(glycyl-L-leucyl-mescaline) bound to blood plasma expander (polyvinylpyrrolidone) as a new depot form of a biologically active primary amine (mescaline), (in German), *Z. Naturforsch.*, 10 b, 27, 1955.
106. Fu, T. Y. and Morawetz, H., Enzymatic attack on side chains of synthetic polymers, *J. Biol. Chem.*, 251, 2083, 1976.
107. Drobník, J., Kopeček, J., Labský, J., Rejmanová, P., Exner, J., Saudek, V., and Kálal, J., Enzymatic hydrolysis of side chains in synthetic water soluble polymers, *Makromol. Chem.*, 177, 2833, 1976.
108. Schechter, I. and Berger, A., On the size of the active site in proteases. I. Papain, *Biochem. Biophys. Res. Comm.*, 27, 157, 1967.
109. Abramowitz, N., Schechter, I., and Berger, A., On the size of the active site in proteases. II. Carboxypeptidase-A, *Biochem. Biophys. Res. Comm.*, 29, 862, 1967.
110. Schechter, I. and Berger, A., On the size of the active site in proteases. III. Mapping of the active site of papain; specific peptide inhibitors of papain, *Biochem. Biophys. Res. Comm.*, 32, 898, 1968.
111. Berger, A. and Schechter, I., Mapping the active site of papain with the aid of peptide substrates and inhibitors, *Philos. Trans. R. Soc. London*, B 257, 249, 1970.
112. Morihara, K. and Oka, T., The complex active sites of bacterial proteases in relation to their specificities, *Biochem. Biophys. Res. Comm.*, 30, 625, 1968.
113. Kurachi, K., Powers, J. C., and Wilcox, P. E., Kinetics of the reaction of chymotrypsin-A_a with peptide chloromethyl ketones in relation to its subsite specificity, *Biochemistry*, 12, 771, 1973.
114. Morihara, K., Oka, T., and Tsuzuki, H., Comparison of α-chymotrypsin and subtilisin BPN: size and specificity of the active site, *Biochem. Biophys. Res. Comm.*, 35, 210, 1969.
115. Kopeček, J., Rejmanová, P., and Chytrý, V., Polymers containing enzymatically degradable bonds. I. Chymotrypsin catalyzed hydrolysis of p-nitroanilides of phenylalanine and tyrosine attached to side chains of copolymers of N-(2-hydroxy-propyl)methacrylamide, *Makromol. Chem.*, in press.
116. Rejmanová, P., Obereigner, B., and Kopeček, J., Polymers containing enzymatically degradable bonds. II. Poly [N-(2-hydroxypropyl)methacrylamide] chains connected by oligopeptide sequences cleavable by chymotrypsin, *Makromol. Chem.*, in press.
117. Ulbrich, K., Strohalm, J., and Kopeček, J., Polymers containing enzymatically degradable bonds. III. Poly[N-(2-hydroxy-propyl)methacrylamide] chains connected by oligopeptide sequences cleavable by trypsin, *Makromol. Chem.*, in press.

118. Ulbrich, K., Zacharieva, E. I., Obereigner, B., and Kopeček, J., Polymers containing enzymatically degradable bonds. V. Hydrophilic polymers degradable by papain, *Biomaterials*, 1, 199, 1980.
119. Duncan, R., Lloyed, J. B., and Kopeček, J., Degradation of side chains of n-(2-hydroxypropyl)methacrylamide copolymers by lysosomal enzymes, *Biochem. Biophys. Res. Commun.*, 94, 284, 1980.
120. Kopeček, J., Cifková, I., Rejmanová, P., Strohalm, J., Obereigner, B., and Ulbrich, K., Polymers containing enzymatically degradable bonds. IV. Preliminary experiments in vivo, *Makromol. Chem.*, in press.

Chapter 9

BIOCOMPATIBILITY AND EXPERIMENTAL THERAPY OF IMMOBILIZED ENZYMES AND PROTEINS*

Thomas Ming Swi Chang

TABLE OF CONTENTS

I.   Polymer Immobilization of Enzymes and Proteins 180
     A.   Covalent Linkage Method 180
     B.   Physical Adsorption 181
     C.   Matrix Entrapment 181
     D.   Microencapsulation 181

II. Localization of Immobilized Enzyme and Protein for In Vivo Applications 182
     A.   Local Implantation 182
     B.   Intravenous Injection 182
     C.   Extracorporeal Shunt Systems 182
     D.   Local Topical Applications 183
     E.   Administration of Immobilized Enzymes into the Gastrointestinal Tract 183

III. Experimental Therapy 183
     A.   Red Blood Cells 183
     B.   Hereditary Enzyme Deficiency 183
     C.   Substrate Dependent Tumor 183
     D.   Organ Failure and Metabolic Disorders 184

References 184

---

*   The support of the Medical Research Council of Canada (MRC-SP-4) is gratefully acknowledged. The author is an MRC Career Investigator.

I. POLYMER IMMOBILIZATION OF ENZYMES AND PROTEINS

Introduction of heterogeneous and other foreign proteins into the body have potential problems of producing hypersensitivity reactions, immunological reactions, rapid removal, and inactivation of the proteins. Enzymes in solution at body temperature are also unstable and free enzymes cannot be kept at a site where the action is desired. As a result of these problems, despite their extremely important role in the total body metabolism and other functions, enzymes are used very rarely in therapy.

With the rationale that most of the enzymes in the body are immobilized intracellularly, either in free solution or in combination with intracellular membranes, studies have been carried out on the possible use of immobilized enzymes and proteins for therapeutic applications. The first reported study on the use of immobilized enzymes and proteins in experimental therapy was in the form of artificial cells consisting initially of microencapsulated enzymes.[1,2] Since then an increasing number of laboratories have carried out studies on the possible therapeutic application of immobilized enzymes and proteins. These have been reviewed in detail elsewhere.[3-8]

Historically, enzyme research has gone through several different stages. The first stage involved the extraction, purification, and crystallization of enzymes from cells or other sources. The enzymes obtained this way are used in detailed studies of enzyme kinetics. The next stage involved the attachment of enzymes and proteins to a solid support system. This was done with the hope that this would give a better resemblance to the kinetics of membrane-bound enzymes in nature. Stage three involved the incorporation of enzymes and proteins in artificial cells to afford an artificially synthesized intracellular environment.

Immobilized enzymes have been classified into four major types. The first three, representing stage two of this research, are divided into covalently bound, adsorbed, and matrix-entrapped enzymes. The fourth type, representing stage three of the enzyme research, is in the form of microencapsulated enzymes or artificial cells. Detailed description of the different types of immobilized enzyme systems have been reviewed elsewhere.[4,6,8-15]

A. Covalent Linkage Method

This includes the binding of enzymes to water insoluble carriers by covalent binding. In the covalent reaction the amino, carboxyl, sulfhydryl, hydroxyl, phenolic, and possibly other groups of the enzyme molecules, are used to react with carriers containing reactive species such as diazonium, isocyanate, halides, and acid azide. Another form of covalent linkage consists of the use of bifunctional agents, such as glutaraldehyde, for intermolecular linkage whereby enzyme molecules could be linked with one another and with the bifunctional agents.

The major advantage of covalent linkage of enzymes is the increase in thermal and storage stability in some cases. Furthermore, after covalent binding there is little or no leakage of the enzyme. However, the reaction for covalent binding is not particularly mild, and as a result it may affect the conformational structure and perhaps also the active center of the enzyme, resulting in some loss of activity. Insofar as biomedical applications are concerned, since a significant proportion of covalently-bound enzyme is exposed to the surface, the enzyme would come in contact with other proteins, antibodies, or cells when placed in the body, possibly resulting in immunological problems. On the other hand, in the use of immobilized antibodies or antigens for the removal of circulating antigens or antibodies, covalent binding would be an advantage. Covalently-linked enzymes could also be microencapsulated in artificial cells to prevent their immunological reaction and at the same time taking advantage of their increased stability.

B. Physical Adsorption

Immobilization can be carried out by adsorption of enzymes or proteins onto the surface of water-insoluble carriers. Organic materials such as activated carbon, porous glass, acid clay, kaolinide, clay, silica, alumina, and polymers like starch and gluten have been used. Ionic binding of enzymes and proteins to water-soluble ion-exchange carriers can also be used with, for instance, polysaccharides and synthetic polymers with ion-exchange groups. These adsorption methods involve very mild reactions with very little conformational or structural changes taking place in the proteins and very little destruction of the enzyme activity. Enzymes immobilized by this approach also have higher stability. The problem here, however, is that in this form, changes in pH, electrolyte concentrations, or temperature may result in the desorption of enzymes. As in the case of covalent linkage, adsorbed enzymes are exposed on the surface and as a result, immunological and hypersensitivity reactions can occur when used as such in the body. In therapeutic application, adsorbed urease has been microencapsulated to make use of its stability and ease of preparation to avoid the adsorbed material leaking out.

C. Matrix Entrapment

This method is based on entrapping molecules of enzymes in the lattice of solid polymer matrix. Synthetic polymers such as polyacrylamide and natural polymers like starch have all been used for the immobilization of enzymes by matrix-entrapment. In this method, the ratio of enzyme to matrix polymer is rather low, often in the region of 1 part of enzyme to 100 parts of matrix polymer. Immobilized enzyme of this type, when introduced into the body, will result in a very large polymer component in comparison to enzyme. Furthermore, much of the enzyme is located deep inside a solid polymer matrix, limiting the free diffusion of substrates. There is also some leakage of enzyme from the polymeric material because of the method of molecular entrapment. Increasing the cross-linking of the matrix material will decrease leakage, but this tends to greatly reduce the substrate permeability.[53] As a result, this type of matrix-entrapped enzyme has found much greater applications in analytical methods, preventive medicine, and industry.

D. Microencapsulation

In the use of microencapsulation of enzymes as artificial cells, the enzymes are enveloped with spherical, ultrathin (200 Å) membranes. This way, high concentrations of enzyme could be used and a ratio of enzyme to membrane could be more than 100,000 parts of enzyme to 1 part of membrane. The reason for this is that the microcapsule or artificial cell acts as a microscopic envelope inside which is enclosed a concentrated solution of enzyme. The small dimensions of the microcapsule and the ultrathin membrane allow for a very fast permeation of substrate and product. Furthermore, the enzymes are contained within a membrane which is not permeable to proteins and therefore do not come in direct contact with external proteins or cells and do not result in any immunological reaction. The stability of microencapsulated enzymes can be further increased by enclosing them together with a high concentration of another protein, microencapsulation of insolubilized enzymes, or cross-linking the enzyme after microencapsulation. The membrane material can be made nonthrombogenic, biocompatible, and even biodegradable when required.

The original methods of artificial cells for microencapsulation of enzymes,[2] have now been extended to a very large number of other systems.[3,4] Liposomes are lipid microspheres with concentric, onionskin-like lipid membranes where enzymes and proteins are trapped between the membranes.[16] These have the advantage that they can be made extremely small and could be especially useful for i.v. injection for localization at intracellular sites, and the lipid carrier material of liposomes could easily be metabolized in the body. The amount of enzymes carried by the liposomes is rather small compared to the carrier material, and the enzymes have to be released to carry out their action, since the carrier is impermeable to substrates. Another approach is the encapsulation of enzymes within erythrocytes.[17] This is carried out by the reversed hemolysis of erythrocyte and the trapping of external enzyme during the hemolysis phase. The advantage here is that the membrane will be completely biocompatible. Injection of erythrocyte-encapsulated enzymes this way would allow the enzyme to be removed and located in the reticuloendothelial cells. However, since the permeability of the erythrocyte membrane is rather limited, the use of erythrocyte-encapsulated enzymes is more appropriate for those enzymes that require to be released in the reticuloendothelial system. Another approach involves the dialysis principle,[29] where an enzyme system is enclosed, allowing substrate to diffuse in from outside to be acted on by the encapsulated enzyme. The ease of preparation using the dialysis system is achieved at the expense of the substrate equilibration rate, which is about 400 times slower than that of artificial cells.

II. LOCALIZATION OF IMMOBILIZED ENZYME AND PROTEIN FOR IN VIVO APPLICATIONS

Immobilized enzymes have to be located where they can come in contact with the substrate. A number of routes of administration have been studied.

A. Local Implantation

Suspensions of immobilized enzyme can be injected intramuscularly, subcutaneously, intraperitoneally, or elsewhere. Although a very large amount of immobilized enzymes could be introduced into the peritoneal cavity, this way of introduction carries a risk of peritonitis. However, for animal experimentation, the intraperitoneal route is an easy way of administering a large amount of immobilized enzymes into a location where they can come in contact with those substrates which can equilibrate rapidly across the peritoneal cavity.

B. Intravenous Injection

The fate and sites of distribution of intravenously injected microencapsulated enzymes, other particles, and red blood cells have been studied.[2,3,18] It was found that artificial particulates larger than 2 μm in diameter are filtered out to a great extent by the pulmonary capillaries, while those that passed through the capillaries were removed mainly by the reticuloendothelial system of the liver and spleen. By variations in the surface properties of microcapsules, one can vary the distribution of microcapsules to different organs. Thus, negatively charged particles of small diameters are less liable to be trapped in the lung capillaries, but are removed more by the reticuloendothelial system of the liver and spleen.[3,18] Other surface properties studied included the treatment of red blood cells with neuraminidase, and the resulting rapid removal of these treated hemologous red blood cells by the liver and spleen.[3,18] Intravenously injected liposomes are also removed by the liver and spleen where their contents can enter the cells and even the intracellular organelles.[16] Erythrocyte-encapsulated enzymes can also be introduced intravenously to be removed by the reticuloendothelial system.[17]

C. Extracorporeal Shunt Systems

Immobilized enzymes can be used in an extracorporeal system to act on substrates of blood recirculating through the system.[19] An extracorporeal shunt system containing microencapsulated urease was used in this way to remove blood urea.[19] The additional use of heparin attached to microcapsule membranes, shunt chamber, and tubings have been used to prevent clotting of blood and to avoid the necessity of systemic heparinization.[20] Extracorporeal shunts containing microencapsulated catalase have also been used to recirculate peritoneal fluid for the removal of perborate in acatalasemic mice.[21] A large number of other extracorporeal immobilized enzyme systems were also studied. These include immobilizing enzymes on glass,[22,23] polymethylmethacrylate,[24,25] nylon tubings,[26] gel attached to nylon tubings,[27] collagen systems,[28] dialysis systems,[29] fibers,[30] and combined artificial cell synthetic capillary systems.[31] Albumin immobilized on the surface of microencapsulated adsorbents has resulted in a blood-compatible extracorporeal system being used for the treatment of patients with renal failure, liver failure, and drug intoxication.[3,7,32-40]

Immobilized material in an extracorporeal system does not enter the body at any time and has, therefore, become the obvious candidate for early applications in patients.

D. Local Topical Applications

Immobilized enzymes have been applied directly to local lesions to prevent absorption of the enzyme into the body or to prevent immunological or hypersensitivity reactions. Thus, microencapsulated catalase has been applied directly to oral lesions in mice with hereditary catalase deficiency to replace the enzyme deficiency locally.[41]

E. Administration of Immobilized Enzymes into the Gastrointestinal Tract

Microencapsulated urease was used to act on urea either by direct introduction into the intestine or by oral administration into animals.[21,42-44]

III. EXPERIMENTAL THERAPY

A. Red Blood Cells

Artificial cells containing red blood cell hemolysate have been investigated for possible use as red blood cell substitutes.[1-3,18] However, there is a major problem of rapid removal from the circulation by the reticuloendothelial system. Immobilized urease has been used as a model system for experimental enzyme therapy.[2,45] The basic result obtained paves the way for other types of enzyme replacement therapy.

B. Hereditary Enzyme Deficiency

Microencapsulated catalase has been shown to effectively replace a hereditary catalase deficiency in acatalasemic mice.[21,46] Subsequently, liposome microencapsulated enzymes have been used for replacement in hereditary enzyme-deficiency conditions related to storage diseases.[16] More recently, red-blood-cell-entrapped enzymes and liposome-entrapped enzymes have also been tested for possible use in storage diseases.[17,47]

C. Substrate Dependent Tumor

With the demonstration of the effectiveness of microencapsulated asparaginase for experimental tumor suppression,[48-52] a large amount of work is being carried out in many centers using all available types of immobilized asparaginase. These include matrix entrapment of asparaginase in polyacrylamide gel which is fragmented to fine particles for peritoneal or i.v. injection.[53] Another group of investigators[54] used matrix-entrapped L-asparaginase in poly-2-hydroxyethylmethacrylate which was frozen, dried, crushed, and sieved through a 38-μm screen, giving a mean particle size of 12 μm for injection intraperitoneally into C3H mice for tumor suppression. A number of extracorporeal shunt approaches include the immobilization of asparaginase to glass plates by covalent linkage,[22] to polymethylmethacrylate by covalent linkage,[24,25] to nylon tubing by covalent linkage,[26] to a polycarboxylic gel layer attached to the inner wall of a nylon tube,[27] and others. L-Asparaginase has also been bound covalently to Dacron® vascular prosthesis for installation in the inferior vena cava, or as circular Dacron® grafts for implantation in the peritoneal cavity.[55] Another material investigated for extracorporeal hemoperfusion is asparaginase complexed to collagen wound to form a spiral multipore reactor,[28] while artificial cells have been combined with capillary dialysis membranes.[31]

D. Organ Failure and Metabolic Disorders

Immobilized enzymes and proteins have been used for the construction of artificial kidneys, artificial livers, and detoxifiers.[3,7,32-39] Other types of immobilized enzymes have been studied for use in oxygenators,[56,57] artificial pancreas,[58] and biocompatible surfaces.[32,59] An artificial organ based on albumin immobilized on microencapsulated adsorbent has been used for a number of years for the treatment of patients with renal failure, liver failure, and acute intoxication.[3,7,32-40] Antigens or antibodies have been immobilized and used in an extracorporeal shunt system for the removal of antibodies and antigens, respectively.[60] More recently, artificial cells containing multienzyme systems with cofactor recycling mechanisms have been used to convert unwanted metabolites into useful metabolites,[61] for instance, urea and ammonium could be converted into amino acids.[61]

REFERENCES

1. **Chang, T. M. S.**, Hemoglobin Corpuscles, report of research project for B.Sc. Honours, McGill University, Montreal, 1957.
2. **Chang, T. M. S.**, Semipermeable microcapsules, *Science*, 146, 524, 1964.
3. **Chang, T. M. S.**, *Artificial Cells*, Charles C Thomas, Springfield, Ill, 1972.
4. **Chang, T. M. S.**, *Biomedical Applications of Immobilized Enzymes and Proteins*, Vol. 1 and 2, Plenum Press, New York, 1977.
5. **Zaborsky, O. R.**, *Immobilized Enzymes*, CRC Press, Cleveland, 1973.
6. **Hasselberger, F. X.**, *Uses of Enzymes and Immobilized Enzymes*, Nelson-Hall, Chicago, 1978, 219.
7. **Chang, T. M. S.**, *Artificial Kidney, Artificial Liver and Artificial Cells*, Plenum Press, New York, 1978.
8. **Salmona, M., Saronio, C., and Garanttini, S.**, Eds., *Insolubilized Enzymes*, Raven Press, New York, 1974.
9. **Dunlop, R. B.**, Ed., *Immobilized Biochemicals and Affinity Chromatography*, Plenum Press, New York, 1974.
10. **Stark, G. R.**, Ed., *Biochemical Aspects of Reactions on Solid Surfaces*, Academic Press, New York, 1971.
11. **Mosbach, K.**, Ed., *Immobilized Enzymes (Methods in Enzymology Series)*, Academic Press, New York, 1976.
12. **Pye, E. K. and Wingard, L. B., Jr.**, Eds., *Enzyme Engineering*, Vol. 2, Plenum Press, New York, 1974.
13. **Weetall, H. H.**, Ed., *Immobilized Enzymes, Antigens, Antibodies, and Peptides*, Marcel Dekker, New York, 1975.
14. **Wingard, L. B., Jr.**, Ed., *Enzyme Engineering*, Vol. 1, John Wiley & Sons, New York, 1972.
15. **Broun, G. B., Manecke, G., and Wingard, L. B., Jr.**, Eds., *Enzyme Enginnering*, Vol. 4, John Wiley & Sons, New York, 1978.
16. **Gregoriadis, G. and Ryman, B. E.**, Fate of protein-containing liposomes injected into rats: an approach to the treatment of storage diseases, *Eur. J. Biochem.*, 24, 485, 1972.
17. **Ihler, G. M., Glew, R. H., and Schnure, F. W.**, Enzyme loading of erythrocytes, *Proc. Natl. Acad. Sci. U.S.*, 70, 2663, 1973.

18. Chang, T. M. S., Semipermeable Aqueous Microcapsules, Ph.D. thesis, McGill University, Montreal, Canada, 1965.
19. Chang, T. M. S., Semipermeable aqueous microcapsules ("artificial cells"): with emphasis on experiments in an extracorporeal shunt system, *Trans. Am. Soc. Artif. Intern. Organs*, 12, 13, 1966.
20. Chang, T. M. S., Johnson, L. J., and Ransome, O., Semipermeable aqueous microcapsules. IV. Nonthrombogenic microcapsules with heparin-complex membranes, *Can. J. Physiol. Pharmacol.*, 45, 705, 1967.
21. Chang, T. M. S. and Poznansky, M. J., Semipermeable microcapsules containing catalase for enzyme replacement in acatalasemic mice, *Nature (London)*, 218, 243, 1968.
22. Hyden, H., An extracorporeal shunt apparatus for blood detoxification, *Arzneim. Forschung*, 21, 1671, 1970.
23. Venter, J. C., Venter, B. R., Dixon, J. E., and Kaplan, N. O., Extracorporeal immobilized enzyme reactors, *Biochem. Med.*, 12, 79, 1975.
24. Sampson, D., Hersh, L. D., Cooney, D., and Murphy, G. P., A new technique employing extracorporeal chemotherapy, *Trans. Am. Soc. Artif. Intern. Organs*, 18, 54, 1972.
25. Hersh, L. S., L-Asparaginase from *Escherichia coli* II and *Erwinia carotovora* bound to poly-(methyl methacrylate), *J. Polymer Sci.*, 47, 55, 1974.
26. Allison, J. P., Davidson, L., Gutierrer-Hartman, A., and Kitto, G. B., Insolubilization of L-asparaginase by covalent attachment to nylon tubing, *Biochem. Biophys. Res. Commun.*, 47, 66, 1972.
27. Horvath, C., Sardi, A., and Woods, J. S., L-Asparaginase tubes: kinetic behavior and application in physiological studies, *J. Appl. Physiol.*, 34, 181, 1973.
28. Venkatasubramanian, K., Vieth, W. R., and Bernath, F. R., Use of collagen immobilized enzymes in blood treatment, in *Enzyme Engineering*, Vol. 2, Pye, E. K. and Wingard, L. B., Jr., Eds., Plenum Press, New York, 1974, 439.
29. Apple, M., Hemodialysis against enzymes as a method of "gene replacement" in cases of inherited metabolic diseases, *Proc. West. Pharmacol. Soc.*, 14, 125, 1971.
30. Salmona, M., Saronio, C., Bartosek, I., Vecchi, A., and Mussini, E., Fibre-entrapped enzymes, in *Insolubilized Enzymes*, Salmona, M., Saronia, C., and Garattini, S., Eds., Raven Press, New York, 1974, 189.
31. Chang, T. M. S., Biomedical applications of artificial cells, *Bio-med. Eng.*, 8, 334, 1973.
32. Chang, T. M. S., Removal of endogenous and exogenous toxins by a microencapsulated adsorbent, *Can. J. Physiol. Pharmacol.*, 47, 1043, 1969.
33. Chang, T. M. S. and Malave, N., The development and first clinical use of semipermeable microcapsules (artificial cells) as a compact artificial kidney, *Trans. Am. Soc. Artif. Intern. Organs*, 16, 141, 1970.
34. Chang, T. M. S., Gonda, A., Dirks, J. H., and Malave, N., Clinical evaluation of chronic, intermittent, and short-term hemoperfusions in patients with chronic renal failure using semipermeable microcapsules (artificial cells) formed from coated activated charcoal, *Trans. Am. Soc. Artif. Intern. Organs*, 17, 246, 1971.
35. Chang, T. M. S., Gonda, A., Dirks, J. H., Coffey, J. F., and Lee-Burns, T., ACAC microcapsule artificial kidney for the long-term and short-term management of eleven patients with chronic renal failure, *Trans. Am. Soc. Artif. Intern. Organs*, 18, 465, 1972.
36. Chang, T. M. S., Coffey, J. F., Lister, C., Taroy, E., and Stark, A., Methaqualone, methyprylon, and glutethimide clearance by the ACAC microcapsule artificial kidney: *in-vitro* and in patients with acute intoxication, *Trans. Am. Soc. Artif. Intern. Organs*, 19, 87, 1973.
37. Chang, T. M. S., Coffey, J. F., Barre, P., Gonda, A., Dirks, J. H., Levy, M., and Lister, C., Microcapsule artificial kidney: treatment of patients with acute drug intoxication, *Can. Med. Assoc. J.*, 108, 429, 1973.
38. Chang, T. M. S., Migchelsen, M., Coffey, J. F., and Stark, A., Serum middle molecule levels in uremia during long-term intermittent hemoperfusions with the ACAC (coated charcoal) microcapsule artificial kidney, *Trans. Am. Soc. Artif. Intern. Organs*, 20, 364, 1974.
39. Chang, T. M. S., Chirito, E., Barre, B., Cole, C., and Hewish, M., Clinical performance characteristics of a new combined system for simultaneous hemoperfusion-hemodialysis-ultrafiltration in series, *Trans. Am. Soc. Artif. Intern. Organs*, 21, 502, 1975.
40. Chang, T. M. S., Hemoperfusion alone or in series with ultrafiltration or dialysis for uremia, poisoning, and liver failure, *Kidney Int.*, 10, S305, 1976.
41. Chang, T. M. S., Effects of local applications of microencapsulated catalase on the response of oral lesions to hydrogen peroxide in acatalasemia, *J. Dental Res.*, 51, 319, 1972.
42. Chang, T. M. S. and Loa, S. K., Urea removal of urease and ammonia adsorbents in the intestine, *Physiologist*, 13, 70, 1970.
43. Gordon, A., Greenbaum, M. A., Marantz, L. B., McArthur, M. J., and Maxwell, M. H., Adsorbent based low volume recirculating dialysate, *Trans. Am. Soc. Artif. Intern. Organs*, 15, 347, 1969.

44. Asher, W. J., Bovee, K. C., Frankenfeld, J. W., Hamilton, R. W., Henderson, L. W., Hotzapple, P. G., and Li, N. N., Liquid membrane system directed toward chronic uremia, *Kidney Int.*, 7, S409, 1975.
45. Sekiguchi, W. and Kondo, A., Studies on microencapsulated hemoglobin, *J. Japan Soc. Blood Transfusion*, 13, 153, 1966.
46. Poznansky, M. J. and Chang, T. M. S., Comparison of the enzyme kinetics and immunological properties of catalase immobilized by microencapsulation and catalase in free solution for enzyme replacement, *Biochim. Biophys. Acta*, 334, 103, 1974.
47. Thorne, S. R., Fiddler, M. D., and Desnick, R. J., Enzyme therapy. V. *in-vivo* fate of β-glucuronidase deficient mice, *Pediatr. Res.*, 9, 918, 1975.
48. Chang, T. M. S., Asparaginase-loaded semipermeable microcapsules for mouse lymphomas, *Proc. Can. Fed. Biol. Sci.*, 12, 62, 1969.
49. Chang, T. M. S., The *in-vivo* effects of semipermeable microcapsules containing L-asparaginase on 6C3HED lymphosarcoma, *Nature (London)*, 229, 117, 1971.
50. Siu Chong, E. D. and Chang, T. M. S., *In-vivo* effects of intraperitoneally injected L-asparaginase solution and L-asparaginase immobilized within semipermeable nylon microcapsules with emphasis on blood L-asparaginase, "body" L-asparaginase, and plasma L-asparagine levels, *Enzyme*, 18, 218, 1974.
51. Mori, T., Sato, T., Matuo, Y., Tosa, T., and Chibata, I., Preparation and characteristics of microcapsules containing asparaginase, *Biotechnol. Bioeng.*, 14, 663, 1972.
52. Mori, T., Tosa, T., and Chibata, I., Enzymatic properties of microcapsules containing asparaginase, *Biochim. Biophys. Acta*, 321, 653, 1973.
53. Updike, S. J., Prieve, C., and Magnuson, J., Immobilization in hypoallergenic gel, a method of protecting enzymes from proteolysis and antibody complexing, in *Enzyme Therapy in Genetic Diseases, Birth Defects*, Vol. 9, (Original Article Series), Williams & Wilkins, Baltimore, 1973, 77.
54. Ohnuma, T., O'Driscoll, K. F., Korus, R., and Walczak, I. A., Pharmacological studies and antitumor activity of *E. coli* asparaginase immobilized in 2-hydroxyethylmethacrylate, *Abstr. Papers Amer. Chem. Soc.*, 174, 81, 1974.
55. Cooney, D. A., Weetall, H. H., and Long, E., Biochemical and pharmacologic properties of L-asparaginase bonded to dacron vascular prostheses, *Biochem. Pharmacol.*, 24, 503, 1975.
56. Broun, G., Selegny, E., Minh, C. T., and Thomas, D., The use of proteic and enzymatic coatings and/or membranes for oxygenators, *FEBS Lett.*, 7, 223, 1970.
57. Updike, S. J., Shults, M. C. N., and Magnuson, J., Catalase for oxygenator membranes, *J. Appl. Physiol.*, 34, 271, 1973.
58. Bessman, S. P. and Schultz, R. D., Prototype glucose oxygen sensor for the artificial pancreas, *Trans. Am. Soc. Artificial Internal Organs*, 19, 361, 1973.
59. Kusserow, B., Larrow, R., and Nichols, J., Surface bonded, covalently cross-linked urokinase synthetic surfaces, *Trans. Am. Soc. Artificial Internal Organs*, 19, 8, 1973.
60. Terman, D. S. and Buffaloe, G., Extracorporeal immunoadsorbents for specific extraction of circulating immune reactants, in *Artificial Kidney, Artificial Liver, and Artificial Cells*, Chang, T. M. S., Ed., Plenum Press, New York, 1978, 99.
61. Chang, T. M. S. and Malouf, C., Artificial cells microencapsulated multienzyme system for converting urea and ammonia to amino acid using α-ketoglutarate and glucose as substrate, *Trans. Am. Soc. Artificial Internal Organs*, 24, 18, 1978.

Chapter 10

THE PRINCIPLES OF CONTROLLED DRUG RELEASE

R. D. Bagnall

TABLE OF CONTENTS

I.	Foreword	188
II.	Introduction	188
III.	The Meaning of Control	188
IV.	Controlled Drug Release	190
V.	Controlled Drug Release Systems	192
	A. Controlling the Release Profile	192
	B. Directing the Released Drug to Its Target	193
VI.	Future Problems	195
References		195

I. FOREWORD

In recent years, there has been considerable interest in designing new drug formulations which release therapeutic agents in a controlled fashion from what are sometimes called depot devices. Various terms have been used to describe such release, including sustained, delayed, controlled, and prolonged, and these often seem to be used interchangeably by different authors. The problem for the new research scientist in this area is that there is no clear guideline as to what does and does not constitute genuine control. If he examines the literature, he will find a plethora of release devices in which an extremely large number of polymeric carriers appear to have been used, but there does not at first examination seem to be any pattern to the published works on which the selection of release type and polymer for any new drug problem can be made. This chapter has been deliberately written to provide such a pattern rather than simply catalogue a list of published devices, and it is hoped that it will serve as a basic guide for those researchers concerned with this relatively new area of pharmaceutical science.

II. INTRODUCTION

In the search for new therapeutic systems, it has hitherto been assumed that most drugs will be administered systemically, that is, they will not be delivered directly to their target organs, but will be presented in a convenient form such as a tablet or injection from which they pass rapidly into the bloodstream and are carried to their intended sites of action. Unfortunately, this means that organs other than the target site will also be exposed to the drug, thus raising the possibility of unwanted side effects. This loss of drug to nontarget organs affects potency and unnecessarily exposes the patient to far more of the therapeutic agent than he actually needs for his disorder.

Traditionally, the solution to this has been the province of the synthetic chemist, who by skilled molecular manipulation hopes to provide new drugs which retain their biological activity but partition preferentially from the blood into the target organ only. This is the basis of much current interest in structure/activity relationships of pharmacologically active compounds.

However, there are some problems that the chemist cannot solve. For example, it is becoming harder for potential new drugs to reach the clinic because of the increasingly stringent toxicological tests required and the time that such tests inevitably take. When a drug does reach the market, its efficacy depends largely on the patient administering to himself the prescribed drug at repeating intervals, and patients can be unreliable as dosage regulators, particularly when some modern drugs must be taken for extremely long periods. Even then, the need to maintain an effective concentration of drug at the target organ between doses means that at the start of any one dose period the target site concentration must be much higher than required so that it can decay, through metabolism, without reaching a subeffective level. Inevitably, this increases the risk of side effects and uses the drug inefficiently.

In recent years, therefore, there has been increasing interest in diverting attention away from the discovery of novel active compounds, and considering instead new formulations which deliver already available drugs at a precisely controlled rate directly to the target organ. This new approach has become generally known as controlled drug release.

III. THE MEANING OF CONTROL

Controlled drug release is concerned with delivering to a patient the minimum quantity of drug required for just the period necessary to combat his disorder, exposing

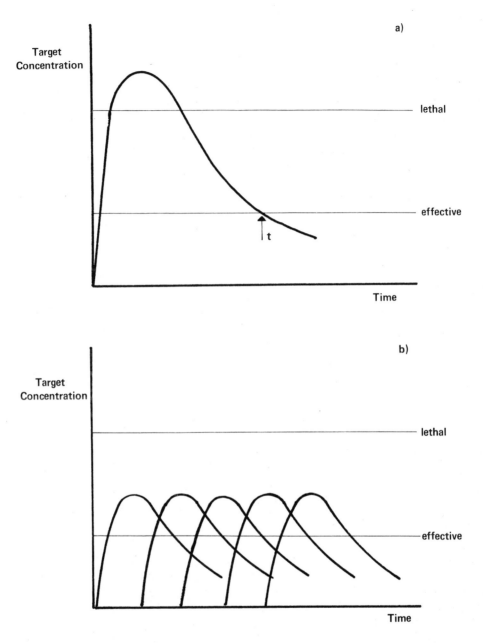

FIGURE 1. (a) Schematic profile of drug level at target organ from normal release system, with single large dose. (b) Schematic effect of repeated small doses of normally released drug.

only the target organ to the therapeutic agent. It is generally accepted in therapeutic research that every drug has a minimum effective concentration at its site of action, and it is the purpose of controlled drug release to just maintain this concentration. It is important to realize that in the most general case the threshold target concentration may vary as a result of the treatment, so that an ideal controlled drug release device should be capable of varying both release rate and release time to create any desired release profile. A suitable definition of controlled drug release is then, 'the delivery of drug at a predetermined rate over a specified time to a preselected target organ.'

What if this control is not present? Suppose that a drug is administered normally so that all of the active component is available immediately. What happens next is shown in Figure 1a. The blood level rises rapidly, and with it the level of drug in the target organ. After the maximum concentration has been reached the level of drug decays in an approximately exponential manner through metabolism until it becomes ineffective after time t. To increase the time of effectiveness from a single dose would mean giving more drug, and this could well push the maximum concentration up beyond the threshold lethal value. It would also expose the patient to a prolonged late period during which the drug was present at subeffective concentrations but could still cause side effects.

The usual answer is to give the drug as a series of doses, producing several overlapping release profiles as shown in Figure 1b. The result is a fluctuating but effective target concentration, with a final subeffective decay due mainly to the final dose only. The fluctuations are still marked, however, and a good profile relies heavily on the cooperation of the patient.

Of late, therefore, there has been some consideration given to removing or drastically reducing the patient variable by providing large doses of drug in a form from which the agent is not all available at one time but leaches slowly over a period. The initial surge of drug is thereby lost, but the benefit of a long effective period from a single dose is retained as shown in Figure 2a. A typical example is a drug dispersed in a polymer matrix, and devices of this type are variously called one of the following:

- prolonged release
- delayed release
- timed release
- sustained release
- slow release
- retarded release
- continuous release
- long acting dosage
- delayed action preparation
- depot drug form
- long-term agent

To the extent that such delivery systems are deliberately designed to alter normal release profiles, they are often described in the literature as controlled drug release. However, it is the belief of this author that this is an erroneous description because the release rate and release time cannot be independently varied to produce any desired release profile. For example, in the case of the polymer matrix the initial release rate is a function of $t^{1/2}$ and the later rate of e^{-kt}. The release profile is always of the same mathematical type, and the only control is in the selection of systems with different rate constants.

True control, on the other hand, involves the separate manipulation of release rate and release time to match the needs of the patient, and includes not only long-term but also short-term drug administration. In this chapter, therefore, controlled drug release will be limited only to those systems capable of producing any desired release profile.

IV. CONTROLLED DRUG RELEASE

Controlled drug release implies that the needs of the patient are considered first, and the release profile is then designed accordingly to maintain the minimum effective

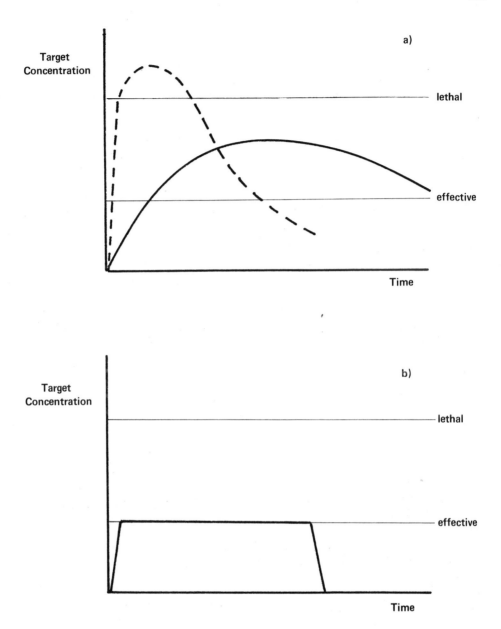

FIGURE 2. (a) Schematic effect of sustained release preparation (solid line) against same dose given normally (broken line). (b) Idealized effect of controlled drug release device.

target concentration. Since any controlled device carries only a finite quantity of drug which is gradually consumed by the patient, the basic problem of controlled drug release is how to maintain the minimum effective concentration from a device which is depleting with time.

What then are the needs of the patient? Although in principle the minimum therapeutic level could vary during treatment, for most practical purposes it may be regarded as constant, and the problem reduces to the maintenance of a constant active site concentration from a device which is depleting with time.

Suppose that a controlled release device delivers drug directly to the target organ, and it is desired to maintain a particular active site concentration. The rate of removal of drug by processes such as metabolism is related to the concentration of drug, and is constant so long as the concentration is constant. To maintain the latter, it is then only necessary to supply drug at a rate equal to the rate of loss, that is at a constant rate.

Suppose, on the other hand, that delivery directly to the target organ is not possible and systemic administration must be used. In this case, it is the blood which supplies the target organ with drug, and it is the rate at which drug passes from blood to target which must remain constant. This requires a constant blood concentration, and since the device delivers drug to the blood, the requirement once again is that the release rate from the device must be constant.

For the normal case of a constant minimum effective drug concentration at the target site, therefore, the release rate from the controlled release formulation must be constant, regardless of the administration site. The problem of controlled drug release reduces finally to how to maintain a constant release rate from a device which is depleting with time. Since the release rate is independent of the quantity of drug remaining in the device at any time, many authors refer to this as zero order release.

In this chapter, the basic principles by which such zero order release can be achieved are outlined with some examples, but it should be remembered that although these relate only to the normal case of a constant threshold effective level, the principle of matching a depleting source to a desired target site concentration remains the same even when the latter varies.

V. CONTROLLED DRUG RELEASE SYSTEMS

The problems of (1) controlling the release profile and (2) directing the drug to its intended target, are essentially independent and will be discussed separately.

A. Controlling the Release Profile

As explained above, the problem is to maintain a constant release rate from a depleting device. There are three basic release systems which may themselves be further subdivided. Disintegrating systems release drug by some form of erosion, whereas intact devices remain essentially intact during the release period. Multiple release devices are pseudo-zero order release methods in which many overlapping sustained release profiles, some of which are short acting and some of which are long acting, are blended in such a way that the final release profile approximates to a constant release rate (Figure 2b).

Disintegrating devices can use biodegradation, dissolution, hydrolysis, or mechanical erosion, but the release mechanism is really not involved in the control process. The erosion rate is governed by the surface area of the device, and the problem is one of geometry to maintain constant surface area. As a general rule, convex surfaces decrease in area by erosion while concave surfaces increase in area, and the problem can be solved by a suitable combination of both in one device. Typical examples are apertured tablets such as hollow cones or rings,[1] or concave/convex tablets with random projections and depressions.[2] A special case is the thin erodable film where edge effects are negligible, a typical example being biodegradable polylactic acid films for local drug delivery to wounds and burns.[3,4]

Intact devices may be further subdivided into passive devices which release drug by diffusion through the material of the device, and active devices which release drug by means of an applied force.

Passive devices are of two types, each being governed by Fick's law of diffusion,

$$J = -D \frac{dC}{dx} \tag{1}$$

where J is the flux of material across plane x in the diffusion field at time t, D is the diffusion coefficient of the field, and $\frac{dC}{dx}$ is the concentration gradient of the drug at x. For constant release therefore,

$$\left(D \frac{dC}{dx}\right)_{t_1} = \left(D \frac{dC}{dx}\right)_{t_2} \tag{2}$$

Devices which do not alter during release, i.e., where $D_1 = D_2$, must, therefore, be constant concentration gradient devices. Alternatively, increasing permeability devices compensate for a falling gradient with an increasing diffusion coefficient.

Constant gradient devices require constant boundary concentrations and a diffusion field of constant thickness. This is best achieved by membrane-bound delivery systems. Since the desired aim is a constant physiological concentration, the requirement is for a constant inner drug concentration for the device, and the most favored way of achieving this is with a solid core of drug which either forms a saturated solid solution with the material of the membrane or is poorly soluble in physiological fluids which diffuse in through the membrane. The choice of membrane material is limited only by permeability considerations. A somewhat more sophisticated version of this system is a drug-filled matrix surrounded by a membrane, such that diffusion through the matrix is faster than through the membrane. Potential buildup of drug at the inner membrane boundary is compensated for by a decreasing output from the matrix, and with suitable choice of materials a constant release rate can be obtained.[6,7]

Increasing permeability devices include both leaching systems in which a membrane- or matrix-bound drug has as part of the structural material of the device, a component which is slowly soluble in physiological fluids,[8-10] and swelling devices in which the structural material of the device gradually becomes more permeable by swelling in body fluids.[11-13]

Active devices are of two types. Externally activated devices expel drug by means of an externally applied force, and include the standard diffusion pump which can be miniaturized for implantation and operated remotely by magnets if required.[14] Internally activated devices generate their own expulsion pressure on contact with body fluids. Gas pressure can be generated by bicarbonate/acid systems,[15,16] and liquid pressure by osmosis.[17,18]

Multiple release systems are of three main types. The blended release system is the simplest, and usually consists of a multiparticulate formulation where each particle has its own sustained release profile, slow and fast types being blended to give the desired overall release profile. The multiple bursting device, on the other hand, is a blend of drug-containing membrane-bound particles which swell and rupture on contact with body fluids. Variable membrane thickness ensures variable bursting times, and the overall release profile is a summation of the individual bursts.[19] Finally, the sequential laminate consists of successively eroding layers of matrix-bound drug in a tablet, such that each layer is a sustained release system and the concentration of drug in each layer can be adjusted to compensate for the decreasing size of the device.[20]

B. Directing the Released Drug to Its Target

There are three main methods for drug targetting, although undoubtedly others are possible and will be reported from time to time in the literature. Implants are self-evident and include not only surgically placed devices, but also externally positioned

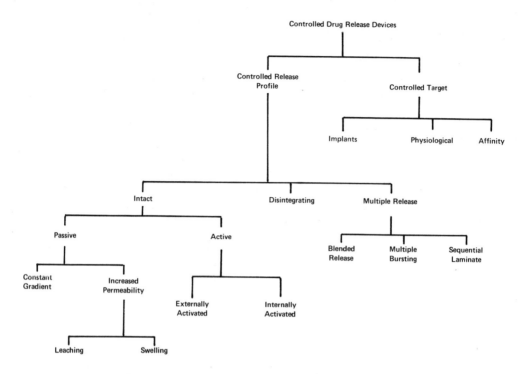

FIGURE 3. A classification of controlled drug release systems.

systems such as intrauterine devices, ocular inserts, dermal patches, and buccal strips. Physiological devices wander through the body until a relatively abrupt change in physiological conditions initiates drug release. Perhaps the most common example is the enteric coated tablet where the coating is soluble only at certain pH values and is designed to protect the drug from the action of gastric juices, but release it in the more alkaline small intestine. Human enteric coatings are acid-insoluble, whereas ruminant enteric coatings are base-insoluble. High density pellets have been claimed to sink in the rumen of farm animals and thereby be retained,[21] and swelling tablets which cannot pass the pylorus when swollen have been described for gastric retention in humans.[22] Drugs specifically designed for use in birds have also been coated with highly cross-linked epoxy resin, the principle being that the coating is totally resistant to human digestive tract fluids, but can be fractured by the mechanical action of the avian gizzard.[23]

Finally, affinity devices are designed to wander through the body until they locate an organ for which they have a particular affinity. Currently there is great interest in the use of liposomes for such purposes, these being minute drug-carrying bodies formed by the emulsification of aqueous drug solutions with materials resembling natural cell membrane components. Each liposome thus resembles a drug-filled cell, the hope being that by suitable modification of the emulsifying agent, the affinity of the liposome for different organs will be altered.[24] On a more limited scale, it has been claimed that liposomes will penetrate the intestinal wall and can be used for oral administration of large molecules such as insulin which are not normally absorbed from the gastrointestinal tract and must instead be given by injection.[25]

These then are the basic principles of controlled drug release, and they are summarized for clarity in Figure 3. It is hoped that these principles provide at least a first guide for new workers in this area in which the choice of materials is really secondary to a recognition of the underlying principles of zero order release.

VI. FUTURE PROBLEMS

Unfortunately, although the description of controlled drug release given in this chapter works very well in vitro, matters are not really that simple when devices are tried in vivo. Implanted or swallowed devices become rapidly coated with adsorbed layers of natural macromolecules, such as proteins, which can act as an extra diffusion barrier which rather inconveniently varies with time. In the case of implants the situation is even more complicated because of the gradual formation of a scar tissue capsule around the device which acts as a further variable diffusion barrier. Even when a drug gets through these extra barriers, it may fail to reach the target because of some other tissue which has a particularly avid affinity for the therapeutic agent. Fatty tissue is a notorious problem, soaking up the drug and then releasing it at some future time so that the blood level rises and falls unpredictably, and this unpredictability is made the more difficult by the inherent variation in fatty tissue content in any population.

There is, therefore, a degree of randomness about the recipient which tends to reduce the intended degree of control in any controlled release formulation, and until problems of this type can be routinely overcome, controlled drug release will inevitably remain more the province of the academic scientist than of the practicing pharmacist.

REFERENCES

1. Jacobs, H. R., Medicinal tablets, U.S. Patent 3,113,076, 1956.
2. Reid, A. F., Shaped pellets, U.S. Patent 3,279,955, 1963.
3. E. I. DuPont de Nemours, Polylactidhaltige pharmazeutische zubereitungen, DT. Patent 2,207,635, 1971.
4. Schribner, R. M., Polylactide-drug mixtures for topical application, *U.S. Patent 3,755, 558, 1971.*
5. Shippy, R. L., Hwang, S. T., and Bunge, R. G., Controlled release of testosterone using silicone rubber, *J. Biomed. Mater. Res.,* 7, 95, 1973.
6. Zaffaroni, A., Drug delivery system, ZA Patent 7,007,144, 1970.
7. Zaffaroni, A., Drug delivery system, DT Patent 2,054,488, 1970.
8. Tansey, R. P., Sustained release therapeutic tablet compositions comprising organic solvent-gelled gums, U.S. Patent 3,133,863, 1961.
9. Nikken Chemical Co., Delayed release medicament, JA Patent 7,118,151, 1969.
10. Nikken Chemical Co., Sustained release drug production, JA Patent 7,343,848, 1970.
11. Klimet, K., Protracted activity oral hydrogel bead, U.S. Patent 3,689,634, 1966.
12. Klimet, K., Carriers for biologically active substances, U.S. Patent 3,551,556, 1967.
13. Cesk. Akad, Ved., Improvements in or relating to sustained release medicaments, U.K. Patent 1,135,966, 1966.
14. Charles River Foundation, Magnetically operated capsule for administering drugs, U.S. Patent 3,659,600, 1970.
15. Place, V. A., Device for dispensing product with directional guidance member U.S. Patent 3,797,492, 1972.
16. Michaels, A. S., Drug delivery device with self actuated mechanism for retaining device in selected area, U.S. Patent 3,786,813, 1972.
17. Higuchi, T. and Leeper, H. M., Osmotic dispenser, U.S. Patent 3,732,865, 1971.
18. Alza Corporation, Osmotic pressure chambers for continuous and controlled release of active materials, NL Patent 7,307,699, 1972.
19. National Patent Development Corporation, Drogenfreisetzungssystem, DT Patent 2,311,843, 1972.
20. Stephenson, D., Prolonged action medicinal tablets, U.S. Patent 3,184,386, 1962.
21. Marston, H. R., Therapeutic pellet for ruminants, U.S. Patent 3,056,724, 1958.

22. Purdue Research Foundation, Arzneimittelformulierung mit kontrollierter verweilzeit im gastrischen bereich, DT Patent 2,328,580, 1973.
23. U.S. Sec. Interior, Avian specific bio-affecting preparation and treatment method, U.S. Patent 3,629,390, 1969.
24. **Inchema S. A.**, Verfahren zur inkapsulation von insbesondere wasserlöslichen, DT Patent 2,249,552, 1971.
25. Bayer, Nouveaux liposomes contenant des composés chimiques et leur procédé de preparation, BE Patent 796,610, 1973.

INDEX

A

AAS, see Atomic absorption spectrophotometry
Absorption, and corrosion process, see also
    Metabolism, II: 7
Acetates, nickel, I: 116
Acetylcholinesterase, effects of aluminum on, I;
    195
Acetyl-D,L-penicillamine, interactions of, I; 91
Acroosteolysis, in vinyl chloride diseases, II:
    65—66, 75
Acrylate monomers, LD_{50} values for, II: 54
Acrylic acid-maleic anhydride copolymer, II: 167
Acrylic cement, see also Bone cement
    and arterial blood pressure changes, II: 135
    hypersensitivity to, II: 56
Active devices, in release profile, II: 195
Addison's disease, copper concentration in, I: 225
Additives (in medical plastics)
    nature of, II: 146
    rate of leaching out of plastics, II: 147
    toxicology of, II: 147—153
Adenomas, and vinyl chloride exposure, II: 70
Adhesives, see Cyanoacrylic adhesives
Adipates, toxicology of, II: 148
Adrenal cortical carcinomas, copper
    concentrations in, I: 225
Adrenal gland, aluminum in, I: 190—191
Agarose, cell migration in, II: 38
Age
    and copper concentrations, I: 218
    and metabolic interactions, I: 94—95
Air
    aluminum in, I: 188
    chromium in, I: 145
    nickel in, I: 116
    vinyl chloride in, II: 75
Albumin
    cobalt binding to, I: 136
    nickel binding and, I: 118—119
Alcoholics, copper concentrations in, I: 228
Aldehyde oxidase
    and iron, I: 28
    and molybdenum, I: 30, 165
Aliphatic acids, toxicology of, esters of, II: 148
Alkaline phosphatase, cadmium substituted for,
    I: 33
Alkylchlorosilanes
    in animal experiments, II: 85
    irritation caused by, II: 84
Alkyl mercury compounds, I: 243—244
Allergy, see also Dermatitis; Sensitivity
    chromium, I: 156—157
    nickel, I: 123, 124, 125
    vinyl chloride, II: 65
ALMA, see Methacrylic acid, allylester of
Aluminum
    absorption of, I: 197
    carcinogenicity of, I: 191

clinical toxicity of, I: 199—201
excretion of, I: 197
in hemodialysis, II: 26—28
in nuclei, I: 42—43
metabolism of, I: 197
occurrence of, I: 188—189
    home food preparation and storage, I:
        189—190
    in adrenal gland, I: 190—191
    in blood, I: 191—193
    in bone, I: 193
    in cardiovascular tissue, I: 193
    in eye, I: 194
    in food, I: 190
    in kidney, I: 194
    in liver, I: 194
    in lung, I: 194—195
    in male genital tissue, I: 195
    in milk, I: 195
    in muscles and tendons, I: 195
    in nerve tissue, I: 195
    in pancreas, I: 195
    in skin, I: 196
    in spleen, I: 196
    in tooth, I: 196
    in uterus, I: 196
    treatment of drinking water, I: 190
    water storage, I: 190
physiology of, I: 197
role in immune response, I: 199
toxicity of, I: 197—199
Aluminum hydroxide, toxicities of, I: 198
Aluminum sulfate, for treatment of drinking
    water, I: 190
Alzheimer's disease, I: 199
Amalgams, dental, I: 244—245
Amenorrhoea, and copper concentrations, I: 299
Amine oxidases, I: 27
Analgesia, MMA-caused, II: 118
Angiosarcomas, and vinyl chloride exposure, see
    also Carcinogenesis, II: 70, 74, 76—77
Animal studies, see also specific animal; specific
    metal
    for chromium, I: 148
    on sensitivity to metallic implants, II: 45
Animal tissues, wound healing in, I: 70
Anorexia nervosa
    copper levels in, I: 215
    vinyl chloride-induced, II: 64
Antagonism, biochemical mechanism of, I: 88, 90
Antibiotics, antitumoral, in ligand-DNA
    interactions I: 62
Antibodies
    immobilization of, II: 186
    tissue damage by, II: 35—36
Anticomplement effect, of vinyl chloride, II: 63
Antigens
    immobilization of, II: 186
    in host defense response, II: 34—35

Antimony, interaction with BAL, I: 90
Antineoplastic compounds, I: 58—62
Antioxidants
  additives as, II: 146
  toxicology of, II: 150—152
Antitumoral substances, I: 59—60, 62—63
Apnoea, MMA-induced, II: 125
Argyria
  clinical manifestations of, I: 252—254
  definition of, I: 252
  differential diagnosis of, I: 254, 256
  generalized, I: 253—254
  history of, I: 252
  pathology of, I: 254, 255
  treatment for, I: 256
Agyrosis, I: 252
Arsenic, interaction with selenium, I: 94
Arthritis
  aluminum levels in, I: 193
  silicone fluid injection for, II: 87
Artificial cells
  composition of, II: 182, 183
  experimental therapy with, II: 186
  red blood cell substitutes, II: 185
Asbestos, in peridontal dressings, II: 152
Ascorbic acid
  carcinogenicity of, I: 57
  mutagenicity of, I: 57
Ashing, for hair analysis, I: 105
Asparaginase, immobilized, II: 185
Asthma, chromium-induced, I: 152
Ataxia, vinyl chloride-induced, II: 63
Atomic absorption spectroscopy (AAS)
  for metal determination in hair, I: 105
  of aluminum detection, I: 200
  of cobalt, I: 135
  of copper, I: 213
  of nickel, I: 117
ATPase, interaction with lead, I: 34

B

BAL, see Dimercaptoropanol Bases, metal ions
    binding to, I: 45—47
Bentonite
  toxicities of, I: 198
  treatment of drinking water, I: 190
Beryliosis, aluminum levels in, I: 192
Beryllium
  carcinogenicity of, I: 57
  genetically controlled immune responses with,
    I: 95
  in nuclei, I: 42—43
  mutagenicity of, I: 57
Biocompatibility
  systemic, II: 5—6
  testing for, II: 34
Biodegradation, in release profile, II: 194
Biomaterials, sensitivity to, see also Corrosion;
    specific material
  dental surgery, II: 41—42

immune responses to metal salts, II: 37—40
induction of, II: 40—45
orthopedic surgery, II: 42—45
sensitivity to surgical reaction, II: 40—41
Biometals, see also Metals
  adverse effects of chromium in, I: 155—157
  uses of chromium in, I: 154—155
Bis-methylmercury selenide, I: 92
Bismuth
  in nuclei, I: 42—43
  interaction with BAL, I: 90
Blastogenic factor, in sensitivity reactions, II: 36
Bleomycin, in ligand-DNA interactions, I: 62
Blood
  aluminum in, I: 191—193
  chromium in, I: 150
  mercury levels in, I: 239
  MMA in, II: 116
  molybdenum in, I: 164
  rate of MMA accumulation in, II: 110
  vinyl chloride in, II: 68
Blood-brain barrier
  aluminum interaction with, I: 199
  interaction with mercury, I: 240
  soluble polymers and, II: 161
Blood plasma expanders
  conditions for, II: 160
  dextrans, II: 161—163
  gelatin as, II: 163—164
  hydroxyethyl starch as, II: 163
  kinds of, II: 166—167
  polyvinylpyrrolidone, II: 164—166
Blood pressure, see also Hypertension;
    Hypotension
  and acrylic cement, II: 131—132
  effect of vanadium on, I: 183
  n-BMA see Methacrylic acid, n-butylester of
Body fluids, simulation of, I: 16
Body metal burden
  assessment of, I: 102
  hair metal determinations as measure of, I: 110
Bone, aluminum levels in, I: 193, 198
Bone cement, see also Methylmethacrylate
  harm from, II: 133
  pMMA in, II: 108
Bradypnoea, MMA-induced, II: 125
Brain, aluminum in, see also Central nervous
    system, I: 191, 195; II: 27
Breast cancer
  copper concentrations and, I: 229
  role of silicones in, II: 90
  vinyl chloride exposure and II: 70, 74
Bronchitis, vinyl chloride-induced, II: 64
Bronchoconstriction, in MMA experiments, II:
    129
Bronchospasm, MMA-induced, II: 125
Burn therapy, silicone in, II: 87
Butyl 2-cyanoacrylate, experimental studies with,
    II: 91, 92
Butyl oleate, toxicology of, II: 63
Butylphthalyl butyl glycolate, II: 153

C

Cadmium
  antagonisms of, I: 88
  carcinogenicity of, I: 57; II: 10
  competition for carriers and, I: 92
  in enzyme interactions, I: 33
  in hair analysis, I: 106—107
  in hemodialysis, II: 29
  low protein diet and, I: 89
  mutagenicity of, I: 57
  pretreatment with, I: 94
  synergism of, I: 88
Cadmium intoxication, I: 91
Calcium
  deficiency in, I: 94
  hemodialysis, and, II: 24
  in body metabolism, I: 22
  in dialysate, II: 26
Camphor, toxicology of, II: 149
Carbonic anhydrase, I: 32
Carboxypeptidase, I: 32
Carcinogenicity, I: 35
  of aluminum, I: 191
  of chromium, I: 57, 151—152, 154, 157
  of cobalt, I: 139
  of cyanoacrylic adhesives, I: 94
  of nickel, I: 120—122
  of pMMA implantations, II: 113—114
  of silicones, II: 90
  of vinyl chloride, II: 69—75
  susceptibility to, I: 34
Carcinogens, see also specific carcinogen
  characteristics of, I: 57
  in modification of reiterative DNA sequences, I: 55—58
Cardiac arrhythmias, vinyl chloride-induced, II: 63
Cardiomyopathy, cobalt-induced, I: 138
Cardiovascular tissue, aluminum in, I: 193—194
Castor oil, II: 149
Catabolite gene activator protein (CAP), I: 67
Catalase, interaction with mercury, I: 240
Catalysts, toxicology of, II: 153
Catheters, silicone rubber, II: 87
Central nervous system, see also Blood-brain barrier
  and mercury toxicity, I: 244
  and toxic actions of MMA, II: 121—124
Central venous pressure, effect of MMA on, II: 132
Ceramic whiskers, toxicology of, II: 152
Cerebral infarction, copper elevations in, I: 227
Cerebrocuprein, I: 221
Ceruloplasmin, I: 217
  circadian rhythm, I: 220
  deficiency of, I: 16
  ferroxidase activity of, I: 218
  molybdenum toxicity and, I: 166
  oxidase activity of, I: 218
  regulatory mechanism for, I: 223
Chelate effect, I: 14

Chelating agents, I: 56
  dimercaptopropanol, I: 90
  DNA-metal ligand complex and, I: 63
  effect on copper, I: 16—17
Chelation, I: 90—91
Chemotactic factor, in sensitivity reactions, II: 36
Chemotherapy, and copper concentrations, I: 229
Chloral acid, II: 68
Chlorinated aromatic compounds, toxicology of, II: 148
Chlorinated diphenyls, toxicology of, II: 150
2-Chloro acetaldehyde, and mutagenesis, II: 67
Chloroacetic acid, II: 68, 69
Chloroethylene oxide and mutagenesis, II: 67
Cholinesterase
  effect of aluminum on, I: 195
  molybdenum toxicity and, I: 166
Chromates, I: 150
  carcinogenic effect of, I: 152—153
  hexavalent state of, I: 144
Chrome holes, I: 151
Chromite, I: 144
Chromium, see also Cobalt-chromium alloys
  allergic reaction to, I: 156—157
  biometals, I: 154—155
  carcinogenicity of, I: 57, 151—152, 154, 157
  clinical toxicology of
    acute effects, I: 150—151
    biochemical mechanisms of, I: 152—154
    chronic effects, I: 151—152
    subacute effects, I: 151
  compounds, I: 144
  corrosive release of, I: 155—156
  enzyme interactions and, I: 25—32
  excretion of, I: 145
  experimental sensitivity to, II: 39
  genetically controlled immune responses with, I: 95
  hair analysis, I: 107
  human exposure to, I: 145—146
  interaction with BAL, I: 90
  interaction with iron, II: 19
  intracellular distribution of, I: 149—150
  methods of analysis and, I: 144
  mutagenicity of, I: 57
  normal levels, I: 150
  organ distribution of, I: 149
  physical and chemical properties of, I: 144
  sensitivity to, II: 37
  skin reactions to, I: 152
  storage and distribution of, I: 148
  transplacental transfer of, I: 148—149
  uptake and excretion of, I: 146—148
Chromium pigments, I: 151—152
Chromocenters, formation of, I: 66
Chromomere, activity of, I: 68
Circadian rhythm, for copper, I: 220
Cirrhosis, copper elevatsio in, I: 227
Coal, nickel in, I: 116
Cobalt
  absorption of, I: 136—137
  biochemistry of, I: 137—138

carcinogenicity of, I: 57; II: 10
concentrations in humans, I: 136
enzyme interactions and, I: 25—26
excretion of, I: 136—137
in hemodialysis, II: 29
inorganic chemistry of, I: 134—135
interaction with zinc, I: 135
interaction with zinc metalloenzymes, I: 138
intracellular distribution of, I: 135—136
mutagenicity of, I: 57
oxidation states of, I: 10
sensitivity to, II: 37
toxicology of, I: 138—139
vitamin B_{12} and, I: 8
Cobalt-chromium alloys, I: 8, 152
chromium in, I: 154—155, 156
corrision predicted for, II: 3
in dentistry, I: 156
in vivo corrosion rates for, II: 2
sensitivity reactions to, II: 43, 44
Cobalt salts, in vitro testing of, II: 38
Computer programs, for stability constants, I: 15
Constant gradient devices, in release profile, II: 195
Copolymers, II: 160
Copper
analytical determination of, I: 212—214
competition for carriers and, I: 92
complexes of, I: 16
diet and, I: 214—215
enzyme interactions and, I: 26—27
erythrocyte concentration of, I: 220—221
hemodialysis and, II: 28
intestinal absorption of, I: 216—217
metabolism of, I: 165
plasma and, I: 217
serum concentrations for, I: 219
Wilson's disease, I:225
Copper concentrations
hormonal influences on, I: 224
in contraceptive therapy, I: 220
in hair analysis, I: 103, 104, 107
in human tissues, I: 221
extrahepatic, I: 223—224
hepatic, I: 222—223
in Menkes' kinky hair syndrome, I: 226—227
in pregnancy, II: 220
in various diseases, I: 227—229
Copper deficiencies
in diet, I: 214
in livestock, I: 215
Cornea, in silicone experiments, II: 84
Coronary atherosclerosis, aluminum levels in, I: 192
Coronary atherosclerosis, aluminum levels in, I: 192
Coronary heart disease, copper concentrations in, I: 228
Corrosion (from metallic implants)
carcinogenic effects of
clinical carcinogenicity, II: 11—12
industrial exposure and experimental

investigations, II: 10—11
experimental induction of, II: 40
general principles, II: 4
immunologic considerations
iron and susceptibility to infectious disease, II: 8—9
metal hypersensitivity, II: 9—10
measurement of, II: 4
metabolic effects of, II: 12
iron metabolism, II: 13—15, 17
iron overload, II: 15—17
pharmacotoxicological considerations, II: 6—7
process of II: 2—4
role of trace elements in normal metabolism, II: 5—6
systemic effects of, II: 7—8
Cotton, as filler, II: 151
Covalent linkage method, of polymer immobilization, II: 182
Cushing's syndrome, copper concentrations in, I: 225
Cyanoacrylic adhesives, II: 61
analysis of, II: 90
carcinogenicity of, II: 94
mutagenicity of, I: 94
physical properties of, II: 90
teratogenesis and, II: 94
toxicology of
in of
in animals, II: 90—92
in humans, II: 93
Cysteine
antagonism of, II: 88
biliary excretion of $MeHg^{2+}$ and, I: 94
synergism of, I: 88
Cystic fibrosis, aluminum levels in, I: 193
Cysts, in silicone experiments, II: 83
Cytochrome, interaction with aluminum, I: 193
cytochrome oxidase
copper in, I: 27
copper-dependent, I: 217
interaction with copper, I: 222, 223
molybdenum toxicity and, I: 166
oxygen reduction and, I: 8
Cytocuprein, I: 221

D

Deacon catalyst, II: 61
Deafness, associated with long-term vinyl chloride exposure, II: 66
DEHP, see Di-2-ethylhexyl phthalate
Dehydrogenases, and zinc, I: 32
Denaturation of DNA, I: 43
Dental alloys, cobalt-based, I: 135
Dental calculus, and aluminum concentrations, I: 196
Dental caries
and molybdenum, I: 166
and titanium, I: 175
Dental fillings, MMA in, II: 108
Dental plaque, and aluminum concentrations, I: 196

Dental prostheses, and chromium allergy, I: 156
Dental surgery
  polymeric materials used in, II: 55
  sensitivity to biomaterials and, II: 41—42
Dentistry
  mercury hazards in, I: 244—246
  metallic implants in, II: 41—42
Dermatitis, see also Allergy; Skin
  aluminum levels and, I: 193
  associated with metal implants, II: 40
  chromium, I: 156
  from dentures, II: 41
  metallic implants associated with, II: 44
  MMA-caused, II: 115
  nickel-caused, I: 123
  vinyl chloride and, II: 65
Detoxifiers, II: 186
Dextrans, II: 161—163
Diabetes mellitus, aluminum levels in, I: 192
Dialysate, see also Hemodialysis, II: 24
  potential toxic concentrations in, II: 25
  removal of metal ions from, II: 30
  sodium and potassium in, II: 25—26
Dialysis encephalopathy syndrome, I: 199
Dialysis osteodystrophy, osteomalacic, II: 27—28
Dialysis patients, aluminum hydroxide in, see also Hemodialysis, I: 200
Diazepam, II: 27
Dibutylphthalate
  allergic response to, II: 65
  released from PVC film, II: 62
Dibutyl sebacate, toxicology of, II: 63
Diet, see Food
Diethyldithiocarbamate, for nickel carbonyl poisoning, I: 120
Di-2-ethylhexyl phthalate (DEHP), II: 153
  extraction of, II: 155
  structure of, II: 154
  toxicology of, II: 154—156
Diethylmaleate, and biliary excretion of MeHg^{2+}, I: 94
Diethyl phthalyl, II: 153
Digestion of tissues, for hair analysis, I: 105
Diisooctyl phthalate, II: 153
Dimercaptopropane sulfonate (DMPS), interactions of, I: 91
Dimercaptopropanol (BAL), interactions of, I: 90—91
Dimercaptosuccinic acid, interactions of, I: 91
Dimethylglyoxime test, for sensitivity, II: 41
Dimethyl phthalate, II: 63
Dimethylpolysiloxane implants, as contraceptive devices, II: 87
Dimethyl sebacate, toxicology of, II: 63
1,5-Diphenylcarbazide, for analysis of chromium, I: 144
Di-n-propyl phthalate, toxicology of, II: 63
Disease, see also Allergy: Toxicity
  copper homeostasis in, I: 225—229
  copper levels in, I: 220
  nickel concentrations and, I: 118
Dissolution, in release profile, II: 194
Distribution, intracellulaar, essential vs. nonessential metals in, II: 5
Divalent metals, I: 24—25
DNA (deoxyribonucleic acid)
  aluminum interaction and, I: 197
  double-helix scheme, I: 44
  interaction of carcinogens with, I: 34
  metal-binding sites in I: 46, 48
  metal ion interaction with, I: 43—44
  metal occurrences in, I: 41—42
  primary structure of, I: 49
  quaternary structure, I: 50
  secondary structure of, I: 49—50
  structural modifications induced by metal ions, I: 47—51
  tertiary structure of, I: 50
  vinyl chloride and, II: 69
DNA-metal complexes, I: 51—54
  constitutive heterochromatin areas and, I: 66—68
  interactions with various ligands, I: 58—65
  metallo-DNA and, I: 65
Dopamine hydroxylase, copper in, I: 224
Dose-effect relationships, I: 89
Dose-response relationship, I: 89
  curve for, II: 6
  for vinyl chloride carcinogenesis, II: 70
Dow Corning medical antifoam A compound, II: 81
Dow Corning 360 fluid
  experimental studies with, II: 88, 89
  physical properties of, II: 80
Dressing, peridontal, asbestos in, II: 152
Drug release, controlled
  definition of, II: 190—191
  future of, II: 197
  principle of, II: 192—194
  systems for, II: 194—197
Drugs
  synthetic polymers as, II: 167
  target organs for, II: 190

E

ECG, MMA-induced changes in, II: 131
Eczematization, MMA-caused, II: 115
Edema
  MMA-caused, II: 115
  silicone-induced, II: 82
EDTA, interactions of, I: 91
Elastomers, development of, II: 60
Electron orbitals, effect of ligands on, I: 11
Electrophoresis, of copper, I: 213
Electroretinogram (ERG), influence of MMA on, II: 123
Electrothermal atomization, of copper, I: 213
EMA, see Ethylmethacrylate
Emission spectroscopy, for metal determination in hair, I: 105
Encephalopathy, dialysis, II: 26—27
Endoanaesthetic effect, of MMA, II: 126
Enteric coated tablet, see also Drug release, II: 196

Enzyme(s), see also Metalloenzymes; specific
    enzyme, I: 4
  adsorption of, II: 183
  immobilized, II: 182
    experimental therapy with, II: 185—186
    in gastrointestinal tract, II: 185
    in shunt systems, II: 184—195
    localization of, II: 184—195
  matrix entrapment of, II: 183
  metal-activated, I: 22, 29
  microencapsulation of, II: 183
  topical applications, II: 185
  zinc-containing, I: 137
Enzyme deficiency, experimental therapy for, II: 185
Enzyme interactions
  cobalt in, I: 137
  copper and, I: 218, 221
  divalent metals and, I: 24—25
  heavy metals in, I: 32—34
  molybdenum and, I: 165—166
  monovalent metals and, I: 24
  nickel and, I: 118
  vinyl chloride and, II: 68
epidemiologic study, of vinyl chloride problem, II: 74
Epoxy, allergy to, II: 56
Erythema
  cyanoacrylate adhesives causing, II: 93
  MMA-caused, II: 115
Erythrocuprein, I: 221
Erythrocytes
  chromium tagging of, I: 149
  copper in, I: 220—221
Erythropoietic effect, of cobalt, I: 137, 138
esophageal reconstruction, silicone for, II: 84
Essential metals, see also Metals; Transient metals, I: 3, 8, 14
  aluminum as, I: 191, 195, 196
  copper, I: 212
  in implant alloys, II: 5—6
  interaction with toxic metals, I: 89—90
  molybdenum as, I: 164
  nickel as, I: 31, 119
  quantity in body, I: 9
  titanium as, I: 172
  vanadium as, I: 180
Estrogen, interaction with copper, I: 224
Ethoxysilanes, experimental studies with, II: 85, 86
Ethylene chlorohydrin, II: 68
Ethylene diamine tetra-acetate, see EDTA
Ethylene oxide, in sterilization process, II: 56, 57
Ethylmethacrylate (EMA), 107, 126
Ethylphthalyl ethyl glycolate, II: 153
Excretion, and corrosion process, see also Metabolism, II: 7
Excretion rate, vs. rate of metallic release, II: 3
Experimental supplementation, metal occurrence after, I: 42
Experiments, see Animal studies
Eye, aluminum in, I: 194
Eye irritation, and silicone, II: 82

F

Farr technique, for testing sensitivity, II: 39
Feces
  chromium excreted in, I: 146, 147
  nickel excreted in, I: 118
Ferritin, I: 9; II: 15
Ferrochromium, I: 144
Fetal loss rate, and vinyl chloride, II: 67
Fetus, and transplacental transfer of chromium, I: 148
Fillers, toxicology of, II: 151—152
Flavoprotein, molybdenum in, I: 165
Fluoride, in hemodialysis, II: 29—30
Foley catheters, siliconizing of, II: 87
Food
  aluminum compounds in, I: 190, 197
  chromium in, I: 145
  copper in, I: 214
  molybdenum in, I: 165
  nickel in, I: 116
Formaldehyde, allergy to, II: 56
Formication, vinyl chloride-induced, II: 64

G

Galactoseoxidase, copper in, I: 224
Gamma radiation, II: 56
Gastrointestinal system
  aluminum levels in, I: 192
  chromium in, I: 146, 150—151
  copper in, I: 216
  vanadium in, I: 183
Gelatin, as blood plasma expander, II: 163—164
Genital tissue, aluminum in, I: 195
Giddiness, associated with long-term vinyl chloride exposure, II: 66
Glass, toxicology of, II: 152
Glucose-6-phosphate dehydrogenase (G6PD), and titanium, I: 174
Glucose tolerance factor (GTF), and chromium exposure, I: 145
Glutaminase, and molybdenum toxicity, I: 166
Glycerine, toxicology of, II: 149
Glycols, toxicology of, II: 148, 149
Glycolysis, and reactions of aluminum-containing compounds, I: 197
Golden hamster, vinyl chloride inhalation experiments in, II: 73, 74
Granuloma formation, in silicone experiments, II: 82
GSH_{130}, and biliary excretion of $MeHg^{2+}$, I: 94
Guinea pig
  for allergy testing, II: 39
  MMA studies with, II: 126

H

Hair (as biopsy material)
  cadmium detected in, I: 106—107
  chromium in, I: 107

copper in, I: 107
dynamics of growth and, I: 102
mercury in, I: 109
methods used, I: 105
nickel in, I: 107, 117—118
normal values for copper in, I: 104
sampling techniques for, I: 103—104
trace metals in, I: 102
washing, I: 104—105
zinc in, I: 107, 108
Hair pool, for quality control of metal determinations, I: 106
Hamster, vinyl chloride inhalation experiments in, II: 73, 74
Heavy metals, in enzyme interactions, I: 32—34
Heme production, effect of cobalt on, I: 137
Hemin biosynthesis, and aluminum interaction, I: 201
Hemodialysis, see also Dialysate
aluminum in, II: 26—28
copper in, II: 28
fluoride in, II: 29—30
principle of, II: 24—25
zinc in, II: 28
Hemoglobin, iron as, I: 9
Hemolysis, ethylene oxide-produced, II: 57
Hemosiderrin, iron as, I: 9; II: 15
Hepatitis, vinyl chloride-induced, see also Liver dysfunction, II: 64, 73
Hepatocuprein, I: 221
Hepatolenticular degeneration, I: 16
Hepatomegaly, vinyl chloride-induced, II: 64
HES, see Hydroxyetheyl starch
Heterochromatic areas, constitutive
application of telestability to, I: 67—68
properties of, I: 66—68
Hetero-polyacids, oxidation states of, I: 10
Hip arthroplasty
acrylic stabilization of, II: 130
hypotension associated with, II: 134
MMA blood levels after, II: 110, 117
Hip prostheses
and cardiorespiratory changes, II: 108
and corrosion, II: 3
Hodgkin's disease
copper concentrations in, I: 228
elevation of serum aluminum in, I: 191
Holo-ceruloplasmin, copper interaction with, I: 218
Hormones, influences on copper homeostasis of, I: 224
Host defense mechanisms
immunologic, II: 34—37
nonspecific, II: 34
Host susceptibility, specificity of, I: 34
Hydration energy, I: 11
Hydrolysis
in release profile, II: 194
of metal complexes, I: 12
Hydroxyethyl starch (HES), as blood plasma expander, II: 163
N-(2-Hydroxypropyl)methacrylamide, copolymers of, II: 171

8-Hydroxyquinoline (8-HG), in ligand-DNA interactions, I: 61
Hypercalcemia, II: 26
Hyperkalemia, acute, II: 26
Hyperkeratosis, MMA associated with, II: 113
Hypermanesemia, during hemodialysis, II: 26
Hyperphosphatemia, prophylaxis of, I: 198
Hypersensitivity, to polymers, see also Sensitivity, II: 55—56
Hypertension
in MMA experiments, II: 131—132
vinyl chloride-induced, II: 63
Hypopituitarism, copper concentrations in, I: 225
Hypotension
associated with hip arthroplasty, II: 134
bone cement and, II: 135
Hypotony, in silicone experiments, II: 83
Hypoxemia, and bone cement, II: 135

I

i-BMA, see Methacrylic acid, isobutylester of
Immune response, role of aluminum in, I: 199
Immune system
antigens in, II: 34—35
cell-mediated, II: 36
pathways of immune response in, II: 35
sensitivity reactions, II: 35—37
Implant(s)
chromium-containing, I: 157
hair metals commonly found in, I: 108
metallic, II: 40—45
silicone, II: 84
titanium, I: 174—175
Implant alloys, essential metals in, see also specific alloys, II: 5
Implant loosening, reactions to, II: 10
Infants, copper concentrations in, I: 218
Infectious disease, and transferrin (TR), II: 9
Inflammation
cyanoacrylic adhesives and, II: 91
silicone-induced, II: 82
Interactions, see also Enzyme interactions
by association, I: 90
by association of metals and metalloids, I: 91—92
by metabolic interference, I: 93
competition for carriers and, I: 92
pharmacodynamic, I: 90
Intercalation, and chelation, I: 46
Intestinal bypass surgery, and copper concentrations, I: 228
Intrauterine devices, copper-containing, I: 227—228
Ionic strength, variations of, I: 69
Iron
absorption of, I: 9; II: 14
binding capacity of, II: 16
cadmium toxicity and, I: 94
carcinogenicity of, I: 57
competition for carriers and, I: 92
enzyme interactions and, I: 27—28

excretion of, II: 15
interaction with chromium, II: 19
metabolism of, II: 13
mutagenicity of, I: 57
nutritional immunity and, II: 8
overload, II: 15—17
storage of, II: 15
systemic sequellae for, II: 17—18
transport of, II: 14—15
"Irritation theory", I: 153
Irving Williams order, I: 12
Isobutyl 2-cyanoacrylate adhesives, II: 93
Isoferritins, I: 33
Isotactic polyacrylic acids, of antiviral activity, II: 167
Itching, cyanoacrylate adhesives causing, see also Dermatitis, II: 93

K

Δ^5-3 Ketosteroid isomerase, interaction with copper, I: 225
Kidney, see also Hemodialysis
aluminum concentrations in, I: 194
artificial, II: 186
as target organ for inorganic mercury, I: 242
Kinetics, of metal complexes, I: 12
Kwashiorkor, copper deficiency in, I: 214

L

Laccase, copper in, I: 224
Lacrimal glands, vinyl chloride in, II: 68
Lactate dehydrogenase, and titanium, I: 174
Lead
carcinogenicity of, I: 57
dose-effect relationships of, I: 89
in enzyme interactions, I: 33—34
in hemodialysis, II: 29
interaction with D-penicillamine, I: 91
low protein diet and, I: 89
mutagenicity of, I: 57
protective effect of selenite against, I: 92
synergism of, I: 88
Lead intoxication, EDTA for, I: 91
Lesion processes, ternary complexes during, I: 35
Leukemia, and copper concentrations, I: 229
Leukocytes, titanium in, I: 174
Leukocyte endogenous mediators (LEM), and copper concentrations, I: 229
Leukocyte migration inhibition test (LMI), for nickel hypersensitivity, I: 124
Leukoplasia, MMA associated with, II: 113
Lewis acids
cobalt as, I: 134
defined, I: 22
Ligands
interaction with DNA molecules, I: 58—65
metal and, in complexes, I: 9
modification or exchange of, I: 58

Ligand field stabilization energy (LFSE), I: 11, 13
Ligand field theory, I: 10—12
Ligandin, I: 94
Liposomes, II: 183, 196
Liver
artificial, II: 186
copper in, I: 222
effect of aluminum on, I: 194
vinyl chloride exposure and, II: 74, 75
Liver dysfunction, and vinyl chloride exposure, II: 73, 74
Lubricants, additives as, II: 146
Lung
aluminum concentrations in, I: 194—195
effect of chromium on, I: 152
effects of vanadium, I: 182—183
mercury vapor absorbed by, I: 239
Lymphocyte transformation test (LTT), for nickel hypersensitivity, I: 124
Lymphokines
and cell-mediated immunity, II: 42
in sensitivity reactions, II: 36
tests for, II: 38
Lymphotoxin, in sensitivity reactions, II: 36
Lysosomes, binding of mercury to, I: 241
Lysyl oxidase, interaction with copper, I: 223

M

Macrodex, II: 162
Macroglobulin, cobalt binding to, I: 136
Macrophage inhibition tests (MIF), for nickel hypersensitivity, I: 124
Magnesium
deficiency, I: 94
in dialysate, II: 24, 26
interaction with aluminum, I: 197
Maleic anhydride-furan copolymer, II: 167
MAM, see Methacrylamide
Manganese
carcinogencity of, I; 57
enzyme interactions and, I: 28—29
in hair analysis, I: 108
mutagenicity of, I: 57
stability of complexes and, I: 12
Marasmus, copper deficiency in, I: 214
Melanin, in production of argyric color, I: 253—254
Menkes' kinky hair syndrome, copper metabolism in, I: 226-227
Mercury
animal tissues and, I: 238—239
biotransformation of, I: 238, 243
cadmium pretreatment and, I: 94
compounds of, I: 239, 240
dentistry and
as hazard to patient, I: 245
as hazard to personnel, I: 245—246
use of, I: 224—245
EDTA and, I: 91
enzyme interactions and, I: 34

genetically controlled immune responses, I: 95
hair analysis and, I: 108, 109
interaction with dimercaptopropanol, I: 91
interaction with selenium, I: 91
metabolism of
    absorption, I: 239
    excretion, I: 241
    inhalation, I: 239
    transport and distribution, I: 240—241
occurrence of, I: 238
properties of, I: 238
toxicity of
    alkyl mercury compounds, I: 243—244
    general, I: 241
    inorganic mercury, I: 242
    metallic mercury, I: 241—242
    unstable organomercurials, I: 243
Mercury vapor, I: 239, 242
Metabolism
    corrosion process and, II: 7
    iron, II: 13—15
    molybdenum, I: 164—165
    nickel, I: 117
    physiological role of metals in, I: 24
    titanium, I: 171—172
    vanadium, I: 181—182
Metal(s)
    biological effects of, I: 48
    heavy, I: 32—34
    transuranic, I: 3
Metal-activated enzymes
    manganese-activated, I: 29
    vs. metalloenzymes, I: 22
Metal balance, variations of, I: 68—69
Metal chelating substance, see also Chelating agents
    metal-transfer chains and, I: 64—65
    nature of, I: 63—64
Metal chelation therapy, I: 90
Metal complexes, see also Transition metals, I: 12
    enzyme catalysis, I: 23
    formation of, I: 12
Metal deficiency, I: 41
Metal-DNA interactions, in vitro studies, see also DNA-metal complexes, I: 43—51
Metal-enzyme interactions, see also Enzyme interactions, I: 4
Metal ion(s)
    calculating distribution of, I: 15—16
    defined, I: 22
    different possible cellular localization, of, I: 52
    DNA structural modifications induced by, I: 47—51
    hydrolysis and, I: 12
    in vitro studies on, I: 44—45
    reduction potential of, I: 13
Metal ion-DNA complexes, I: 41—51
    binding sites and, I: 45—48
    in experimental animals, II: 45
    metal ion concentration and, I: 47
    specificity of, I: 50
Metallic implants, sensitivity to

general surgery, II: 40—41
in orthopedic surgery, II: 42—45
Metallo-DNA
    as active site for metal binding, I: 53
    concept of, I: 65
Metalloenzymes
    as active sites for metal binding, I: 52—53
    copper-containing, I: 26
    iron containing, I: 28
    manganese-containing, I: 29
    metal-activated enzymes and, I: 22
    molybdenum-containing, I: 30, 165
    zinc-containing, I: 33, 138
Metalloprotein, intestinal, I: 216
Metallothioneins, I: 52, 216
    characteristics of, I: 93
    cobalt absorption and, I: 137
    mercury bound to, I: 94
Metal salts, animal sensitivity to, II: 39—40
Metal-transfer chain, functioning of, I: 64—65
Methacrylic acid
    allylester of (ALMA), II: 107, 126
    isobutylester of (i-BMA), II: 107, 126
    n-butylester of (n-BMA), II: 107, 126
Methoxyethyl-mercury compounds, I: 240
Methyl 2-cyanoacrylate, experimental studies with, II: 92—94
Methylmercury
    biliary excretion of, I: 94
    interactions
        with BAL, I: 90
        with DMPS, I: 91, 92
        with DMSA, I: 91
    metabolism of, I: 240
    toxicity of, I: 89, 244
Methylmethacrylate (MMA)
    acute toxicity of, II: 111—112
    allergic reactions to, II: 115
    cardiovascular effects of, II: 130—133
    chronic toxicity of, II: 112—113
    clinical observations with, II: 133—135
    effect on nervous system of, II: 118
    effects on respiration of, II: 124—130
    intracellular distribution of, II: 116
    liberation from polymer, II: 109—111
    localized actions of, II: 114—115
    pharmacokinetics of, II: 115—117
    physicochemical properties of, II: 108—111
    polymerization of, II: 112
    side effects of, II: 109
    teratology of, II: 113
Mice, vinyl chloride inhalation experiments in, II: 70, 72, 73
Microembolism, MMA-induced, II: 129
Microenvironment, modification of composition of, I: 68—70
Migration inhibition factor (MIF)
    in sensitivity reactions, II: 36
    tests for, II: 38
Milk
    aluminum in, I: 195
    nickel concentrations in, I: 117

Minamata epidemic, I: 244
Mineral deficiencies, see also specific minerals, I: 94
Mitochondria, chromium in, I: 149
MMA, see Methylmethacrylate
Molecular weight
  of soluble polymers, II: 161
  of synthetic polymers, II: 169
  urinary excretion and, II: 164
Molybdates, absorption of, I: 164—165
Molybdenum
  biological activity of, I: 165—166
  dental caries and, I: 166
  enzyme interactions and, I: 29—31
  metabolism of
    absorption, I: 164—165
    effects of copper and sulfate on, I: 165
    excretion of, I: 165
  metalloenzymes and, I: 165—166
  oxidation states of, I: 10
  tissues and fluids and, I: 164
  toxicity of, I: 166
Monoamine oxidases, interaction with copper, I: 223
Mononuclear phagocytic system (MPS), deposition of PVP in, II: 165
Monovalent metals, and enzyme activity, I: 24
Moore's prosthesis, II: 130
MPS, see Mononuclear phagocytic system
Multiple element analysis, I: 214
Multiple release systems, II: 195
Multiple sclerosis, copper concentrations in, I: 228
Musculature, aluminum in, I: 195
Mutagenesis
  of cyanoacrylic adhesives, II: 94
  of silicones, II: 89
  susceptibility to, I: 34
  vinyl-chloride and, II: 67
Mutagens
  characteristics of, I: 57
  frameshift, I: 56
  metals, I: 54—55
  modification of reiterative DNA sequences, I: 55—58
Mutations, induced, I: 55
Myocardial infarction, aluminum levels in, I: 192

N

NADH dehydrogenase, and role of iron, I: 27
NaMa, see Sodium methacrylate
Narcosis, vinyl chloride-induced, II: 63
n-BMA, see Methacrylic acid, n-butylester of
Necroses
  MMA-induced, II: 114
  silicone-induced, II: 82
Necrotizing enterocolitis, role of DEHP in, II: 156
Neoplastic cells, see also Carcinogenesis
  metal content in, I: 33
  metal requirements of, I: 33
  metal transport in, I: 33
Nephritis, aluminum in, I: 192
Nervous system, see also Central nervous system
  aluminum in, I: 195
  effect of MMA on, II: 118
Neuritis, MMA-induced, II: 114
Neurofibrillary degeneration (NFD), aluminum-induced, I: 199
Neurological asthenia, vinyl chloride-induced, II: 64
Neutron activation analysis, I: 106
  for copper determination, I: 214
  for metal determination in hair, I: 105
N-(2-hydroxypropyl)methacrylamide, copolymers of, II: 171
Nickel, I: 43
  carcinogenesis, I: 120
    experimental data on, I: 121—122
    occupational aspects of, I: 121
  carcinogenicity of, I: 57, II: 10
  characteristics of, I: 116
  compounds, I: 116
  environmental, I: 116
  enzyme interactions and, I: 31
  experimental sensitivity to, II: 39
  hair analysis and, I: 107
  hemodialysis and, II: 29
  hypersensitivity to
    allergic reactions to "internal" nickel, I: 123—124
    clinical manifestations of, I: 123
    diagnosis of, I: 124
    incidence of, I: 122
    in vitro tests, I: 124
    sensitizing forms, I: 122—123
    treatment of, I: 125
  interaction with BAL, I: 90
  metabolism
    binding, I: 118—119
    in vitro observations, I: 120
    nutritional essentiality in, I: 119
    routes of uptake, I: 117
    tissue distribution and elimination, I: 117—118
  monitoring of, I: 117
  mutagenicity of, I: 57
  sensitivity to, II: 37, 41
  toxicity of
    epidemiological data on, I: 120
    experimental data on, I: 119—120
    nickel carbonyl poisoning, I: 120
Nickel oxides, I: 116
Nicotinamide adenine dinucleotide (NADH), and chromium reduction, I: 153
Nitrate reductase, molybdenum in, I: 165
p-Nitroanilides, as drug models, II: 172
Nitrogenase, molybdenum in, I: 165
Nuclei
  cobalt concentrated in, I: 139
  metal occurrence in, I: 41—42
Nutritional immunity, II: 8

O

OPG, see oxypolygelatin
Oral contraceptives, and copper levels, I: 218, 220, 224
Oral mucosa, metal sensitivity of, II: 41
Orthopedic surgery, metallic implants in, II: 42—45
Osteoclastic activating factor (OAF), in sensitivity reactions, II: 36
Osteodystrophy, osteomalacic dialysis, II: 27—28
Osteosarcoma, and copper concentrations, see also Carcinogenesis, I: 229
Oxidation-reduction, in cobalt, I: 134
Oxidation states
   for cobalt, I: 134
   of transition elements, I: 10
Oxygen transport, I: 8
Oxypolygelatin (OPG), II: 163

P

Pancreas, aluminum in I: 195—196
Pancreas tumor, and vinyl chloride exposure, II: 75
Paraffin wax, and arterial blood pressure changes, II: 135
Parethesia, MMA-caused, II: 115, 118
Passive devices, in release profile, II: 194—195
Pathology
   ionic strength in, I: 69
   of mi-roenvironment, I: 68—70
   of pH variations, I: 70
   of transmembrane potential, I: 69—70
   of wound healing, I: 70
D-Penicillamine, interactions of, I: 91
Petroleum, nickel in, I: 116
Pharmacodynamic interaction, I: 90
Pharmacokinetic interaction, I: 90
Phenobarbitone, effect of, I: 94
Phenylmercury, toxicity of, I: 243
Phosphates
   metal ions binding to, I: 45, 47
   toxicology of, II: 149
Phosphoric esters, toxicology of, II: 148
Phosphorylation, and reactions of aluminum-containing compounds, I: 197
Phthalates, toxicology of, II: 149
Phthalic acid esters, toxicity of, II: 153
Phthalic anhydride, structure of, II: 154
Phthalic esters, toxicology of, II: 148, 149
α-Picolonic acid, in ligand-DNA interactions, I: 61
Plant tissues, lesion process in, I: 70
Plasma, see also Blood plasma expanders
   copper in, I: 217
   stability of ternary complexes and, I: 14
Plasma ashing, for hair analysis, I: 105
Plasticine
   animal experiments with, II: 133
   arterial blood pressure changes and, II: 135

Plasticizers
   additives as, II: 146
   toxicology of, II: 148, 149, 153, 156
Plastics
   copper in, I: 214
   introduction of, II: 61
   major additives in, II: 147
   medical applications of, II: 146
   toxicology of, II: 52
Platelets, chromium tagging of, I: 149
Platinum compounds, in ligand-DNA interactions, I: 61
pMMA, see Polymeric methylmethacrylate
Poisoning, see also Toxicity
   mercury, I: 246
   nickel carbonyl, I: 120
Polyacetic acid, degradation products of, II: 55
Polyacrylic acid, II: 167
Polycaprolactam, oral administration of, II: 53
Polycarbonate, oral administration of, II: 53
Polyelectrolytes, biological activity of, II: 167
Polyester, allergy to, II: 56
Polyethylene, oral administration of, II: 53
Polyglycolic acid, II: 55
Poly[N-(2-hydroxypropyl)methacrylamide, structure of, II: 166
Polymer(s)
   degradation products of, II: 54—55
   development of, II: 60
   hypersensitivity to, II: 55—56
   inert high, II: 53
   particles derived from, II: 55
   soluble, II: 55
      as blood plasma expanders, II: 161—167
      as carriers of biologically active compounds, II: 168
      biocompatibility of, II: 168—169
      destructibility of cross-links and, II: 173—175
      therapeutical uses of, II: 160
   sterilization of, II: 56—57
   synthetic
      as drugs, II: 167
      degradability of bonds in side chains of, II: 172—173
      enzymatically degradable bonds, II: 170—172
      preparation of, II: 169—175
   toxicology of, II: 52
Polymer chains, steric hindrance of, II: 174, 175
Polymeric methylmethacrylate (pMMA)
   carcinogenicity of, II: 113—114
   localized actions of, II: 114—115
   physiochemical properties of, II: 108—111
Polymerization, process of, II: 53—54
Polymer macromolecules, toxicology of, II: 53—55
Polymethacrylic acid, II: 167
Polymethylmethacrylate, II: 54
Polyperoxide content, in vinyl chloride, II: 62
   oral administration of, II: 53
   particles derived from, II: 55
Polyurethane, allergy to, II: 56

Poly(vinyl alcohol), II: 161
Polyvinyl chloride (PVC), see also Vinyl chloride
  additives in, II: 147
  degradation of, II: 150
  leaching of plasticizers from, II: 155
  tissue responses to, II: 155
Polyvinylpyridine-N-oxide, II: 160, 161
Poly(2-vinyl-pyridine-l-oxide), as drug, II: 167
Polyvinylpyrrolidone (PVP), as blood plasma
    extender, II: 161, 164—167
Porphyria, aluminum-caused, I: 197
Porphyrin complexes, interaction with cobalt, I: 135
Potassium
  in body metabolism, I: 22
  in dialysate, II: 24, 25—26
Potentiation, I: 88
Pregnancy, and copper concentrations in, I: 218, 220, 224
Premature infants, copper levels in, 215
Prostheses, see also Implants
  corrosion and, II: 4
  nickel reactions and, I: 123—124
  pMMA, II: 115
Proteins
  adsorption of, II: 183
  localization of immobilized, II: 184—185
  polymer immobilization of, II: 182
Pulmonary arterial pressure
  effect of MMA prosthesis, II: 133
  influence of MMA on, II: 129
Pulmonary diseases, aluminum levels in, see also
    Lung, I: 192
Pulmonary hemorrhages, and MMA, II: 111, 130
PVP, see Polyvinylpyrrolidone
Pyran copolymer, II: 167
Pyruvate carboxylase (PC), I: 28

Q

Quartz, toxicology of, II: 152

R

Rabbit, as experimental animal, II: 39
Radioactive metals, and nuclear localization, I: 42
Rat
  alkylchlorosilanes in, II: 85
  Dow Corning 360 fluid in, II: 88, 89
  ethoxysilanes in, II: 85, 86
  MMA in, II: 116
  tetraethoxysilanes in, II: 86
  vinyl chloride inhalation experiments in, II: 70, 71
Raynaud's syndrome, vinyl chloride-induced, II: 64, 66, 74
Reduction potentials
  and chelation, I: 63
  for chromium, I: 153
  of metal ions, I: 13
  range of, I: 8

Release profile
  controlling, II: 194—195
  target organ and, II: 195—196
Respiration
  effects of MMA on, II: 124—130
  silicone inhalation and, II: 82
Reticuloendothelial system
  chromium accumulation and, I: 148
  polymers deposited in, II: 165
Retina, in silicone experiments, II: 83
Retinal surgery
  cyanoacrylate for, II: 93
  silicone buckling procedures for, II: 87
Rhabdosarcomata, cobalt-induced, I: 139
Rheomacrodex, II: 162
Ricinoleates, toxicology of, II: 148, 149
RNA (ribonucleic acid), and aluminum
    interaction, I: 197

S

Salivary glands, vinyl chloride in, II: 68
Sampling techniques, for hair biopsy, I: 103—104
Schizophrenia, copper elevations in, I: 227
Scleroderma, and vinyl chloride exposure, II: 64, 75
Seawater, nickel in, I: 116
Sebacates, toxicology of, II: 148, 149
Selenite, I: 88
Selenium
  interaction with arsenic, I: 94
  interaction with mercury, I: 91
Senile dementia, and aluminum toxicity, I: 199
Sensitivity, see also Allergy
  induction of, II: 40
  testing for, II: 37—39
Sensitivity reactions, in host defense response, II: 35—37
Sequestration, and corrosion process, II: 7
Sexual differences
  in chromium allergy, I: 156
  in copper concentrations, I: 218
  metabolic interactions and, I: 94—95
Shunt system, for immobilized enzymes, II: 184—185
Silastic® 382 medical grade elastomer, II: 79
Silastic® medical-grade tubing, II: 79
Silicone(s), II: 61
  absorption of, II: 89
  analysis of, II: 78
  carcinogenicity of, II: 90
  intracellular distribution of, II: 89
  metabolism of, II: 89—90
  mutagenicity of, II: 89
  physical properties of, II: 75, 78
  structural formula of, II: 78
  teratogenesis of, II: 88
  toxicology of
    in animal experiments, II: 79—85
    in humans, I: 85—88
Silicone rubber, additives in, II: 146—147

experiments with, II: 84
oral administration of, II: 53
Siliconomas, II: 87
Silver, see Argyria
Sisal, as filler, II: 151
Skin, see also Dermatitis
   aluminum in, I: 196
   chromium uptake through, I: 148
   effect of chromium on, I: 156
   effect of titanium on, I: 175
   mercury in, I: 239
   metal implant hypersensitivity and, II: 10
   reaction to chromium in, I: 152
   vinyl chloride in, II: 66, 68
Sleeping time, influence of MMA on, II: 124
SOD, see Superoxide dismutase
Sodium
   hemodialysis and, II: 24
   in body metabolism, I: 22
   in dialysate, II: 25—26
Sodium methacrylate, II: 107, 128
Soya bean oil, toxicology of, II: 150
Spectro chemical series, I: 11
Spectrophotometric methods
   copper analysis, I: 212
   silicone determination, II: 79
Spleen, aluminum in, I: 196
Splenomegaly, vinyl chloride-induced, II: 64
Sprague-Dawley rats, vinyl chloride inhalation experiments in, II: 71
Stability constants, I: 13—17
Stabilizers, toxicology of, II: 148, 150
Stainless steel alloys, I: 8
   chromium in, I: 154, 156
   clinical carcinogenicity and, II:11
   for dental prostheses, I: 156—157
   in vivo corrosion rates for, II: 2
   nickel in, I: 122; II: 18
   sensitivity reactions to, II: 43—44
   316 stainless steel
      carcinogenicity associated with, II: 11, 12
      corrosion of, II: 3, 17
      infection and, II: 9
Starr Edwards prosthetic aortic valves, II: 87
Sterilization
   chemical, II: 56
   of metal vs. polymers, II: 52
   polymers, II: 56—57
Stomatitis, MMA-caused, II: 115
Stress, copper elevations in, I: 227
Subsite interactions, II: 175
Succinate dehydrogenase, iron in, I: 28
Succinate oxidase, interaction with aluminum, I: 193
Sulfhydryl groups, binding of mercury to, I: 124
Sulfite oxidase
   and molybdenum, I: 30, 165, 166
   and tungsten, I: 31
Superoxice dismutases, I: 221
   as manganese metalloenzyme, I: 29
   copper in, I: 26—27
Surgery, see also Dental surgery
   metallic implants in, II: 40—41
   polymeric materials in, II: 55
Surgical alloys, cobalt-based, I: 135
Sweat, nickel concentrations in, I: 117
Swiss mice, vinyl chloride inhalation experiments in, II: 70, 72, 73
Synergism, biochemical mechanisms of, I: 88, 90
Synovial membrane, silicone in, II: 83

T

Tachypnoea, MMA-induced, II: 125
Tattooing, silver-amalgam, I: 252
Teeth, aluminum levels in, see also Dentistry, I: 196
Telestability, concept of, I: 67
Tellurium, interaction with mercury, I: 92
Teratogenesis
   from exposure to vinyl chloride, II: 66—67
   of cyanoacrylic adhesives, II: 94
   of MMA, II:113
   of silicones, II: 88
Ternary complexes
   enzyme catalysis and, I: 23
   formation of, I: 34
   ligand modification or exchange, I: 58
   low molecular weight, I: 14
   modifications induced in, I: 54—58
   reiterative DNA sequence-metal-ligand and, I: 54
Testicular damage, cadmium caused, I: 89
Tetrachlorosilane, exposure to, II: 88
Tetraethoxysilane, experimental studies with, II: 86
Thiosemicarbazones, in ligand-DNA interactions, I: 58, 61
Thompson's arthroplasty, II: 130
Thrombocytopenia, vinyl chloride-induced, II: 64, 66
Thromboplastic activity, MMA-induced, II:129
Thymus, vinyl chloride in, II: 68
Thyroid gland, in copper metabolism, I: 225
Titanium, I: 8
   animal tissues and, I: 171
   biological activity of, I: 172—175
   biologically inert behavior of, I: 175
   biomedical uses of, I: 170
   carcinogenicity of, I: 57, 175
   chromium in, I: 154
   dental caries prevention, I: 175
   effects of
      after implanted, I: 174—175
      after ingestion, I: 173
      after inhalation, I: 173
      on cells, I: 174
      on skin, I: 175
   essentiality of, I: 172
   human tissues and, I: 171
   industrial uses of, I: 170
   metabolism of, I: 171
   mutagenicity of, I: 57

occurrence of, I: 170
properties of, I: 170
reduction of folate, I: 175
Toxicity
   aluminum, I: 197—201
   antagonism and, I: 88
   of molybdenum, I: 166
   of plastics vs. metals, II: 52
   shift in dose-effects and, I: 89
   synergism and, I: 88
Trace elements
   in hair, I: 102, 104—105
   in normal metabolism, II: 5—6
   interaction of corrosion products with, II: 2
Transferrin, I: 9; II: 8—9
Transition metals, see also specific metals
   definition of, I: 8
   d electron distribution of, I: 10
   enzyme interactions and, I: 25—32
   ligand field theory and, I: 10—12
   occurrence of, I: 9
   oxidation-reduction and, I: 13
   stability constants for, I: 13
Transmembrane potential, variations of, I: 69
Trien, I: 16, 17
Triresylphosphate, allergic response to, II: 65
Triton®, II: 167
Tubing, silicone rubber, II: 84
Tumor, substrate dependent, see also specific tumor, II: 185—186
Tumor formation, latency periods for, see also Carcinogenesis, II: 12
Tungsten
   in enzyme interactions, I: 31
   oxidation states of, I: 10
Tyrosinase, I: 27, 224

U

Ulcers, chromium-induced, I: 151
Uricase, copper in, I: 27, 224
Urinary tract, vinyl chloride in, II: 68, 74
Urine, chromium in, I: 146, 147, 150
Uterus, aluminum levels in, I: 196

V

Vagal activity, MMA-induced, II: 125
Vanadium
   blood pressure and, I: 183
   dietary levels of, I: 180
   differential diagnosis of, I: 183
   effect on gastrointestinal tract of, I: 183
   effect on various organs, I: 183
   enzyme interactions, I: 31
   essentiality of, I: 180
   implantation of, I: 180
   injection of, I: 180
   interactions of, I: 182

   iron metabolism and, I: 181
   lipid metabolism and, I: 181
   metabolism of
      absorption of, I: 180
      excretion of, I: 182
      transport of, I: 181
   metabolism of various substances and, I: 181—182
   pulmonary effects, I: 182—183
   sulfur amino acid metabolism and, I: 181—182
   tolerance levels for, I: 184—185
   toxicity, I: 184—185
Vinyl chloride, II: 61
   absorption of, II: 67—68
   analysis of, II: 62
   angiosarcoma associated with, II: 76—77
   biotransformation of, II: 69
   carcinogenesis of, II: 69—75
   chronic exposure to, II: 65
   intracellular distribution of, II: 68
   metabolism of, II: 68—69
   mutagenesis and, II: 67
   synthesis and physical properties of, II: 61—62
   teratogenesis associated with, II: 66—67
   toxicology of
      in animals, II: 63—64
      in humans, II: 64—66
Vinyl chloride-vinylidene chloride copolymers, II: 53
Vision failure, associated with long-term vinyl chloride exposure, II: 66
Vitallium
   carcinogenicity of, II: 11
   composition of, I: 155
Vitamin B_{12}, see also Cobalt, I: 8
   and enzyme interactions, I: 25—26, 31
   in human tissue, I: 136
Voltametric techniques, for cobalt detection, I: 135

W

Washing, of hair for biopsy, I: 104—105
Wear
   generation of sensitizing complexes by, II: 40
   implanted polymers, II: 55
Welders, chromium exposure of, I: 145, 148
Wilson's disease
   copper metabolism in, I: 212, 225
   treatment of, I: 16
Wood chips, as fillers, II: 151
Working environment, chromium in, I: 145—146
Wound healing, I: 70
   cyanoacrylate adhesive in, II: 92
   DNA-metal complexes and, I: 94
Wounding, ternary complexes during, I: 35

X

Xanthine oxidase
   iron in, I: 28

molybdenum in, I: 30, 165
tungsten and, I: 31

Z

Zinc
 cadmium substituted for, I: 33
 cadmium toxicity and, I: 94
 carcinogenicity of, I: 57
 competition for carriers and, I: 92
 in enzyme interactions, I: 32
 in hair analysis, I: 107, 108
 in hemodialysis, II: 28
 interaction with cobalt, I: 135
 mutagenicity of, I: 57
Zinc deficiency, I: 94, II: 24